Asymptotic Distribution of Eigenvalues
of Differential Operators

Mathematics and Its Applications (*Soviet Series*)

Volume 53

Asymptotic Distribution of Eigenvalues of Differential Operators

by

Serge Levendorskiĭ
*Institute of National Economy,
Rostov-on-Don, U.S.S.R.*

KLUWER ACADEMIC PUBLISHERS
DORDRECHT / BOSTON / LONDON

Library of Congress Cataloging-in-Publication Data

Levendorskiĭ, Serge, 1951–
    Asymptotic distribution of eigenvalues of differential operators /
  Serge Levendorskiĭ.
        p.   cm. -- (Mathematics and its applications. Soviet Series)
    Translated from the Russian.
    Includes bibliographical references and indexes.
    ISBN 0-7923-0539-6
        1. Differential operators.   2. Eigenvalues.   3. Asymptotic
  distribution (Probability theory) 4. Distributions, Theory of
  (Functional analysis)   I. Title.   II. Series: Mathematics and its
  applications (Kluwer Academic Publishers).   Soviet Series.
  QA329.4.L48   1990
  515'.7242--dc20   90-40287

ISBN 0-7923-0539-6

Published by Kluwer Academic Publishers,
P.O. Box 17, 3300 AA Dordrecht, The Netherlands.

Kluwer Academic Publishers incorporates
the publishing programmes of
D. Reidel, Martinus Nijhoff, Dr W. Junk and MTP Press.

Sold and distributed in the U.S.A. and Canada
by Kluwer Academic Publishers,
101 Philip Drive, Norwell, MA 02061, U.S.A.

In all other countries, sold and distributed
by Kluwer Academic Publishers Group,
P.O. Box 322, 3300 AH Dordrecht, The Netherlands.

Printed on acid-free paper

Printed in the Netherlands

SERIES EDITOR'S PREFACE

'Et moi, ..., si j'avait su comment en revenir,
je n'y serais point allé.'

Jules Verne

The series is divergent; therefore we may be
able to do something with it.

O. Heaviside

One service mathematics has rendered the
human race. It has put common sense back
where it belongs, on the topmost shelf next
to the dusty canister labelled 'discarded non-
sense'.

Eric T. Bell

Mathematics is a tool for thought. A highly necessary tool in a world where both feedback and non-linearities abound. Similarly, all kinds of parts of mathematics serve as tools for other parts and for other sciences.

Applying a simple rewriting rule to the quote on the right above one finds such statements as: 'One service topology has rendered mathematical physics ...'; 'One service logic has rendered computer science ...'; 'One service category theory has rendered mathematics ...'. All arguably true. And all statements obtainable this way form part of the raison d'être of this series.

This series, *Mathematics and Its Applications*, started in 1977. Now that over one hundred volumes have appeared it seems opportune to reexamine its scope. At the time I wrote

> "Growing specialization and diversification have brought a host of monographs and textbooks on increasingly specialized topics. However, the 'tree' of knowledge of mathematics and related fields does not grow only by putting forth new branches. It also happens, quite often in fact, that branches which were thought to be completely disparate are suddenly seen to be related. Further, the kind and level of sophistication of mathematics applied in various sciences has changed drastically in recent years: measure theory is used (non-trivially) in regional and theoretical economics; algebraic geometry interacts with physics; the Minkowsky lemma, coding theory and the structure of water meet one another in packing and covering theory; quantum fields, crystal defects and mathematical programming profit from homotopy theory; Lie algebras are relevant to filtering; and prediction and electrical engineering can use Stein spaces. And in addition to this there are such new emerging subdisciplines as 'experimental mathematics', 'CFD', 'completely integrable systems', 'chaos, synergetics and large-scale order', which are almost impossible to fit into the existing classification schemes. They draw upon widely different sections of mathematics."

By and large, all this still applies today. It is still true that at first sight mathematics seems rather fragmented and that to find, see, and exploit the deeper underlying interrelations more effort is needed and so are books that can help mathematicians and scientists do so. Accordingly MIA will continue to try to make such books available.

If anything, the description I gave in 1977 is now an understatement. To the examples of interaction areas one should add string theory where Riemann surfaces, algebraic geometry, modular functions, knots, quantum field theory, Kac-Moody algebras, monstrous moonshine (and more) all come together. And to the examples of things which can be usefully applied let me add the topic 'finite geometry'; a combination of words which sounds like it might not even exist, let alone be applicable. And yet it is being applied: to statistics via designs, to radar/sonar detection arrays (via finite projective planes), and to bus connections of VLSI chips (via difference sets). There seems to be no part of (so-called pure) mathematics that is not in immediate danger of being applied. And, accordingly, the applied mathematician needs to be aware of much more. Besides analysis and numerics, the traditional workhorses, he may need all kinds of combinatorics, algebra, probability, and so on.

In addition, the applied scientist needs to cope increasingly with the nonlinear world and the extra mathematical sophistication that this requires. For that is where the rewards are. Linear models are honest and a bit sad and depressing: proportional efforts and results. It is in the non-linear world that infinitesimal inputs may result in macroscopic outputs (or vice versa). To appreciate what I am hinting at: if electronics were linear we would have no fun with transistors and computers; we would have no TV; in fact you would not be reading these lines.

There is also no safety in ignoring such outlandish things as nonstandard analysis, superspace and anticommuting integration, p-adic and ultrametric space. All three have applications in both electrical engineering and physics. Once, complex numbers were equally outlandish, but they frequently proved the shortest path between 'real' results. Similarly, the first two topics named have already provided a number of 'wormhole' paths. There is no telling where all this is leading - fortunately.

Thus the original scope of the series, which for various (sound) reasons now comprises five sub-series: white (Japan), yellow (China), red (USSR), blue (Eastern Europe), and green (everything else), still applies. It has been enlarged a bit to include books treating of the tools from one subdiscipline which are used in others. Thus the series still aims at books dealing with:

- a central concept which plays an important role in several different mathematical and/or scientific specialization areas;
- new applications of the results and ideas from one area of scientific endeavour into another;
- influences which the results, problems and concepts of one field of enquiry have, and have had, on the development of another.

Eigenvalues of all kinds of operators are important. Mostly (all) eigenvalues cannot be calculated exactly. Then various kinds of asymptotic data on the eigenvalues become most important. There have been developed a variety of techniques to obtain such asymptotic data. One of these, to which contributions have been made by many mathematicians, and among them most definitely the present author, is the approximate spectral projection method. It is a systematic method to analyze spectral asymptotics which yields good remainder estimates in quite a variety of different classes of problems and sometimes the sharpest ones known so far, for instance. The method is in particular suited to the case of pseudo-differential operators.

The ASP method is clearly a powerful one and destined to be developed further and to be applied on a wide scale. This is the first systematic book devoted to it and thus a welcome addition to this series.

The shortest path between two truths in the real domain passes through the complex domain.

J. Hadamard

La physique ne nous donne pas seulement l'occasion de résoudre des problèmes ... elle nous fait pressentir la solution.

H. Poincaré

Never lend books, for no one ever returns them; the only books I have in my library are books that other folk have lent me.

Anatole France

The function of an expert is not to be more right than other people, but to be wrong for more sophisticated reasons.

David Butler

Bussum, July 1990   Michiel Hazewinkel

CONTENTS

Introduction

0.1. Eigenvalue problems for self-adjoint operators appear in many fields of mathematical physics and mechanics. For example, eigenvalues of the Schrödinger operator are permissible energy levels for a quantum-mechanical particle, and eigenvalues of Maxwell's and Navier–Stokes's systems of operators in elasticity theory, membrane theory, etc., are squares of eigenfrequencies of respective bodies.

It is clear that calculating the eigenvalues of self-adjoint operators constitutes quite an important problem; however, its precise solution, except for certain simple cases, is impossible. Therefore, the methods of approximate calculation of eigenvalues are of great importance. One of these methods involves calculating the asymptotics of certain sequences of eigenvalues λ_n of an operator A as $n \to \infty$.

If A is an operator with discrete spectrum, then instead of an asymptotics of positive (negative) eigenvalues $\lambda_n^{\pm}$ of an operator A, it is usually more convenient to calculate an asymptotics of the distribution function of the sequences $\{\lambda_n^{\pm}\}$:

$$N_{\pm}(t, A) = \operatorname{card}\{i \mid 0 < \pm\lambda_i^{\pm} < t\}, \quad \text{as} \quad t \to \infty;$$

and if the sequence $\{\lambda_n\} \subset [\alpha, \beta)$ becomes more dense near β and, in addition, if $[\alpha, \beta)$ contains only the points of the discrete spectrum, then the asymptotics of the function

$$N([\alpha, \beta - \tau), A) = \operatorname{card}\{i \mid \alpha \leqslant \lambda_i < \beta - \tau\}, \quad \text{as} \quad \tau \to +0.$$

Classical examples of operators with discrete spectrum are the operator for a boundary value problem in a bounded domain for a system in elasticity theory and the operator for harmonic oscillator; examples of operators with nonempty essential spectrum are Schrödinger and Dirac operators with potentials decreasing at infinity and the operator for the multigrouped diffusion model.

If an operator depends on a small parameter h (Planck's constant in Schrödinger and Dirac operators; the relative thickness of thin elastic shells), then information about eigenvalues can be obtained by analyzing the asymptotics of $N((a, b), A_h)$ as $h \to 0$.

Besides calculating the principal term of the asymptotics of the spectrum distribution function, of great interest is the estimate for the remainder (in cases where it is impossible to calculate the second term of the asymptotic function). For instance, an estimate for the remainder is important in substantiating certain methods for the expansion by eigenvectors of weakly non-self-adjoint operators (see Agranovich [1]).

We shall develop a method allowing us to use a single scheme to analyze the spectral asymptotics in different problems. In many cases, this method yields good estimates for a remainder (in a number of cases, they are sharper than those obtained by other methods).

The basic theorems of our method allow us to analyze the following asymptotics.

1) Asymptotics as $t \to \infty$ of the distribution functions $N_\pm(t)$ of positive and negative spectra of the linear bundle

$$(0.1) \qquad\qquad Au = tBu,$$

where the operator A is of the higher order and one of the operators A, B is positive-definite. Equation (0.1) is considered in both bounded and unbounded domains. The operators A, B can be elliptic or Douglis–Nirenberg elliptic and can degenerate. They are defined by differential operators (DO), by pseudodifferential operators (PDO), and by quadratic forms with continuous coefficients.

Examples of problems of the type in (0.1) are the system in elasticity theory and the Schrödinger operator with a potential increasing near infinity. Potentials that in some directions do not increase could also be considered.

2) Spectral asymptotics in problems with constraints (problems on fundamental frequencies for piezoactive bodies and electromagnetic resonators (wave conductors); linearized Navier–Stokes's systems).

3) Asymptotics of discrete spectrum accumulating near the boundary of the essential spectrum (Schrödinger and Dirac operators with potentials decreasing near infinity and a system for a multigrouped diffusion model).

4) Quasi-classical spectral asymptotics; i.e., asymptotics of the distribution function $N((\alpha,\beta)A_h)$ as $h \to 0$ for an operator of the type $A_h = A(x, hD)$ (Schrödinger operators and Dirac operators). Problems in strongly anisotropic domains also reduce to this type; in these cases, the operator A_h has an operator-valued symbol.

5) Asymptotic functions with respect to a small parameter h (relative thickness) of the distribution function for eigenfrequencies of a thin elastic shell—in a vacuum or containing liquid.

6) Asymptotics of spectra with respect to several parameters.

In addition, by using the general results of Marcus and Matsaev [1–3] and Michaileč [4], one can also obtain asymptotic formulas for weakly non-self-adjoint operators.

0.2. The study of spectral asymptotics has an old tradition. First, the asymptotic function of $N(t, A)$ was studied by Weyl [1, 2]. In implicit form, he applied the now well-known variational minimax principle, which Courant later extracted from Weyl's proof. Methods based on variational principles are also called variational.

The second large group of methods is based on the application of Tauberian theorems to suitable functions of operators (for example, the kernel of a resolvent, ζ-functions, and spectral functions). The Tauberian technique was first applied by Carleman [1]. Tauberian methods, unlike variational ones, apply in calculating asymptotics of spectral functions. In addition, these very methods (mainly the hyperbolic operator method) recently produced sharp estimates for the remainder in a number of problems and, with global assumptions, yielded the second term in an asymptotic function.

The hyperbolic operator method was first used by Levitan [1] and Avakumovic [1]. By this method, the asymptotics for a spectral function is calculated, and from that, the asymptotics is derived for the distribution function of the spectrum. The general method for estimating the remainder in asymptotic formulas for spectral functions of elliptic PDO on a closed manifold was developed by Hörmander [2], based on the technique of Fourier integral operators. This method was further developed in works by Levitan [2], Duistermaat and Guillemin [1], Ivriĭ [1–10], Heffer and Robert [1–6], Heffer [1], Feigin [6–8], Vasilev [2, 3], and Safarov [1–3] (see also the monograph of Ivriĭ and its literature review).

In particular, we should mention the method developed by Ivriĭ, who succeeded in expanding the applications of the hyperbolic operator method and in obtaining the second term of the asymptotics in a number of problems. The second term of the asymptotics was obtained by other authors, too (see the Review of the Bibliography).

The Tauberian methods are more sensitive to the smoothness of a problem's data than are variational methods, and they allow only regular degeneration; however, the results of variational methods are not as accurate.

Weyl's variational methods were advanced by Metivier [1], Kozlov [2, 3], and Andreev [1–8], allowing us to improve the estimate for remainders in nonsmooth problems. Earlier, an important variant of variational method was developed by Birman and Solomak [1–3], based on abstract operator perturbation theory. It was precisely in this way that the classical formula was established for spectral asymptotics under minimal conditions of smoothness of the coefficients and the boundary of a domain. For further results in this direction, see their review [4] and the works by Rosenblum [1–5].

The variational methods mentioned above either do not apply to the analysis of PDO or only yield a relatively poor estimate for their remainders. This shortcoming is not characteristic of the approximate spectral projector (ASP) method, introduced by Tulovsky and Shubin [1] and further developed in works by Roytburd [1], Bezyaev [1, 2], Feigin [1–5], Hörmander [4], and this author [1–16]. The ASP method is grouped with variational methods but, according to its characteristics, takes an intermediate place between other variational methods and Tauberian methods. In sufficiently smooth cases, this method usually gives better estimates for remainders than do other variational methods; and in nonsmooth cases, worse estimates. And unlike Tauberian methods, it does not allow to obtain accurate results, although in nonregular cases it gives better estimates for remainders than do Tauberian methods. The ASP method has a broader scope of application than the Tauberian methods, but it does not work in the case of discontinuous coefficients.

0.3. Before presenting the main idea of the ASP method, let us recall the quasi-classical Weyl formula. In (0.1) let A be an elliptic positive-definite operator in $L_2(\Omega)^m$, where $\Omega \subset R^n$ is a bounded open set and B is an operator subordinate to A. Then under certain regularity conditions, the spectrum of the bundle (0.1) is discrete, and the asymptotics for the distribution functions for this spectrum is given by the formula

$$(0.2) \qquad N_{\pm}(t) \sim (2\pi)^{-n} \iint\limits_{\Omega \times R^n} N(0, a(x,\xi) \mp tb(x,\xi)) dx\, d\xi$$

as $t \to \infty$, where a and b are principal symbols of operators A and B, respectively, and where $N(\lambda, a)$ on the right-hand side is a distribution function for the spectrum of the operator in C^m. If Ω is unbounded, then the discreteness of spectrum and its asymptotic function usually are not determined by principal symbols; therefore, full Weyl symbols should generally be used in (0.2) (since only they are always symmetric, provided that the operators are). We recall that a is a Weyl symbol of an operator A, if

$$(Au)(x) = (a_w u)(x) = (2\pi)^{-n} \int d\xi \int dy\, e^{i(x-y,\xi)} a\left(\frac{x+y}{2}, \xi\right) u(y).$$

In what follows, we assume that a and b, in (0.2), are Weyl symbols. Then formula (0.2) is also valid for sufficiently regular elliptic operators with weight in an unbounded domain, as well as for operators with operator-valued symbols.

For quasi-classical asymptotics for the spectrum of the operator A_h with the Weyl symbols $a_h(x, \xi) = a(x, h\xi)$, the formula analogous to (2) is valid:

$$(0.3) \qquad N((\alpha, \beta), A_h) \sim (2\pi)^{-n} \iint\limits_{\Omega \times R^n} N((\alpha, \beta), a_h(x,\xi)) dx\, d\xi$$

as $h \to 0$. Formula (0.3) is also valid for operators in shell theory, whose dependence on h is more complicated.

Finally, for the asymptotic function for a set of eigenvalues accumulating near $\beta - 0$, we have as $\tau \to +0$,

$$(0.4) \qquad N((\alpha, \beta - \tau), A) \sim (2\pi)^{-n} \iint\limits_{\Omega \times R^n} N((\alpha, \beta - \tau), a(x, \xi)) dx \, d\xi.$$

In (0.3) and (0.4), if $\alpha = -\infty$, then (0.2)–(0.4) can be written uniformly under the assumptions that in the first case, $A_t = A \mp tB$, $a_t = a \mp tb$; in the second case, $t = h^{-1}$, $A_t = A_h - \beta$, $a_t = a_h - \beta$; and in the third case, $t = \tau^{-1}$, $A_t = A - \beta + \tau$, $a_t = a - \beta + \tau$. Then according to the minimax principle, formulas (0.2)–(0.4) are reduced to one formula,

$$(0.5) \qquad N(0, A_t) \sim (2\pi)^{-n} \iint\limits_{\Omega \times R^n} N(0, a_t(x, \xi)) dx \, d\xi.$$

In the case where $\alpha > -\infty$ in (0.3), we can select a suitable rational function f such that $N((\alpha, \beta), A_h) = N(0, f(A_h))$, and can assume that $A_t = f(A_h)$, $a_t = f(a_h)$, in order to arrive to the same formula (0.5). Formula (0.4) with $\alpha > -\infty$ is transformed similarly.

This transformation generally leads us to operators containing inverses of operators of boundary value problems, but this difficulty can be overcome within the framework of the ASP method.

Thus, we reduced the problem indicated above to a single one—namely, the analysis of an asymptotic function of the number of eigenvalues of the operator A_t depending on t as $t \to \infty$. For this problem, we have the generalized Weyl formula (0.5).

0.4. In the scheme of the ASP method, it is most important to prove formula (0.5) [more precisely, some two-sided estimates for the function $N(0, A_t)$] for the Friedrichs extension of the operator

$$r_+ a_{t,w} e_+ : C_0^\infty(\Omega; H) \to L_2(\Omega; H),$$

where H is the Hilbert space, and r_+ and e_+ are the standard operators of contraction of generalized functions onto Ω and of continuation by zero of functions from $L_2(\Omega; H)$, respectively. The analysis of the operators of general boundary value problems is reduced to the analysis of the Friedrichs extension by means of standard variational arguments.

We start with Glazman's Lemma, according to which, for a self-adjoint operator A semi-bounded from below in the Hilbert space H,

$$(0.6) \qquad \dim L_\lambda^- \leqslant N(\lambda, A) \leqslant \dim L_\lambda^+,$$

if subspaces $L_\lambda^\pm \subset D(A)$ are such that

$$(0.7) \qquad \langle Au, u \rangle < \lambda \|u\|^2 \quad \forall (0 \neq) u \in L_\lambda^-,$$

$$(0.8) \qquad \langle Au, u \rangle \geqslant \lambda \|u\|^2 \quad \forall u \in {}^\perp L_\lambda^+ \cap D(A),$$

where $^\perp L$ is an orthogonal complement of L. Inequalities (0.6) become equalities if

$$L_\lambda^- = L_\lambda^+ = E_\lambda H = \text{span}\{u_i \mid \nu_i = 1\} = \text{span}\{u_i \mid \nu_i \geqslant \tfrac{1}{2}\}$$

(where ν_i and u_i are, respectively, the eigenvalues and their corresponding eigenvectors of a spectral projector E_λ of the operator A), but in the general case, we cannot construct it. With the ASP method, we construct the operators $\mathcal{E}_\lambda^\pm$, approximating E_λ, such that the subspaces $L_i^\pm = \text{span}\{u_i^\pm \mid \nu_i^\pm \geqslant \tfrac{1}{2}\}$, where $\nu_i^\pm$ and $u_i^\pm$ are eigenvalues and eigenvectors of operators $\mathcal{E}_i^\pm$; satisfy (0.7) and (0.8), and estimate $\dim L_\lambda^\pm$. From (0.6) we obtain estimates for $N(\lambda, A)$.

In order to clarify the main idea of constructing $\mathcal{E}_\lambda^\pm$, we assume temporarily that the symbol for the product of the PDO is the product of the symbols: $(ab)_w = a_w b_w$ (this is not far from the truth, since for PDO from "nice classes," the order of $(ab)_w - a_w b_w$ is smaller than that of $a_w b_w$), and we suppose that one can treat PDO with discontinuous symbols the same as PDO with smooth symbols. Then in particular, for finite a,

$$(0.9) \qquad \text{Tr}\, a_w = (2\pi)^{-n} \iint \text{Tr}\, a(x, \xi)dx\, d\xi,$$

if $a(x, \xi)$ are finite-dimensional operators for all (x, ξ); and if $a(x, \xi) > 0$ for all (x, ξ), then

$$(0.10) \qquad \langle a_w u, u \rangle = \|(a^{1/2})_w u\|^2 \geqslant 0 \quad \forall u \in \mathcal{S}(R^n; H).$$

It could be shown that if a_w is a hypoelliptic PDO, then (0.10) is valid for $u \in L \subset \mathcal{S}(R^n; C^m)$, where $\text{codim}\, L \leqslant C$.

Our assumptions allow us to apply formulas (0.6)–(0.8) in the following way. Let ψ, χ be characteristic functions of the set Ω and of the interval $(-\infty, 0)$, and let $d_t > 0$ be some auxiliary symbol such that

$$N(0, (a_t - d_t)(x, \xi)) < \infty \quad \forall (x, \xi) \in \Omega \times R^n.$$

Let $\overset{\circ}{e}_t^\pm = \chi(d_t^{-1/2} a_t d_t^{-1/2} \mp 1)$, $e_t^\pm = \psi_\Omega \overset{\circ}{e}_t^\pm \psi_\Omega$,

$$\mathcal{E}_t^\pm = e_{t,w}^\pm, \qquad L_t^\pm = \text{span}\{u_i^\pm \mid \nu_i^\pm \geqslant \tfrac{1}{2}\}.$$

Then (0.9) yields

$$\dim L_t^\pm = \text{Tr}\, \mathcal{E}_t^\pm = (2\pi)^{-n} \iint\limits_{\Omega \times R^n} N(0, (a_t \mp d_t)(x, \xi))dx\, d\xi.$$

Let $\hat{L}_t^- = (d_t^{-1/2})_w L_t^-$. Then $\dim \hat{L}_t^- = \dim L_t^-$; and for all $v \in L_t^-$ and $u = (d_t^{-1/2})_w \mathcal{E}_t^- v \in \hat{L}_t^-$, we have

$$\langle A_t u, u \rangle = \langle (a_{t,w} + d_{t,w})(d_t^{-1/2})_w \mathcal{E}_t^- v, (d_t^{-1/2})_w \mathcal{E}_t^- v \rangle - \|\mathcal{E}_{1,t}^- v\|^2$$
$$\leqslant -\langle q_{t,w} v, v \rangle,$$

where

$$q_t = -e_t^-(d_t^{-1/2} a_t d_t^{-1/2} + 1) e_t^- + \tfrac{1}{4} I$$

is positive by construction. (0.10) yields $\langle A_t u, u \rangle < 0 \quad \forall (0 \neq) u \in \hat{L}_t^-$, and (0.6) yields an estimate from below for the function $N(0, A_t)$.

In order to obtain an estimate from above, we assume that

$$A_t^+ = \psi_\Omega((d_t^{-1/2} a_t d_t^{-/12})_w - I)\psi_\Omega + I$$

and notice that

$$\forall u \in C_0^\infty(\Omega; H), \quad \langle A_t u, u \rangle = \langle A_t^+ (d_t^{1/2})_w u, (d_t^{1/2})_w u \rangle.$$

Therefore, $N(0, A_t) \leqslant N(0, A_t^+)$, where A_t^+ is an operator in $L_2(R^n; H)$. Let $\hat{L}_t^+ = {}^\perp L_t^+$. Then for $v \in \hat{L}_t^+$, $u = (I - \mathcal{E}_t^+) v \in \hat{L}_t^+$, we have $\|u\| \geqslant \tfrac{1}{2}\|v\|$ and

$$\langle A_t^+ u, u \rangle = \langle (I - \mathcal{E}_t^+)\psi_\Omega((d_t^{-1/2} a_t d_t^{-1/2})_w - I)\psi_\Omega(I - \mathcal{E}_t^+) v, v \rangle + \|u\|^2$$
$$\geqslant \langle q'_{t,w} v, v \rangle,$$

where

$$q'_t = (I - e_t^+)(d_t^{-1/2} a_t d_t^{-1/2} - I)(I - e_t^+) + \tfrac{1}{4} I$$

is a positive symbol. Thus, $\langle A_t^+ u, u \rangle \geqslant 0 \quad \forall u \in \hat{L}_t^+$, and an estimate from above follows from (0.6).

Thus, we obtain the estimate
(0.11)

$$N(0, A_t) = (2\pi)^{-n} \iint\limits_{\Omega \times R^n} N(0, a_t(x, \xi)) dx \, d\xi$$

$$+ O(\iint\limits_{\Omega \times R^n} (N(0, (a_t - d_t)(x, \xi)) - N(0, (a_t + d_t)(x, \xi)) dx \, d\xi),$$

where d_t is an arbitrary positive symbol, which is valid under the simplifying assumptions mentioned above.

In order to carry out a rigorous proof, we must smoothe the functions $\overset{\circ}{e}_t^\pm$, ψ_Ω and choose the function d_t (it cannot be arbitrarily small) such that all operators that we have to consider belong to some sufficiently nice algebra of

PDO. As a result, a summand related to the boundary will be added to the O-term of formula (0.11), and if the operator A_t is not scalar, another summand will follow. In addition, the closer the approximation of the functions $\overset{\circ}{e}_t^{\pm}$, ψ_Ω by smooth functions, the sharper the estimate that we obtain for a remainder. The estimate for a remainder turns out to be good in the scalar case, as well as in the matrix case, provided that the symbol is microlocally smoothly diagonalizable.

In conclusion, we point out that for hypoelliptic PDO on the entire space, the ASP method allows simple formalizations. The first one, introduced by Tulovsky and Shubin [1] (see also Hörmander [4]), utilizes estimates for the kernel norm of the operator $(\mathcal{E}_t^{\pm})^2 - \mathcal{E}_t^{\pm}$; the second, introduced by Feigin [2] and explicitly formulated by Beziaev [2], is considerably more convenient, since it utilizes only formula (0.9). Here the idea of Feigin [2] is generalized; the idea of Hörmander [4], allowing an improved estimate for a remainder, is also used. Of course, the generalization of the ASP method proposed here could also be formalized, but the corresponding algorithm would be too complex.

0.5. We shall schematically describe the conditions under which the application of the ASP method is possible.

1) The integrals in (0.5) converge.

2) The symbol a_t is a hypoelliptic symbol from some algebra of PDO, which generally depends on t. In most applications, this means that there exist scalar functions $\Psi_{t,j}$, $\psi_{t,j}$ and an operator-valued function q_t such that for all multiindices α, β,

$$(0.12) \qquad \|(q_t^{-1})^* a_{(\beta)}^{(\alpha)} q_t^{-1}\| \leqslant C_{\alpha\beta} \Psi_t^{-\alpha} \psi_t^{-\beta},$$

where

$$a_{(\beta)}^{(\alpha)}(x,\xi) = \partial_\xi^\alpha D_x^\beta a(x,\xi), \qquad \Psi^\alpha = \Psi_1^{\alpha_1} \cdots \Psi_n^{\alpha_n},$$

and

$$(0.13) \qquad a_t(x,\xi) \geqslant c q_t(x,\xi)^* q_t(x,\xi), \quad \text{if } |x| + |\xi| > C(t).$$

The weight functions $\Psi_{t,j}$, $\psi_{t,j}$, q_t must satisfy certain conditions (see §4 and §7), among the more important being the strong uncertainty principle, as follows: either

$$(0.14) \qquad h_t(x,\xi) \overset{\text{def.}}{=} \max_{1 \leqslant j \leqslant n} \Psi_t(x,\xi)^{-1} \psi_t(x,\xi)^{-1} \leqslant C(1 + |x| + |\xi|)^{-c},$$

uniformly with respect to $t \geqslant 1$, where $c > 0$, or

$$(0.15) \qquad h_t(x,\xi) \to 0 \quad \text{as} \quad t \to \infty.$$

uniformly with respect to x, ξ.

In most cases, the remaining conditions are satisfied as long as one of the conditions (0.15) and (0.14) and both conditions (0.12) and (0.13) are satisfied.

When 1) and 2) hold, the ASP method allows us to obtain an asymptotic formula for $N(0, A_t)$ in the following form. The principal term of the asymptotic function is defined by formula (0.5), and the remaining term is estimated by the measures of sets associated with a boundary and with level surfaces $\lambda_{t,j}(x, \xi) = 0$ of eigenvalues of the symbol $a_t(x, \xi)$; if the operator is not scalar, then another summand is added.

Thus, the problem of calculating the asymptotics for a spectrum and of evaluating a remainder generally reduces to the following problems:

1) Selecting weight functions that would allow us to evaluate derivatives of the symbol, i.e., describing the algebra of PDO to which the operator belongs; and

2) Calculating the eigenvalues of the symbol and evaluating the measure of sets on which they satisfy certain conditions.

The first problem is rather simple, since in most cases, such classes of PDO are well-known, whereas in other cases, measure functions are introduced in a natural way. If the first problem is solved, the second one does not entail great difficulties, but it often requires extensive calculation.

Once again, the direct scheme described above applies only to Friedrichs extensions; in other cases, it is necessary to draw on additional variational arguments.

0.6. In some cases—for example, hypoelliptic PDO with multiple characteristics or degenerate elliptic DO—formula (0.2) [and hence, formula (0.5)] predicts $N_{\pm}(t, A) = \infty$ for operators with discrete spectrum or incorrectly predicts asymptotics. In some cases, it is possible for condition 2) in the preceding part not to be satisfied, even though formula (0.2) is valid. The general scheme of analyzing such operators, together with a generalization of Weyl formula, is presented in Chapter 6. We should note that Fefferman [1] recently proposed a conjecture replacing the Weyl formula; however, as far as the author knows, it has not yet been applied to the calculation of spectral asymptotics.

0.7. The fundamental technical apparatus of the ASP method is the Weyl–Hörmander calculus of PDO (see Hörmander [3, 5]). More precisely, in Chapter 1, we develop the generalization of this calculus. Here, as opposed to the work by Hörmander, an order of an operator is determined not by one scalar but by two operator-valued functions. We state only the results necessary for the analysis of spectral asymptotics and do not prove the characteristic property of Weyl PDO, i.e., the invariance of the introduced classes with respect to symplectomorphisms of the space $R_x^n \times R_\xi^n$. We also do not consider a number of applications observed by Hörmander [3].

In addition, we must also construct the theory of Sobolev-type spaces in which the introduced operators act. Here we modify Beals' scheme [2].

0.8. We refer to theorems, formulas, etc., by section number. For example, (2.1) is the first formula in §2.

By c, C, c_1, etc., we denote various positive constants whose precise value is nonessential. Unless stated otherwise, they can be chosen to have equal value for all parameters on which they may depend.

Functions and operators of the form fI and cI, where f is a scalar function and c is a scalar, will often be denoted merely by f, c.

By H, H_j we denote complex Hilbert spaces, and $B_{ij} = \mathrm{Hom}(H_j; H_i)$ denotes the Banach space of continuous linear operators acting from H_j to H_i. The group of continuously invertible operators in H is denoted by $\mathrm{Aut}\, H$; and the identity element of the algebra, $B_{ii} = \mathrm{End}\, H_i$ by $I_{(i)}$, or simply by I if it is clear from context which algebra is being considered. If $H_i = \mathbb{C}^m$, then $I_m = I_{(i)}$.

If A and B are operator functions, then $A \leqslant B$ means $A(x) \leqslant B(x)$ for any x; inequality $\|A\| \leqslant C$ has a similar meaning.

The points in the space $R_X^{2n} = R_x^n \times R_\xi^n$ will be denoted by $X, Y, \ldots$, as well as by $(x, \xi), (y, \eta), \ldots$.

The standard notations used are as follows:

$$\langle x \rangle = (1 + |x|^2)^{1/2}, \qquad \partial_j = \partial/\partial x_j, \qquad D_j = -i\partial_j,$$

$$\partial^\alpha = \partial_1^{\alpha_1} \cdots \partial_n^{\alpha_n}, \qquad a^{(\alpha)} = \partial^\alpha a, \qquad a_{(\beta)}^{(\alpha)}(x, \xi) = \partial_\xi^\alpha D_x^\beta a(x, \xi),$$

$$|\alpha| = \sum \alpha_j, \qquad d'\xi = (2\pi)^{-n} d\xi,$$

where α, β are multiindices and x, $\xi \subset R^n$.

If $\Omega \subset R^n$ is an open set and $S \geqslant 0$ is an integer, then $H^S(\Omega)$ is a standard Sobolev space with the norm

$$\|u\|_S = \left(\sum_{|\alpha| \leqslant S} \|D^\alpha u\|_{L_2}^2 \right)^{1/2},$$

and $\overset{\circ}{H}{}^S(\Omega)$ is the closure of $C_0^\infty(\Omega)$ with respect to the norm $\|\cdot\|_S$.

Acknowledgments. The author is deeply grateful to V. Y. Ivriĭ, V. B. Lidskiĭ, M. Z. Solomiak, and M. A. Shubin for useful and stimulating discussions.

CHAPTER 1

The Weyl–Hörmander Calculus
of Pseudodifferential Operators

§1 Classes of Symbols

1.1. We shall use the standard notation of distribution theory with values in a Banach space (see Schwartz [1]). If $\Omega \subset R^n$ is an open set, then $C^\infty(\Omega; B)$ is a space of infinitely differentiable functions on Ω with values in B, and $C_0^\infty(\Omega; B)$ is a subspace of $C^\infty(\Omega; B)$ consisting of functions f with compact support $\operatorname{supp} f \subset \Omega$. The sequence $\{f_n\} \subset C^\infty(\Omega; B)$ $[\subset C_0^\infty(\Omega; B)]$ converges to $f \in C^\infty(\Omega; B)$ $[\in C_0^\infty(\Omega; B)]$ if for any multiindices α, $\partial^\alpha f_n(x) \to \partial^\alpha f(x)$ and there exists a compact set $K \subset \Omega$ such that $\operatorname{supp} f_n \subset K$ for all n.

$\mathcal{S}(R^n; B)$ is a subspace of $C^\infty(R^n; B)$ consisting of functions that satisfy the condition

$$(1.1) \qquad |u|_k = \max_{|\alpha|+|\beta| \leqslant k} \sup_{R^n} \|x^\alpha \partial^\beta u(x)\|_B < \infty.$$

The set of seminorms $|\cdot|_k$ defines the topology of a Frechet space in $\mathcal{S}(R^n; B)$.

The formulated definitions are also correct for a linear real n-dimensional space V, since the supply of elements and topology in all introduced spaces does not depend on the choice of Euclidean structure and of an orthonormal basis in V, which makes V to be R^n. Therefore, we can write $C^\infty(V; B)$, and so on.

By V' we denote a space dual to V; by $\langle \cdot, \cdot \rangle$, the pairing between V and V'; and by dx and $d\xi$, the Lebesgue measures on V and V' normalized in such a way that the direct and inverse Fourier transforms

$$F : \mathcal{S}(V; H) \to \mathcal{S}(V'; H), \qquad F^{-1} : \mathcal{S}(V'; H) \to \mathcal{S}(V; H)$$

are defined by the equalities

$$(Fu)(\xi) = \hat{u}(\xi) = \int e^{-i\langle x,\xi\rangle} u(x)\,dx,$$

$$(F^{-1}\hat{u})(x) = \int e^{i\langle x,\xi\rangle} \hat{u}(\xi)\,d'\xi.$$

Here the integrals are taken over the entire space which is always implied when the limits of integration are not indicated.

Let $L_2(V;H)$ be a space of strongly measurable vector functions with values in H and with the norm

$$\|u\|_{L_2(V;H)} = \left(\int \|u(x)\|_H^2\,dx\right)^{1/2} < \infty.$$

Then $C_0^\infty(V;H) \subset \mathcal{S}(V;H) \subset L_2(V;H)$ densely in the topology of $L_2(V;H)$, and the operator F extends to an isomorphism $F : L_2(V;H) \to L_2(V';H)$.

From the Sobolev inequality we can conclude that the topology in $\mathcal{S}(V;H)$ can be defined by the system of seminorms

$$(1.2) \qquad |u|'_k = \max_{|\alpha|+|\beta|\leqslant k} \|x^\alpha \partial^\beta u\|_{L_2(V;H)}$$

for any choice of Euclidean structure in V.

1.2. If $a : V \to B_{12}$ is a function of temperate growth such that

$$au \in \mathcal{S}(V;H_2) \quad \forall u \in \mathcal{S}(V;H_1)$$

and $L : V \times V' \to R$ is a linear function, then the equality

$$(L(x,D)au)(x) = F^{-1}L(x,\xi)\widehat{au}(\xi)$$

defines a slowly increasing function $L(x,D)a : V \to B_{12}$.

The kth differential of a slowly increasing function a is defined by the equality

$$d^k a(x;t_1,\ldots,t_k) = (\langle t_1,D\rangle \cdots \langle t_k,D\rangle a)(x).$$

If $P_j \in \operatorname{Aut} H_j$ and if g is a positive-definite quadratic form on V, then we normalize the kth differential by the equality

$$(1.3) \qquad |a|_{g,k}^{p_1,p_2}(x) = \sup_{(0\neq)t_j\in V} \|p_1^{-1}(d^k a)(x;t_1,\ldots,t_k)p_2^{-1}\| \prod_{1\leqslant j\leqslant k} g(t_i)^{-1/2}.$$

If $a \in C^\infty(\Omega;B_{12})$, $b \in C^\infty(\Omega;B_{23})$, and $P_j \in \operatorname{Aut} H_j$, then Leibnitz's rule yields

$$(1.4) \qquad |ab|_{g,k}^{p_1,p_3}(x) \leqslant \sum_{0\leqslant j\leqslant k} \binom{k}{j} |a|_{g,j}^{p_1,p_2^{-1}}(x)|b|_{g,k-j}^{p_2,p_3}(x).$$

If $a(x_0)$ has a left inverse $a(x_0)^{-1}$, then for x near x_0, the series

$$a(x)^{-1} = \sum_{0 \leqslant i < \infty} (-1)^i (a(x_0)^{-1} a(x) - I)^i a(x_0)^{-1}$$

converges, and $a(x)^{-1}$ is a left inverse of $a(x)$. Hence, applying (1.4) yields

$$(1.5) \qquad |a^{-1}|_{g,k}^{p_2^{-1}, p_1^{-1}}(x) \leqslant C_k (|a|_{g,1}^{p_1, p_2}(x) + \cdots + |a|_{g,k}^{p_1, p_2}(x)^{1/k})^k,$$

where C_k depends only on k, $\|p_2 a(x_0)^{-1} p_1\|$, $\|a(x_0)^{-1} a(x) - I\|$.
A similar estimate holds for a right inverse.

1.3. In the definition of classes of symbols, the form g and the operators p_j will depend on x. Therefore, g is a Riemann metric on V; i.e., $\forall x \in V$, $g_x(\cdot)$ is a positive-definite quadratic form on V. We assume that the metric g and the functions p_j vary slowly as x varies.

Definition 1.1. The metric g is said to be slowly varying if there exist $c, C > 0$ such that

$$(1.6) \qquad g_x(x - y) \leqslant c \quad \Rightarrow \quad g_y(\cdot) \leqslant C g_x(\cdot).$$

If $g_x(x - y) \leqslant \min\{c, c/C\}$, then $g_y(x - y) \leqslant c$; and according to (1.6), $g_x \leqslant C g_y$. Therefore, by decreasing c if necessary, we can give (1.6) a symmetric form

$$(1.6') \qquad g_x(x - y) \leqslant c \quad \Rightarrow \quad C^{-1} g_x(\cdot) \leqslant g_y(\cdot) \leqslant C g_x(\cdot).$$

Definition 1.2. A function $p : V \to \text{Aut } H$ is left (right) g-continuous if there exist $c = c(p) > 0$, $C = C(p) > 0$ such that

$$(1.7) \qquad g_x(x - y) \leqslant c \quad \Rightarrow \quad \|p(y)^{-1} p(x)\| \leqslant C \qquad [\|p(x) p(y)^{-1}\| \leqslant C].$$

If the metric g varies slowly (which later on is always assumed), then we can symmetrize (1.7) similarly to how we did (1.6):

$$(1.7') \qquad g_x(x - y) \leqslant c \quad \Rightarrow \quad \|p(x)^{-1} p(y)\| + \|p(y)^{-1} p(x)\| \leqslant C$$
$$[\|p(y) p(x)^{-1}\| + \|p(x) p(y)^{-1}\| \leqslant C].$$

Condition (1.7′) can be conveniently verified as follows:

$$(1.7'') \; g_x(x - y) \leqslant c \quad \Rightarrow \quad C^{-1} p(y) p(y)^* \leqslant p(x) p(x)^* \leqslant C p(y) p(y)^*$$
$$[C^{-1} p(y)^* p(y) \leqslant p(x)^* p(x) \leqslant C p(y)^* p(y)].$$

Functions that are left and right g-continuous at the same time will be called g-continuous. Obviously, a left (right) symmetric g-continuous function is g-continuous.

We say that left (right, positive symmetric) continuous functions p and p_1 are equivalent (denoted $p \asymp p_1$) if there exists $C > 0$ such that

(1.8)
$$\|p^{-1}p_1\| + \|p_1^{-1}p\| \leqslant C$$
$$\left[\|pp_1^{-1}\| + \|p_1p^{-1}\| \leqslant C, \quad C^{-1}p \leqslant p_1 \leqslant Cp\right].$$

Conditions (1.7′) and (1.7″), the polar expansion theorem, and the inequality $\|(p_1 + p_2)^{-1}p_1\| \leqslant C$ for positive-definite operators yield the following proposition:

Proposition 1.1. a) *If p is a right (left) g-continuous function, then p^* and p^{-1} are left (right) g-continuous functions, and p is equivalent to the positive symmetric g-continuous function*

$$p_1 = (p^*p)^{1/2} \qquad \left[= (pp^*)^{1/2}\right].$$

b) *If p_1, p_2 are positive symmetric g-continuous functions and c_1, c_2 are positive constants, then $c_1p_1 + c_2p_2$ is a g-continuous function.*   ∎

1.4. In the following definition of classes of symbols, $|a|_{g,k}^{p_1,p_2}(x)$ is defined by equality (1.3), with $g = g_x$ and $p_j = p_j(x)$.

Definition 1.3.   Let $\Omega \subset V$ and $a \in C^\infty(\Omega; B_{12})$, and $\forall k$, let

(1.9)
$$n_k(g, p_1, p_2, \Omega, a) \stackrel{\text{def.}}{=} \sup_{\Omega} |a|_{g,k}^{p_1,p_2}(x) < \infty.$$

Then we write $a \in S(g, p_1, p_2, \Omega)$.

Writing $a \in S(g, p_1, p_2, \Omega)$ throughout below implies that the metric g varies slowly and that p_1 and p_2 are left and right g-continuous functions, respectively.

Example 1.1.   Let the metric g (on R^n) have the form

$$g_x(y) = \sum_{1 \leqslant i \leqslant n} \Phi_i(x)^{-2}|y_i|^2;$$

and for any multiindex α, let

(1.9′)
$$\|p_1(x)^{-1}(\partial^\alpha a)(x)p_2(x)^{-1}\| \leqslant C_\alpha \Phi(x)^{-\alpha}.$$

Then $a \in S(g, p_1, p_2, \Omega)$.

Example 1.2. Let g be the same as in Example 1.1 and $\forall u \in H$, $\forall \alpha$, $\forall x \in \Omega$, let

$$(1.9'') \qquad |\langle a^{(\alpha)}(x)u, u \rangle_H| \leqslant C_\alpha \langle p(x)^* p(x)u, u \rangle_H \Phi(x)^{-\alpha}.$$

Then $a \in S(g, p^*, p, \Omega)$.

This form allows the simplest verification of the inclusion $a \in S(g, p^*, p, \Omega)$ in applications to problems concerning spectral asymptotics.

Remark 1.1. If $p'_j \asymp p_j$, then $S(g, p_1, p_2, \Omega) \cong S(g, p'_1, p'_2, \Omega)$ as countably normed spaces with seminorms $n_k(g, p_1, p_2, \Omega, \cdot)$ and $n_k(g, p'_1, p'_2, \Omega, \cdot)$; therefore, it follows from part a) of Proposition 1.1 that the classes of symbols could be defined only by positive symmetric g-continuous functions. However, when verifying the conditions of Definition 1.3 in applications, it is convenient to consider nonsymmetric functions as well. ∎

Let $S(g; p; \Omega) = S(g, I; p; \Omega) \cap S(g; p; I; \Omega)$,

$$n_k(g; p; \Omega; a) = \max\{n_k(g, I, p, \Omega, a), n_k(g, p, I, \Omega, a)\},$$
$$S(g, p_1, p_2) = S(g, p_1, p_2, V), \qquad S(g, p) = S(g, p, V),$$
$$n_k(g, p, a) = n_k(g, p, V, a), \qquad n_k(g, p_1, p_2, a) = n_k(g, p_1, p_2, V, a).$$

Then (1.3)–(1.5) and (1.9) yield the following lemma:

Lemma 1.1.
  a) *If* $a \in S(g, p_1, p_2^{-1}\Omega)$ *and* $b \in S(g, p_2, p_3, \Omega)$, *then* $ab \in S(g, p_1, p_3, \Omega)$.
  b) *If* $a \in S(g, p_1, p_2, \Omega)$, *if there exists a left (right) inverse* $a(x)^{-1}$ *for all* $x \in \Omega$, *and if*
$$\|p_2(x)a(x)^{-1}p_1(x)\| \leqslant C \quad \forall x \in \Omega,$$
then $a^{-1} \in S(g, p_2^{-1}, p_1^{-1}, \Omega)$.
  c) *If* $a \in S(g, p_1, p_2, \Omega)$, *then* $a^* \in S(g, p_2^*, p_1^*, \Omega)$.
  d) *If* p *is a scalar function, then* $S(g, pp_1, p_2, \Omega) = S(g, p_1, pp_2, \Omega)$; *and if* p, $a \in S(g, p, \Omega)$ *are scalar functions, then* $\forall p_1$, $a \in S(g, pp_1, p_1^{-1}\Omega)$. ∎

The following lemma allows us to verify the conditions of Definition 1.2.

Lemma 1.2. *Let for all* $x, y \in V$ *and* $u \in H$,

$$(1.10) \qquad |\langle \langle d(pp^*)(x), y \rangle u, u \rangle_H| \leqslant C \langle (pp^*)(x)u, u \rangle_H g_x(y)^{1/2}$$
$$[|\langle \langle d(p^*p)(x), y \rangle u, u \rangle_H| \leqslant C \langle (p^*p)(x)u, u \rangle_H g_x(y)^{1/2}].$$

Then p is a left (right) g-continuous function.

Proof. Let $\tilde{p} = (pp^*)^{1/2} \left[= (p^*p)^{1/2} \right]$. It follows from (1.10) and Lagrange's theorem that $\forall u \in H$,

(1.11)

$$\sup_y |\langle \tilde{p}(y)u, u\rangle_H| \leqslant \langle \tilde{p}(x)u, u\rangle_H + \sup_z \sup_y |\langle \langle d\tilde{p}(z), x - y\rangle u, u\rangle_H|$$

$$\leqslant \langle \tilde{p}(x)u, u\rangle_H + C \sup_{z \in U} |\langle \tilde{p}(z)u, u\rangle_H| \sup_{y \in U} g_z(x - y)^{1/2},$$

where $U = U(x, \varepsilon) = \{y \mid g_x(x - y) < \varepsilon^2\}$. If $\varepsilon > 0$ is sufficiently small, then (1.11) yields (1.7). ∎

Remark 1.2. Sometimes it is more convenient to verify the condition

(1.10′)
$$\|p(x)^{-1}\langle dp(x), y\rangle\|_H \leqslant Cg_x(y)^{1/2}$$
$$\left[\|\langle dp(x), y\rangle p(x)^{-1}\|_H \leqslant Cg_x(y)^{1/2} \right],$$

which yields (1.10). ∎

Lemma 1.3. *For $x, \tau \in V$, let*

$$|\langle dg_x(t), \tau\rangle| \leqslant Cg_x(t)g_x(\tau)^{1/2}$$

uniformly with respect to $0 \neq t \in V$. Then g varies slowly.

Proof. Let $\varepsilon = (2C)^{-1}$ and let $y \in U = U(x, \varepsilon)$. Then for some z,

$$g_x(x - y)^{1/2} g_y(x - y)^{-1/2} \geqslant 1 - \tfrac{C}{2} g_x(x - y)^{1/2} g_z(x - y)^{-1/2} g_z(x - y)^{1/2} \geqslant \tfrac{3}{4};$$

and

$$\mathcal{J}_v \overset{\text{def.}}{=} \sup_{y \in U} g_y(v)^{1/2} g_x(v)^{-1/2} \leqslant 1 + \tfrac{C}{2} \sup_{z \in U} g_z(v)^{1/2} g_x(v)^{-1/2} g_z(x - y)^{1/2}$$

$$\leqslant 1 + \tfrac{2C}{3} \mathcal{J}_v g_x(x - y)^{1/2} \leqslant 1 + \tfrac{1}{3}\mathcal{J}_v.$$

Therefore, $g_x \leqslant \tfrac{2}{3} g_y$ if $g_x(x - y) \leqslant (2C)^{-2}$. ∎

If g is the same as in Example 1.1, then the conditions of Lemmas 1.2 and 1.3 are conveniently verified in the following form: $\forall |\alpha| = 1$, $\forall j$,

$$\|p^{-1}p^{(\alpha)}\| \leqslant c\Phi^{-\alpha} \quad \left[\|p^{(\alpha)}p^{-1}\| \leqslant c\Phi^{-\alpha} \right], \qquad \Phi_j^{(\alpha)} \leqslant c\Phi_j\Phi^{-\alpha}. \quad ∎$$

Lemma 1.4. *Let $0 < \varepsilon < c$, where c is the constant in (1.6′). Then the points $x_i \in V$, $i = 1, 2, \ldots$ can be chosen such that the balls*

$$U_i = U_{i,r} = \{x \mid g_{x_i}(x - x_i) \leqslant r\}$$

cover V if $r \geqslant \varepsilon$, and there exists $C > 0$ such that the multiplicity of the covering $\{U_i\}$ does not exceed C when $r \leqslant c$.

Proof. Let us construct the maximal sequence of points $x_i \in V$ such that

$$(1.12) \qquad g_{x_i}(x_i - x_j) \geqslant \varepsilon, \quad \text{if } j > i.$$

In order to do this, it is sufficient to choose as many such points as possible on a fixed compact set, and then to construct additional points on a larger compact set, and so on.

If $j > i$ and $g_{x_j}(x_i - x_j) \leqslant c$, then $(1.6')$ yields

$$c \leqslant g_{x_i}(x_i - x_j) \leqslant C g_{x_j}(x_i - x_j).$$

Hence,

$$g_{x_j}(x_i - x_j) \geqslant \min\{c, \varepsilon/C\}, \quad \text{if } j > i.$$

Let us find the number of points x_j for which we have

$$(1.13) \qquad g_{x_j}(x_j - x) < r, \quad \text{if } r \leqslant c.$$

(1.13) yields $g_x(x_j - x) < Cr$. If (1.13) is valid with i instead of j, for $i < j$, then

$$\varepsilon \leqslant g_{x_i}(x_i - x_j) \leqslant C g_x(x_i - x_j);$$

i.e., $g_x(x_i - x_j) \geqslant \varepsilon/C$. Obviously, the number of points x_i, x_j that satisfy this condition and belong to the ball $g_x(y - x) < Cr$ is bounded uniformly with respect to x. Now, the balls $U_{i,\varepsilon}$ cover V, since otherwise we could add some point to our sequence $\{x_i\}$ without violating condition (1.12). ∎

Corollary 1.1. *For any $r \in (\varepsilon, c)$, there exists a partition of unity $\Sigma \varphi_i = \Sigma \varphi_{i,r} = 1$ on V with the following properties:*

$$(1.14) \qquad n_k(g, 1, \varphi_i) \leqslant C_k \quad \forall i, k;$$

$$(1.15) \qquad \varphi_i\big|_{U_{i,\varepsilon}} = 1, \qquad \varphi_i\big|_{V \setminus U_{i,r}} = 0, \qquad 0 \leqslant \varphi_i \leqslant 1.$$

Proof. Let us fix $\psi \in C^\infty(R^1)$ such that $0 \leqslant \psi \leqslant 1$, $\psi\big|_{(-\infty, \varepsilon)} = 1$, $\psi\big|_{(r, \infty)} = 0$; and let

$$\psi_i(x) = \psi(g_{x_i}(x - x_i)), \qquad \Psi = \sum_i \psi_i, \qquad \varphi_i = \psi_i \setminus \Psi.$$

Then the functions ψ_i satisfy condition (1.14), since g is equivalent to g_{x_i} on $U_{i,r}$. Since the multiplicity of the covering $\{U_{i,r}\}$ is finite, estimate (1.14) holds for Ψ as well. Finally, the balls $U_{i,\varepsilon}$ cover V. Therefore, $\Psi^{-1} \leqslant 1$; and (1.14) follows from Lemma 1.1 and from the fact that the multiplicity of the covering $\{U_{i,r}\}$ is finite.

We say that $\Sigma\varphi_i = 1$ is the partition of unity associated with g, r, ε.   ∎

Corollary 1.2. *Let $c > 0$ and $U \subset V$. Then there exists a function $\psi_c \in S(g,1)$ such that*

$$\operatorname{supp}\psi_c \subset U(2c,g), \qquad \psi_c\big|_{U(c,g)} = 1,$$

where

$$U(c,g) = \{x \mid \exists y \in U : g_x(x-y) < c\}.$$

Proof. Let $r > 0$ be sufficiently small and let $\Sigma\varphi_{i,r} = 1$ be the partition of unity associated with g, r, $\frac{r}{2}$. Then the function $\psi_c = \sum_i \varphi_{i,r}$ [the summation is extended to the values of i for which $\operatorname{supp}\varphi_i \cap U(c,g) \neq \emptyset$] satisfies the required conditions.

We say that ψ_c is a cutoff function associated with g, c, U.   ∎

Corollary 1.1 allows us to regularize left and right g-continuous functions and the metric g itself.

Lemma 1.5. *Let p be a right (left) g-continuous function. Then there exists a positive symmetric function $p_1 \in S(g,p_1)$ that is equivalent to p.*

Proof. By part a) of Proposition 1.1, we can assume that p is a positive symmetric g-continuous function. We fix a small $\varepsilon > 0$, construct the partition of unity $\sum \varphi_i = 1$ associated with g, ε, $\frac{\varepsilon}{2}$, and introduce

$$\tilde{p}(x) = \sum_i \varphi_i(x)p(x_i)^2.$$

If ε is sufficiently small, then by virtue of (1.7″) and the properties of a partition of unity, we have

(1.16) $$C^{-1}p(x)^2 \leqslant \tilde{p}(x) \leqslant Cp(x)^2, \qquad \tilde{p} \in S(g,p,p).$$

But $(1.16)_2$ yields (1.11); therefore, if r is sufficiently small and $g_x(x-y) < r^2$, then

(1.17) $$(1 - Cr)\tilde{p}(x) \leqslant \tilde{p}(y) \leqslant (1 + Cr)\tilde{p}(x),$$

where C does not depend on $r > 0$. Let $q = \tilde{p}^{1/2}$. Then for the same x, y, the operator $B = B(x,y) = q(y)q(x)^{-1}$ satisfies the condition

(1.18) $$(1 - Cr)I \leqslant B^*B \leqslant (1 + Cr)I.$$

Let $B = UD$ be a polar expansion of operator B. Then (1.18) is valid when D is substituted for B. Hence, $D = I + K$, where $\|K\| \leqslant Cr$. But then

$$(1.19) \qquad\qquad B = U + K', \quad \text{where } \|K'\| \leqslant C'r.$$

We recall that B is a product of positive-definite self-adjoint operators so the spectrum is real and positive. But U is a unitary operator; hence, (1.19) implies that for $r \to 0$, its spectrum must accumulate near 1. Therefore, $\|U - I\| \to 0$ and $\|B - I\| \to 0$ as $r \to 0$. Thus, we have shown that

$$(1.20) \qquad\qquad g_x(y - x) < r \quad \Rightarrow \quad q(y)q(x)^{-1} = I + K,$$

where $\|K\| \to 0$ as $r \to 0$. Let

$$p_1(x) = \sum_i \varphi_i(x)q(x_i).$$

From (1.20), with the symmetry of p_1, q, we obtain $p_1 \in S(g, q)$. Furthermore,

$$\|p_1(x)^{-1}q(x)\| = \|(\sum \varphi_i(x)q(x_i)q(x)^{-1})^{-1}\|$$

$$= \|(\sum_i \varphi_i(x)(I + K_{i,\varepsilon}))^{-1}\|$$

$$= \|(I + K_\varepsilon)^{-1}\| < C$$

for small ε, since $\|K_\varepsilon\| \to 0$ as $\varepsilon \to 0$. Therefore, $p_1 \asymp q$. But from $(1.16)_1$ it follows that $q \asymp p$; hence, $p \asymp p_1$, and $p_1 \in S(g, p_1)$. ∎

We say that the metrics g, g' are equivalent if there exists $C > 0$ such that $C^{-1}g \leqslant g' \leqslant Cg$.

Corollary 1.3. *Let g be a slowly varying metric. Then there exists a slowly varying metric g' equivalent to g such that $g'(t) \in S(g', g'(t))$ uniformly with respect to $(0 \neq)t \in V$.*

Proof. According to condition (1.6), $g(t)$ is a function that is uniformly g-continuous with respect to $t \neq 0$. Hence, it is sufficient to apply the proof of Lemma 1.5 to the function depending on a parameter. ∎

Lemma 1.6. *Let $a = a^* \in S(g, p^*, p)$ and let $a \geqslant cp^*p$. Then there exists $b \in S(g, I, p)$ such that $a = b^*b$.*

Proof. Let $p_1 \in S(g, p)$ be the function constructed in Lemma 1.5. Then $a_1 = p_1^{*-1}ap_1^{-1} \in S(g, 1)$ and $c_1 I \leqslant a \leqslant C_1 I$. In the complex plane, we choose a closed contour Γ containing the interval $[c_1, C_1] \subset \{\lambda \mid \operatorname{Im}\lambda = 0\}$, but not containing zero, and let

$$b_1(x) = -\frac{1}{2\pi i} \int\limits_\Gamma \lambda^{1/2}(a_1(x) - \lambda)^{-1}d\lambda.$$

Then for all $x \in V$ and $\lambda \in \Gamma$, $\|(a_1(x) - \lambda)^{-1}\| \leqslant c_2$, $|\lambda| \leqslant c_2$ uniformly with respect to $\lambda \in \Gamma$, $(a_1 - \lambda)^{-1} \in S(g, 1)$, and $b_1 \in S(g, 1)$. Therefore, we can write $b = b_1 p_1$. ∎

The function constructed in Lemma 1.6 will be denoted by $a^{1/2}$. Throughout its applications, its non-uniqueness will not be essential. If p is a scalar function, then $a^{1/2}$ is a usual square root.

If $\dim H < \infty$, then we can show (see Feigin [2]) that the usual square root $a^{1/2} \in S(g, p)$; but this refinement of Lemma 1.6 will not be used.

§2 Estimates for Solutions of Schrödinger-Type Equations

In this section, we obtain estimates that will enable us to prove fundamental theorems of PDO calculus.

2.1. Let B be real quadratic form in V' and let $a \in \mathcal{S}(V; B_{12})$. Then $\exp(itB(D))a(x)$ is a solution in $\mathcal{S}(V; B_{12})$ of the equation

$$\partial U/\partial t = iB(D)U, \qquad U = a \quad \text{for } t = 0,$$

which is analogous to the Schrödinger equation. For our goals it is sufficient to obtain estimates for the function U for $t = 1$.

We shall start with the remark that for any integer $k \geqslant 0$, according to Taylor's formula,

$$\left| e^{iw} - \sum_{j<k}(iw)^j/j! \right| \leqslant |w|^k/k!.$$

Applying the Fourier transform, we have $\forall u \in H_2$,

$$(2.1) \qquad \left\| (\exp(iB(D))a\sum_{j<k}\frac{1}{j!}(iB(D))^k a)u \right\| \leqslant \|B(D)^k au\|/k!,$$

where $\|\cdot\|$ is a norm in $L_2(V; H_1)$.

Suppose a Euclidean norm is introduced on V by means of a positive-definite quadratic form g. Let $K = \{x \mid g(x) < 1\}$ be a unit ball. We apply (2.1) to the function of the form $\langle t, D\rangle^j a$, where $a \in C_0^\infty(K, B_{12})$, $t \in V$, and $g(t) < 1$. Since a differential operator commutes with a convolution operator in (2.1), the Sobolev inequality yields
(2.2)

$$\left\| \exp(iB(D))a(x) - \sum_{j<k}\frac{1}{j!}(iB(D))^j a(x) \right\|_{12} \leqslant C\sup_{j\leqslant s} n_j(g, I, I, B(D)^k a),$$

where $\|\cdot\|$ is a norm in B_{12}, and s is the smallest integer greater than $\dim V/2$. In particular, for $k = 0$,

$$(2.2')\qquad \|\exp(iB(D))a(x)\|_{12} \leqslant C \sup_{j \leqslant s} n_j(g, I, I, a).$$

2.2. Outside a unit ball K, the sum in (2.2) vanishes; and we can obtain a better estimate if we show that $\exp(iB(D))$ is an "almost local" operator. To prove this, we fix $x \in V$, $\xi \in V'$ and introduce a multiplicative operator L by the function $L(y) = \langle y - x, \xi \rangle$. Then $L(x) = 0$, $L \exp(iB(D))a(x) = 0$, and

$$\exp(iB(D))La(x) = [\exp(iB(D)), L]a(x)$$
$$= \exp(iB(D))\langle B'(D), \xi \rangle a(x),$$

where $\langle B'(D), \xi \rangle = 2B(D, \xi) = 2\langle B\xi, D \rangle$, $B(\cdot, \cdot)$ is a symmetric bilinear form in V' polar to B, and $B : V' \to V$ is a corresponding linear operator.

If $L \neq 0$ on K, then by substituting $L^{-1}a$ for a, we obtain

$$(2.3)\qquad \exp(iB(D))a(x) = 2\exp(iB(D))\langle B\xi, D \rangle(L^{-1}a)(x).$$

Lemma 2.1. *If L is a linear function and $L \neq 0$ on $\{x \mid g(x) < R^2\}$, where $R > 1$, then*

$$(2.4)\qquad n_k(g; 1; L(0)/L; K) < \infty, \quad k = 0, 1, \ldots.$$

Proof. We can assume that $L(0) = 1$. Then $|\langle x, \xi \rangle| \leqslant 1/R$ for all $x \in K$, such that $1 - 1/R \leqslant L(x) \leqslant 1 + 1/R$ and $|L|_{g,1}^{1,1}(x) \leqslant 1/R$ on K. Since L is a linear function, we have $|L|_{g,k}^{1,1}(x) = 0$, $\forall k \geqslant 2$, $\forall x \in V$; therefore, (2.4) follows from (1.5). ∎

By estimating the right-hand side of (2.3)—using (2.2), Lemma 2.1, and (1.4)—we obtain

$$(2.2'')\qquad \|\exp(iB(D))a(x)\|_{12} \leqslant Cg(B\xi)^{1/2}|L(0)|^{-1} \sup_{j \leqslant s+1} n_j(g, I, I, a).$$

Iteration yields

$$(2.2''')\qquad \|\exp(iB(D))a(x)\|_{12} \leqslant Cg(B\xi)^{k/2}|L(0)|^{-k} \sup_{j \leqslant s+k} n_j(g; I; I; a).$$

In order to find out how small the factor $g(B\xi)^{1/2}/|L(0)|$ can be made, we introduce a form that is dual with respect to $\xi \to g(B\xi)$:

$$(2.5)\qquad g^B(x) = \sup_{\xi \in V'} \langle x, \xi \rangle^2 / g(B\xi).$$

(Of course, $g^B = +\infty$ everywhere except at an orthogonal complement to the radical of B.) And suppose that the distance from a point x to RK in the metric g^B is equal to $d > 0$:

$$\inf_{g(y) < R^2} g^B(x - y) = d^2 > 0.$$

By the Hahn–Banach theorem, we can find $\xi \in V'$ such that $g(B\xi) = 1$ and $\langle y - x, \xi \rangle \geqslant d$, $\forall y \in RK$. Let $L(y) = \langle y - x, \xi \rangle$ and note that $L(x) = 0$, $L \neq 0$ in RK and that $L(0) \geqslant d$. Therefore, (2.2''') yields
(2.6)
$$\| \exp(iB(D))a(x) \|_{12} \leqslant C_{k,R}(1 + \inf_{g(y) < R^2} g^B(x - y))^{-k/2} \sup_{j \leqslant s+k} n_j(g, I, I, a).$$

For later reference, we shall summarize our results.

Lemma 2.2. *Let g be a positive-definite quadratic form on V and let B be a real quadratic form on V'; let the form g^B be defined by equality (2.5) and let K be a unit ball with respect to g; let $a \in C_0^\infty(K, B_{12})$ and let $2s > \dim V$. Then for all $k \geqslant 0$ and $R > 1$, estimates (2.2) and (2.6) are valid.* ∎

2.3. In Lemma 2.2 the derivatives of the function a are not mentioned, since estimates for them can be obtained from (2.2) and (2.6) by replacing a with its derivatives. An important detail in estimate (2.6) is the smallness of its right-hand side for large k at large distances from K in the metric g^B. This property allows us to obtain an existence theorem for a solution of a Schrödinger-type equation in the corresponding classes of symbols. The sufficient conditions are given next.

Definition 2.1. A Riemann metric g on V is called B-temperate with respect to $x \in V$ if g varies slowly and if there exist C, N such that for arbitrary $y, t \in V$,

(2.7) $$g_y(t) \leqslant C g_x(t)(1 + g_y^B(x - y))^N.$$

Now (2.7) and (2.5) yield

(2.7') $$g_x^B(t) \leqslant C g_y^B(t)(1 + g_y^B(x - y))^N$$

[and vice versa: (2.7') implies (2.7) if the form B is nondegenerate]. In particular, for $t = x - y$,

(2.7'') $$1 + g_x^B(x - y) \leqslant C(1 + g_y^B(x - y))^{N+1}.$$

Definition 2.2. A function $p : V \to \operatorname{Aut} H$ is said to be left (right) B,g-temperate with respect to $x \in V$ if p is a left (right) g-continuous function and if there exist C, N such that for all $y \in V$,

(2.8)
$$\| p(y)^{-1} p(x) \| \leqslant C(1 + g_y^B(x - y))^N$$
$$[\| p(x) p(y)^{-1} \| \leqslant C(1 + g_y^B(x - y))^N].$$

Estimate $(2.7'')$ allows us to symmetrize (2.8):

$(2.8')$
$$\|p(y)^{-1}p(x)\| + \|p(x)^{-1}p(y)\| \leqslant C_1(1 + g_y^B(x-y))^{N_1}$$
$$[\|p(y)p(x)^{-1}\| + \|p(x)p(y)^{-1}\| \leqslant C_1(1 + g_y^B(x-y))^{N_1}],$$

and from $(2.8')$ we immediately derive the following lemma:

Lemma 2.3. *If p is a left (right) B,g-temperate function with respect to x, then p^* and p^{-1} are right (left) B,g-temperate with respect to x.* ∎

Now we get rid of the restrictions on $\operatorname{supp} a$ in (2.6) with the help of the partition of unity $\sum \varphi_j = 1$, which had been constructed in Corollary 1.1. We fix $r \in (0,c)$, $r_0 \in (r,c)$ and let

$$U_j' = \{x \mid g_{x_j}(x-x_j) \leqslant c\}, \qquad U_j = \{x \mid g_{x_j}(x-x_j) < r_0\}.$$

Lemma 2.4. *Let $g_x \leqslant g_x^B$ and let g be B-moderate with respect to x. Then there exist constants C, N depending only on the constants in (2.7) and (1.6) such that*

(2.9)
$$\sum_j (1 + d_j(x))^{-N} \leqslant C, \quad \text{where } d_j(x) = \inf_{y \in U_j} g_y^B(x-y).$$

Proof. We introduce on V the Euclidean norm $|\cdot| = g_x(\cdot)^{1/2}$ and, having let $M_k = \{j \mid d_j(x) \leqslant k\}$ for $k \geqslant 1$, choose a point $y_j \in U_j$ for $j \in M_k$ such that $g_{y_j}^B(x-y_j) \leqslant k$. Then (2.7) yields

(2.10)
$$g_{y_j}(t) \leqslant C|t|^2 k^N \quad \forall t \in V.$$

Since g varies slowly, (2.10) is valid with $g_{x_j}(t)$ and with a new constant c; therefore, there exists $c_1 > 0$ such that if $|t| < c_1 k^{-N/2}$, then $g_{x_j}(t) \leqslant (\sqrt{c}-\sqrt{r_0})^2$. By the triangle inequality, $g_{x_j}(y_j + t - x_j) < c$; hence, $y_j + t \in U_j'$.

By the same token, we showed that a Euclidean ball V_j of radius $c_1 k^{-N/2}$ with center at y_j is contained in U_j'. Estimate $(2.7'')$ yields

$$|x - y_j|^2 = g_x(x-y_j) \leqslant g_x^B(x-y_j) \leqslant Ck^{N+1};$$

therefore, for $j \in M_k$, the balls V_j are contained in a Euclidean ball of radius $c_1 k^{(N+1)/2}$ with center at x. Since the multiplicity of the covering $\{U_j'\}$ is finite, then $\sup_i \operatorname{card}\{j \mid V_i \cap V_j = \emptyset\} < \infty$, and

$$c' \operatorname{card} M_k k^{-nN/2} \leqslant \sum_{j \in M_k} \operatorname{mes} V_j \leqslant C \operatorname{mes} \bigcup_{j \in M_k} V_j \leqslant C' k^{n(N+1)/2}.$$

Therefore, card $M_k \leqslant C_1 k^{N_1}$; and by successively summing the summands in (2.9) with respect to sets $M_1, M_2 \backslash M_1, \ldots, M_{2^i} \backslash M_{2^{i-1}}$, we obtain estimate (2.9). ∎

Let the metric g be B-temperate, let $p_i : V \to \operatorname{Aut} H_i$ be B, g-temperate with respect to $x \in V$, and let $a \in C_0^\infty(V; B_{12})$, $a_j = \varphi_j a$. We apply estimate (2.6) to $p_1(x_j)^{-1} a_j(x) p_2(x_j)^{-1}$, with $g = r^{-1} g_{x_j}$, $R = r_0/r$; then (1.14) and (1.4) yield

$$(2.11) \quad \|p_1(x_j)^{-1} \exp(iB(D)) a_j(x) p_2(x_j)^{-1}\|_{12}$$
$$\leqslant C_k (1 + d_j(x))^{-k/2} \sup_{m \leqslant s+k} n_m(g_{x_j}; p_1(x_j); p_2(x_j); a).$$

Since $\operatorname{supp} a_j \subset U_j'$, g varies slowly and since p_1 (p_2) is a left (right) g-continuous function, then by decreasing the constant c, if necessary, in the definition of the balls U_j', we can replace g_{x_j} with g and $p_i(x_j)$ with $p_i(x)$ on the right-hand side of estimate (2.11). If we replace $p_i(x_j)$ with $p_i(x)$ on the left-hand side, then by virtue of (2.8), we will have to add the factor $(1 + d_j(x))^N$ to the right-hand side:

$$(2.12) \quad \|p_1(x)^{-1} (\exp(iB(D)) a_j(x)) p_2(x)^{-1}\|_{12}$$
$$\leqslant C_k (1 + d_j(x))^{N - k/2} \sup_{m \leqslant s+k} n_m(g; p_1; p_2; a),$$

where C_k, N depend only on the constants characterizing g, p_1, p_2. If k is sufficiently large, then (2.9) and (2.12) yield the estimate

$$(2.13) \quad \sum_j \|p_1(x)^{-1} (\exp(iB(D)) a_j(x)) p_2(x)^{-1}\|_{12} \leqslant C \sup_{m \leqslant s+k} n_m(g; p_1; p_2; a).$$

Definition 2.3. A linear mapping from $S(g; p_1; p_2)$ to a linear topological space is called weakly continuous if its restriction to each bounded space $M \subset S(g; p_1; p_2)$ is continuous in C^∞ topology. A bilinear mapping from $S(g; p_1; p_2^{-1}) \times S(g; p_2; p_3)$ into a linear topological space is weakly continuous if it is continuous in C^∞ topology on products of bounded sets. ∎

It is sufficient to define weakly continuous, as opposed to continuous, mappings on functions of the class $C_0^\infty(V; B_{12})$, since partial sums of the series $a = \sum a_j$ are bounded in $S(g; p_1; p_2)$, if $a \in S(g; p_1; p_2)$, and converge to a in C^∞ topology, since they are eventually equal to a on any compact set.

The proof of estimate (2.13) gives a convergent majorizing sequence for the series on the left-hand side of (2.13); and, moreover, this sequence is suitable for estimation of every a from a set bounded in $S(g; p_1; p_2)$. Therefore, the sum

$$\exp(iB(D)) a(x) = \sum_j \exp(iB(D)) a_j(x)$$

defines a weakly continuous mapping $S(g, p_1, p_2) \to B_{12}$. By the same token, we have proved the following theorem:

Theorem 2.1. *Let a metric g be B-temperate, let the functions p_1 and p_2 be, respectively, left and right B,g-temperate functions with respect to x, and let $g_x^B \geqslant g_x$. Then the mapping*

$$C_0^\infty(V; B_{12}) \ni a \to \exp(iB(D))a(x) \in B_{12}$$

can be uniquely continued to a weakly continuous linear mapping $S(g; p_1; p_2) \to B_{12}$, and we have

$$(2.14) \qquad \|p_1(x)^{-1} \exp(iB(D))a(x)p_2(x)^{-1}\|_{12} \leqslant C \sup_{j \leqslant N} n_j(g; p_1; p_2; a),$$

where C, N depend not on a, but only on $\dim V$ and on the constants that characterize g, p_1, p_2.

2.4. Under the conditions of Theorem 2.1, we can also estimate the derivatives of the function $\exp(iB(D))a(x)$. First, we assume that $a \in C_0^\infty(V; B_{12})$ so the indicated derivatives exist. Then

$$(2.15) \quad \|p_1(x)^{-1}[\langle D, t_1 \rangle \cdots \langle D, t_k \rangle \exp(iB(D))a(x)]p_2(x)^{-1}\|_{12}$$

$$\leqslant C \prod_{1 \leqslant j \leqslant k} g_x(t_j)^{1/2} \sup_{r \leqslant M} n_r(g; p_1; p_2; a),$$

where C, M depend on k, $\dim V$, and on the constants characterizing g, p_1, p_2. Indeed, if

$$b = \langle D, t_1 \rangle \cdots \langle D, t_k \rangle a, \qquad p_1'(y) = \prod_{1 \leqslant j \leqslant k} g_y(t_j)^{1/2} p_1(y),$$

then p_1' is a function that is B,g-temperate with respect to x, and the constants characterizing it depend only on k, p_1, g. Furthermore,

$$n_m(g; p_1'; p_2; b) \leqslant n_{k+m}(g; p_1; p_2; a);$$

therefore, by applying formula (2.14) to $b \in S(g, p_1', p_2)$ and taking into account that the convolution operator $\exp(iB(D))$ commutes with the differential operator, we obtain (2.15).

Estimate (2.15) yields the following theorem:

Theorem 2.1'. *Let $V_0 \subset V$ be a subspace and let the conditions of Theorem 2.1 be satisfied uniformly with respect to $x \in V_0$. Then the mapping*

$$S(g; p_1; p_2) \ni a \to \exp(iB(D))a\big|_{V_0} \in S(g\big|_{V_0}; p_1\big|_{V_0}; p_2\big|_{V_0})$$

is weakly continuous.

Proof. In (2.15) it is sufficient to consider $t_1, \ldots, t_k \in V_0$. ∎

The following results can be sharpened if

$$(2.16) \qquad h(x)^2 = \sup_{0 \neq t \in V} g_x(t)/g_x^B(t)$$

is not only less than or equal to 1, but is also small.

Lemma 2.5.   *The function h is B,g-temperate with respect to $x \in V_0$.*

Proof.   Obviously, h is g-continuous; and if we write (2.7) in the form $g_y \leqslant M g_x$, then $g_y^B \geqslant g_x^B / M$ and $g_y / g_y^B \leqslant M^2 g_x / g_x^B$. Therefore, $h(y) \leqslant M h(x)$; and as a result of (2.7″), h satisfies condition (2.8).   ∎

Theorem 2.2.   *Let the conditions of Theorem 2.1′ be satisfied and let a function h be defined by equality (2.16). Then the mapping*

$$S(g; p_1; p_2) \ni a \longrightarrow v_N \in S\big(g\big|_{V_0}, (h^N p_1)\big|_{V_0}, p_2\big|_{V_0}\big),$$

where

$$v_N = \exp(iB(D))a - \sum_{j < N} \frac{1}{j!} (iB(D))^j a,$$

is weakly continuous; and for any k,

$$(2.17) \qquad n_k\big(g\big|_{V_0}, (h^N p_1)\big|_{V_0}, p_2\big|_{V_0}, v_N\big) \leqslant C \max_{N \leqslant j \leqslant M} n_j(g, p_1, p_2, a),$$

where the constants C, M depend only on k, N, $\dim V$, and on the constants characterizing g, p_1, p_2.

Proof.   If coordinates are chosen so that g_x is a Euclidean metric and $B(\xi) = \sum b_j \xi_j^2$, then $h(x) = \max |b_j|$. Hence, $B(D)^k a \in S(g, h^k p_1, p_2)$; and if we start with formula (2.2) instead of (2.2′) and do all the constructions in this section with v_N instead of a, then we arrive at the statement of the theorem.   ∎

§3 The Fundamental Theorems of Calculus

3.1. Let V be an n-dimensional real space, let V' be a space dual to V, and let $W = V \oplus V'$. The function $a : W \to B_{12}$ is usually associated with PDO with a left symbol:

$$(3.1) \qquad (a_\ell u)(x) = \int d'\xi \int dy \, \exp(i\langle x - y, \xi \rangle) a(x, \xi) u(y),$$

if the right-hand side is defined—for example, if $a \in S(W; B_{12})$. In the last case, for all $u \in S(V; H_2)$ and $v \in S(V; H_1)$,

$$\langle a_\ell u, v \rangle_{L_2(V; H_1)}$$
$$(3.2) \qquad = \int d'\xi \int dy \, \exp(i\langle x - y, \xi \rangle) \langle a(x, \xi) u(y), v(x) \rangle_{H_1}$$
$$= \int d'\xi \int dx \int dt \, \exp(i\langle t, \xi \rangle) \langle a(x, \xi) u(x - t), v(x) \rangle_{H_1};$$

and a_ℓ is an integral operator with the kernel $A_\ell(x - y)$,

$$(3.3) \qquad A_\ell(x, x - t) = \int d'\xi \exp(i\langle t, \xi \rangle) a(x, \xi).$$

According to the inverse formula for a Fourier transform,

$$(3.4) \qquad a(x, \xi) = \int dt \exp(-i\langle t, \xi \rangle) A_\ell(x, x - t).$$

If a is a polynomial with respect to ξ, then by applying the well-known property of a Fourier transform, we can obtain the operator a_ℓ by replacing ξ with $D = -i\partial/\partial x$ in the expression for $a(x, \xi)$ and by writing down the coefficients on the left side. (This explains the term "left symbol.") If we write down the coefficients on the left, we obtain PDO with the right symbol:

$$(3.5) \qquad (a_r u)(x) = \int d'\xi \int dy \exp(i\langle x - y, \xi \rangle) a(y, \xi) u(y).$$

Similarly to (3.2)–(3.4), we obtain that a_r is an integral operator with the kernel $A_r(x, y)$,

$$(3.6) \qquad A_r(x + t, x) = \int d'\xi \exp(i\langle t, \xi \rangle) a(x, \xi),$$

and

$$(3.7) \qquad a(x, \xi) = \int dt \exp(-i\langle t, \xi \rangle) A_r(x + t, x).$$

The following definition of PDO with the Weyl symbol is a compromise between the definitions in (3.1) and (3.5):

$$(3.8) \qquad (a_w u)(x) = \int d'\xi \int dy \exp(i\langle x - y, \xi \rangle) a(\tfrac{x+y}{2}, \xi) u(y).$$

The kernel of the operator a_w is A_w,

$$(3.9) \qquad A_w(x + \tfrac{1}{2}t, x - \tfrac{1}{2}t) = \int d'\xi \exp(i\langle t, \xi \rangle) a(x, \xi),$$

and

$$(3.10) \qquad a(x, \xi) = \int dt \exp(-i\langle t, \xi \rangle) A_w(x + \tfrac{1}{2}t, x - \tfrac{1}{2}t).$$

The mappings σ^ℓ, σ^r, σ^w denote the inverses of the mappings $a \to a_\ell$, $a \to a_r$, $a \to a_w$, respectively. If a is a linear function, then $a_\ell = a_r = a_w$; the connection between these operators will be clarified in Section 3.5.

3.2. If $a_1 \in \mathcal{S}(V, B_{12})$ and $a_2 \in \mathcal{S}(V, B_{23})$, then it is not difficult to show that the operators $a_{1,w} : \mathcal{S}(V, H_2) \to \mathcal{S}(V; H_1)$ and $a_{2,w} : \mathcal{S}(V, H_3) \to \mathcal{S}(V, H_2)$ are continuous; hence, we can raise the question concerning the symbol of their product $a_{1,w} a_{2,w}$. The kernel of the operator $a_{1,w} a_{2,w}$ is

$$A(x, y) = \iiint a_1(\tfrac{1}{2}(x + z), \eta) a_2(\tfrac{1}{2}(z + y), \tau)$$
$$\times \exp(i[\langle x - z, \eta \rangle + \langle z - y, \tau \rangle]) dz \, d'\eta \, d'\tau;$$

and applying (3.10) and (3.9) yields that the symbol $a = \sigma^w(a_{1,w} a_{2,w})$ is equal to

$$(3.11) \qquad a(x, \xi) = \iiiint a_1(\tfrac{1}{2}x + \tfrac{1}{2}z + \tfrac{1}{4}t, \eta) a_2(\tfrac{1}{2}x + \tfrac{1}{2}z - \tfrac{1}{4}t, \, t)$$
$$\times \exp(iE) d'\eta \, d'\tau \, dt \, dz,$$

where

$$E = \langle x + \tfrac{1}{2}t - z, \eta \rangle + \langle z - x + \tfrac{1}{2}t, \tau \rangle - \langle t, \xi \rangle$$
$$= \langle x + \tfrac{1}{2}t - z, \eta - \xi \rangle + \langle z - x + \tfrac{1}{2}t, \tau - \xi \rangle.$$

Let us change the variables

$$\eta - \xi = \eta_1, \quad \tau - \xi = \tau_1, \quad \tfrac{1}{2}(z - x + \tfrac{1}{2}t) = z_1, \quad \tfrac{1}{2}(z - x - \tfrac{1}{2}t) = t_1.$$

The Jacobian of this change is equal to 2^{2n}; and by omitting index 1 in each equation, we obtain

$$(3.12) \qquad a(x, \xi) = \pi^{-2n} \iiiint a_1(x + z, \xi + \eta) a_2(x + t, \xi + \tau)$$
$$\times \exp(i 2\sigma(t_2 \tau; z, \eta)) dz \, d\eta \, dt \, d\tau,$$

where $\sigma(t, \tau; z, \eta) = \langle t, z \rangle - \langle x, \eta \rangle$ is a simplectic form on W, which we consider here as a quadratic form on $W \oplus W$. All the integrals in (3.11) and (3.12) converge absolutely, since the integrands belong to $\mathcal{S}(\cdot)$.

In a two-dimensional space we have for $f \in \mathcal{S}(R^2, \cdot)$,

$$\iint f(x, y) e^{i 2xy} dx \, dy = \frac{1}{4\pi} \iint \hat{f}(\xi, \eta) e^{-i\xi\eta/2} d\xi \, d\eta.$$

This follows from the inverse formula for a Fourier transform applied to $f(x, y) = g(x) h(y)$; and since

$$\iint \hat{f}(\xi, \eta) e^{-i\xi\eta/2} d\xi \, d\eta = \exp(-\tfrac{1}{2} i D_x D_y) f(x, y) \Big|_{\substack{x=0 \\ y=0}},$$

then (3.12) yields

$$(3.13) \qquad a(x,\xi) = \exp(\tfrac{1}{2}i\sigma(D_x, D_\xi; D_y, D_\eta))a_1(x,\xi)a_2(y,\eta)\Big|_{(y,\eta)=(x,\xi)}.$$

Here we identified W with its dual by means of the simplectic form σ.

3.3. Formula (3.13) allows us to apply results from §2 in order to estimate the function a. Since the right-hand side of (3.13) contains the product $a_1(x,\xi)a_2(y,\eta)$, it is natural to use a quadratic form on $W \oplus W$ of the form $G(w, w') = g_,(w) + g_2(w')$, where g_1, g_2 are forms on W. A bilinear form polar to $B = 2\sigma$ is defined by the expression

$$\sigma(w_1; w_2') - \sigma(w_2; w_1'), \quad (w_1, w_2), (w_1', w_2') \in W \oplus W;$$

and since we have identified w with w' by means of the simplectic form σ, the corresponding mapping B, used in (2.5), here is $\langle w, w' \rangle \to \langle w', -w \rangle$. Therefore,

$$(3.14) \qquad \begin{aligned} G^B(w, w') &= \sup_{W \oplus W} |\sigma(w, w_1) + \sigma(w', w_2)|^2 (g_1(w_2) + g_2(w_1))^{-1} \\ &= g_1^\sigma(w') + g_2^\sigma(w), \end{aligned}$$

where

$$(3.15) \qquad g_j^\sigma(w) = \sup |\sigma(w, w')|^2 / g_j(w')$$

are forms that are dual to g_j with respect to σ.

We note that if $g_1 = g_2$, then $G^B \geqslant G$ is equivalent to $g_1^\sigma \geqslant g_1$, and $\sup G/G^B = \sup g_1/g_1^\sigma$.

If g is a slowly varying metric on W and

$$G_{w,w'}(t, t') = g_w(t) + g_{w'}(t'),$$

then G is a slowly varying metric on $W \oplus W$. Let us find out when G is uniformly B-temperate with respect to each point on a diagonal, i.e., when

$$(3.16) \quad g_w^\sigma(t_1) + g_w^\sigma(t_2)$$
$$\leqslant C(g_{w_1}^\sigma(t_1) + g_{w_2}^\sigma(t_2))(1 + g_{w_1}^\sigma(w_2 - w) + g_{w_2}^\sigma(w_1 - w))^N$$

for all $w, w_1, w_2, t_1, t_2 \in W$. If $w_2 = w_1$, then (3.16) is equivalent to

$$(3.16') \qquad g_w^\sigma(w') \leqslant C g_{w_1}^\sigma(w')(1 + g_{w_1}^\sigma(w_1 - w))^N,$$

for all $w_1, w \in W$. Let us show that, vice versa, (3.16′) implies (3.16). By applying (3.16′), with $w' = w_2 - w$ and $w = w_1 + w_2 - w$, we obtain

$$g_{w_1+w_2-w}^\sigma(w_2 - w) \leqslant C g_{w_1}^\sigma(w_2 - w)(1 + g_{w_1}^\sigma(w_2 - w))^N \leqslant CM^{N+1},$$
$$M = 1 + g_{w_1}^\sigma(w_2 - w) + g_{w_2}^\sigma(w_1 - w).$$

The same estimate is valid for $g^\sigma_{w_1+w_2-w}(w_1-w)$. Therefore,

$$g^\sigma_{w_1+w_2-w}(w_1+w_2-2w) \leqslant 2CM^{N+1};$$

and by applying (3.16′) again, with $w_1 = w_1 + w_2 - w$ and $w' = w_j - w$, we obtain

$$g^\sigma_w(w_j - w) \leqslant Cg^\sigma_{w_1+w_2-w}(w_j-w)(1+g^\sigma_{w_1+w_2-w}(w_1+w_2-2w))^N$$
$$\leqslant C_1 M^{N_1}.$$

By applying (3.16′) once again, we have

(3.17) $$1 + g^\sigma_w(w - w_j) \leqslant C(1+g^\sigma_{w_j}(w-w_j))^{N+1}.$$

Therefore,

$$g^\sigma_w(t) \leqslant Cg^\sigma_{w_j}(w')(1+g^\sigma_{w_j}(w-w_j))^N$$
$$\leqslant C_2 g^\sigma_{w_j}(t)(1+g^\sigma_w(w-w_j))^{N_2}$$
$$\leqslant C_3 g^\sigma_{w_j}(t)M^{N_3},$$

which yields (3.16).

We remark that (3.16′) and (3.17) imply

(3.18) $$C_2^{-1} g^\sigma_{w'}(t)(1+g^\sigma_{w'}(w'-w))^{-N_2} \leqslant g^\sigma_w(t)$$
$$\leqslant C_2 g^\sigma_{w'}(t)(1+g^\sigma_{w'}(w'-w))^{N_2}.$$

The following definition is similar to Definitions 2.1 and 2.2:

Definition 3.1. *A metric g on W is called σ-temperate if g varies slowly, if $g^\sigma \geqslant g$, and if (3.16′) holds. A function $p : W \rightarrow \mathrm{Aut}\, H$ is left (right) σ,g-temperate if p is a left (right) g-continuous function, respectively, and if*

(3.19) $$\|p(w')^{-1}p(w)\| \leqslant C(1+g^\sigma_{w'}(w-w'))^N$$
$$[\|p(w)p(w')^{-1}\| \leqslant C(1+g^\sigma_{w'}(w-w'))^N].$$

We write $p \in O_\ell(H; g)$ $[p \in O_r(H; g)]$ and assume that $O(H; g) = O_\ell(H; g) \cap O_r(H, g)$ [where the metric g is assumed to be σ-temperate].

We note that condition (3.19) can be symmetrized [see (2.8′)] and that the functions from $O_\ell(H, g) \cup O_r(H, g)$ in the definition of classes of symbols can be replaced by functions from $O(H, g)$. [See part a) of Proposition 1.1 and Remark 1.1 following Definition 1.3.]

3.4. From here on, writing $a \in S(g; p_1; p_2)$ implies that g is a σ-temperate metric and that p_1 and p_2 are, respectively, left and right σ,g-temperate functions.

Lemma 3.1. $\mathcal{S}(W; B_{12}) \subset S(g; p_1; p_2)$.

Proof. (3.18), (3.19), and (2.8′) imply the estimates

$$
(3.19') \qquad
\begin{aligned}
C^{-1} g_0(\cdot)(1 + g_0(w))^{-N} &\leqslant g_w(\cdot) \leqslant C g_0(\cdot)(1 + g_0^\sigma(w))^N, \\
\|p_j(w)\| + \|p_j(w)^{-1}\| &\leqslant C(1 + g_0^\sigma(w))^N.
\end{aligned}
$$

Recalling Definitions 1.1 and 1.9 of the seminorms in $\mathcal{S}(W; B_{12})$, $S(g; p_1; p_2)$, we obtain the statement of the lemma. ∎

Now we can prove the fundamental theorem of PDO calculus.

Theorem 3.1. *The composition formula* (3.13) *defines a weakly continuous mapping* $(a_1, a_2) \rightarrow \sigma^w(a_{1,w} a_{2,w})$ *from* $S(g; p_1; p_2^{-1}) \times S(g; p_2; p_3)$ *to* $S(g; p_1; p_3)$.

If

$$
(3.20) \qquad h(w)^2 = \sup_W g_w(t)/g_w^\sigma(t),
$$

then for an arbitrary N,

$$
(3.21) \qquad
\begin{aligned}
r_N(x, \xi) &\overset{\text{def}}{=} \sigma^w(a_{1,w} a_{2,w})(x, \xi) \\
&\quad - \sum_{j < N} \frac{1}{j!} \left(\tfrac{1}{2} i\sigma(D_x, D_\xi; D_y, D_\eta) \right)^j \\
&\quad \times a_1(x, \xi) a_2(y, \eta) \Big|_{(y, \eta) = (x, \xi)} \in S(g; h^N p_1; p_3),
\end{aligned}
$$

where for any k,

$$
(3.22) \qquad n_k(g; h^N p_1; p_3) \leqslant C \sup_{N \leqslant j + i \leqslant M} n_j(g; p_1; p_2^{-1}; a_1) n_i(g; p_2; p_3; a_2),
$$

where the constants C, M depend only on k, N, $\dim V$, and on the constants characterizing g, p_1, p_2, p_3.

Remark 3.1. When applying Theorem 3.1, it is useful to keep in mind that a positive-definite quadratic form g in a simplectic vector space can be written in the form

$$
g(x, \xi) = \sum(\lambda_i x_i^2 + \mu_i \xi_i^2)
$$

if a corresponding simplectic basis—i.e., a basis for which $\sigma = \sum d\xi_i \wedge dx_i$—is properly chosen. (For this purpose, it is sufficient to diagonalize the hermitian forms σ/i and g simultaneously.)

Since

$$
g^\sigma(x, \xi) = \sum(x_i^2/\mu_i + \xi_i^2/\lambda_i),
$$

we obtain

$$h(x,\xi) = \max(\lambda_i \mu_i)^{1/2}$$

(where only the products $\lambda_i \mu_i$ are simplectic invariants!).   ∎

Proof. It is obvious that the functions $p_1'(w,w')=p_1(w)$, $p_3'(w,w')=p_3(w')$ are uniformly B,G-temperate with respect to the points of the diagonal $W \oplus W$; therefore, in the case $p_2 = I$, Theorem 3.1 follows immediately from Theorems 2.1′ and 2.2. If $p_2 \neq I$, we notice that for arbitrary w, w_j, w_j',

$$\|p_2(w_j)p_2(w_j')^{-1}\| \leqslant \|p_2(w_j)p_2(w)^{-1}\| \, \|p_2(w)p_2(w_j')^{-1}\|$$

$$\leqslant C(1 + g_{w_j}^\sigma(w - w_j) + g_{w_j'}^\sigma(w - w_j'))^N$$

$$\leqslant C_1(1 + G_{\overline{w}_j}^B(\overline{w}_j - \overline{w}))^N,$$

where $\overline{w}_j = (w_j, w_j')$, $\overline{w} = (w,w) \in W \oplus W$. Hence, in (2.10), if we replace the function $a_j(x)$ [here, equal to $a_{1,j}(w)a_{2,j}(w')$, $(w,w') \in W \oplus W$] and take into account that the estimates in §2 are needed for the points of the diagonal $W \oplus W$, then the same calculations as in §2 give us the same estimates as in (2.11) and (2.12)—as well as the later ones with $p_1 = p_1'$, $p_2 = p_3'$, and $a(w,w') = a_1(w)a_2(w')$ and with the subspace V_0 as the diagonal in $W \oplus W$.   ∎

3.5. Let us find the relation between the operators a_ℓ, a_r, a_w. If $a \in \mathcal{S}(V; B_{12})$, then $a_\ell = b_w$, where

$$
\begin{aligned}
b(x,\xi) &= \int A_\ell(x + \tfrac{1}{2}t, x - \tfrac{1}{2}t)\exp(-i\langle t,\xi\rangle)dt \\
&= \iint \exp(i\langle t, \eta - \xi\rangle)a(x + \tfrac{1}{2}t, \eta)d'\eta\, dt \\
&= \pi^{-n} \iint a(x + t, \xi + \eta)\exp(2i\langle t, \eta\rangle)d\eta\, dt \\
&= \exp(-i\langle D_x, D_\xi\rangle/2)a(x,\xi),
\end{aligned}
$$

(3.23)

which follows from (3.9) and (3.3). Similarly, applying (3.3), (3.4), (3.6), (3.7), (3.9), and (3.10) yields

$$a_\ell = b_w = c_r$$

if

$$
\begin{aligned}
&(3.24) \\
&b(x,\xi) = \exp(-i\langle D_x, D_\xi\rangle/2)a(x,\xi) = \exp(+i\langle D_x, D_\xi\rangle/2)c(x,\xi), \\
&a(x,\xi) = \exp(+i\langle D_x, D_\xi\rangle/2)b(x,\xi) = \exp(+i\langle D_x, D_\xi\rangle)c(x,\xi), \\
&c(x,\xi) = \exp(-i\langle D_x, D_\xi\rangle)a(x,\xi) = \exp(-i\langle D_x, D_\xi\rangle/2)b(x,\xi).
\end{aligned}
$$

According to the theory in §2, these formulas can be applied to the symbols of the class $S(g; p_1; p_2)$ if the metric g is B-temperate, where $B(x, \xi) = \langle x, \xi \rangle$ is a quadratic form on W. The corresponding bilinear form

$$W \oplus W \in \left(\begin{pmatrix} x \\ \xi \end{pmatrix}, \begin{pmatrix} y \\ \eta \end{pmatrix} \right) \rightarrow \langle x, \eta \rangle + \langle y, \xi \rangle \in R^1$$

differs from the simplectic form only by a sign of one of the summands; therefore, in order to apply Theorems 2.1' and 2.2 to (3.24), we need the conditions

$$(3.25) \qquad g^\sigma_{x,\xi}(t, \tau) \leqslant C g^\sigma_{y,\eta}(t, \tau)(1 + g^\sigma_{y,\eta}(y - x, \xi - \eta))^N,$$

$$(3.26) \qquad h(x, \xi) = \sup_W g_{x,\xi}(t, \tau) / g^\sigma_{x,\xi}(-t, \tau) \leqslant 1,$$

$$(3.27) \quad \|p_1(y, \eta)^{-1} p_1(x, \xi)\| + \|p_2(x, \xi) p_2(y, \eta)^{-1}\| \leqslant C(1 + g^\sigma_{y,\eta}(y - x, \xi - \eta))^N,$$

whereas for the Weyl calculus, we need conditions (3.25) and (3.27) with $x - y$ instead of $y - x$, and condition (3.26) with t instead of $-t$ on the right-hand side.

Theorem 3.2. *Let g vary slowly, let p_1 and p_2 be left and right g-continuous functions, respectively, and let conditions (3.25)–(3.27) be satisfied. Then the classes of operators $\mathcal{L}(g, p_1, p_2) = \mathcal{L}_w(g, p_1, p_2)$, $\mathcal{L}_\ell(g, p_1, p_2)$, $\mathcal{L}_r(g, p_1, p_2)$— with Weyl, left, and right symbols, respectively—coincide; and the relation between the corresponding symbols is established by formulas (3.24).*

If $a \in S(g, p_1, p_2)$, then for any $N > 0$,

$$(3.28) \qquad \sigma^w(a_\ell)(x, \xi) = a(x, \xi) + \sum_{0 < k < N} c_k \langle D_x, D_\xi \rangle^k a(x, \xi) + r_N(x, \xi),$$

where $r_N \in S(g, h^N p_1, p_2)$, and c_k are absolute constants; and for any k,

$$(3.29) \qquad n_k(g, h^N p_1, p_2, r_N) \leqslant C \sup_{N \leqslant j \leqslant M} n_j(g; p_1; p_2; a),$$

where C, M depend only on N, k, $\dim V$, and on the constants characterizing g, p_1, p_2.

The formulas (3.28) and (3.29) are valid for all possible combinations of indices r, ℓ, w (of course, with other constants c_k, C, M). ∎

We note that (3.25)–(3.27) are equivalent to the conditions of Definition 3.1 if the metric g splits, i.e., if

$$g_{x,\xi}(y, \eta) = g'_{x,\xi}(y) + g''_{x,\xi}(\eta).$$

This condition is satisfied in all the applications that we shall consider.

3.6. We shall prove two useful lemmas that will enable us to verify the conditions of Definition 3.1 with ease.

Lemma 3.2.   *Let a metric g be σ-temperate and let a left (right) g-continuous function p satisfy the condition*

(3.29′)
$$p(w)p(w)^* \leqslant C(h(w)^{-N} + h(w')^{-N})p(w')p(w')^*$$
$$[p(w)^*p(w) \leqslant C(h(w)^{-N} + h(w')^{-N})p(w')^*p(w')].$$

Then p is a left (right) σ,g-temperate function.

Proof.   If $c > 0$ is sufficiently small, and either $g_w(w - w') < c$ or $g_{w'}(w - w') < c$, then by the g-continuity of the function p, (3.19) is valid with $N = 0$. If, on the other hand, $g_w(w - w') \geqslant c$ and $g_{w'}(w - w') \geqslant c$, then (3.16) yields

$$h(w')^{-2} \leqslant g^\sigma_{w'}(w - w')/g_{w'}(w - w') \leqslant c^{-1}g^\sigma_{w'}(w - w'),$$
$$h(w)^{-2} \leqslant c^{-1}g^\sigma_w(w - w') \leqslant Cc^{-1}(1 + g^\sigma_{w'}(w - w'))^{N+1};$$

and (3.19) follows from (3.29). ∎

In the applications, we shall deal mainly with metrics of the form

(3.30)
$$g_{x,\xi}(y, \eta) = (|\xi|^2 + \langle x \rangle^{2k})^{-1}|\eta|^2 + \langle x \rangle^{-2m}|y|^2.$$

Lemma 3.3.   *Metric (3.30) is σ-temperate if and only if $m \leqslant 1$, $k > -1$, $k + m \geqslant 0$.*

Proof.   Obviously, g varies slowly if and only if $m \leqslant 1$; and the condition $h \leqslant 1$ becomes $k + m \geqslant 0$, since

$$g^\sigma_{x,\xi}(y, \eta) = \langle x \rangle^{2m}|\eta|^2 + (|\xi|^2 + \langle x \rangle^{2k})|y|^2.$$

In order to verify the last condition of σ-temperateness, we shall show that

(3.31)
$$\langle y \rangle / \langle x \rangle + \langle x \rangle / \langle y \rangle \leqslant C(1 + \langle y \rangle^k |x - y|)^N,$$

(3.32)
$$(|\xi| + \langle x \rangle^k)/(|\eta| + \langle x \rangle^k) \leqslant C(1 + \langle x \rangle^m |\xi - \eta|)^N.$$

If $k \leqslant -1$, then by assuming in (3.31) that $x = 0$ and by taking the limit as $|y| \to +\infty$, we see that (3.31) does not hold. Now let $k > -1$. If $|y| > 2|x|$, then (3.31) holds for $N \geqslant 1/(k + 1)$; if $|y| < |x|/2$, then it holds for $N \geqslant 1$; and if $|x|/2 \leqslant |y| \leqslant 2|x|$, then (3.31) holds for any N. Thus, (3.31) holds if and only if $k > -1$.

By making the substitution $\xi = \langle x \rangle^k \xi'$, $\eta = \langle x \rangle^k \eta'$ in (3.32) and taking into account the fact that $\langle x \rangle^{k+m} \geqslant 1$ for $k + m \geqslant 0$, we see that it suffices to verify the estimate

$$\langle \xi' \rangle / \langle \eta' \rangle \leqslant C(1 + |\xi' - \eta'|)^N,$$

which obviously holds for $N = 1$.

Applying (3.31) and (3.32) successively, we obtain

$$
\begin{aligned}
g_{x,\xi}^{\sigma}(\cdot) &\leqslant C g_{y,\xi}^{\sigma}(\cdot)(1 + \langle y \rangle^{k} |x - y|)^{N'} \leqslant C_1 g_{y,\xi}^{\sigma}(\cdot)(1 + g_{y,\xi}^{\sigma}(x - y, 0))^{N'} \\
&\leqslant C_2 g_{y,\eta}^{\sigma}(\cdot)(1 + g_{y,\eta}^{\sigma}(x - y, 0))^{N'}(1 + \langle x \rangle^{m}|\xi - \eta|)^{N''} \\
&\leqslant C_3 g_{y,\eta}^{\sigma}(\cdot)(1 + g_{y,\eta}^{\sigma}(x - y, \xi - \eta))^{N'''}.
\end{aligned}
$$

3.7.

Example 3.1. $S_{\rho,\delta}^{m}$ usually denotes the Hörmander class of functions $a \in C^{\infty}$ (R^{2n}) such that

$$
\left| a_{(\beta)}^{(\alpha)}(x, \xi) \right| \leqslant c_{\alpha\beta} \langle \xi \rangle^{m - \rho|\alpha| + \delta|\beta|}, \quad |\alpha| + |\beta| = 0, 1, 2, \ldots .
$$

Thus, $S_{\rho,\delta}^{m} = S(g, p^{m})$, where $p(x, \xi) = \langle \xi \rangle$ and

$$
g_{x,\xi}(y, \eta) = \langle \xi \rangle^{2\delta}|y|^{2} + \langle \xi \rangle^{-2\rho}|\eta|^{2}.
$$

Obviously, g varies slowly if and only if $\rho \leqslant 1$ and if the condition $g^{\sigma} \geqslant g$ is equivalent to the condition $\delta \leqslant \rho$. It suffices to verify the condition

$$
\langle \xi \rangle^{-\delta}\langle \eta \rangle^{\delta} + \langle \xi \rangle^{\delta}\langle \eta \rangle^{-\delta} \leqslant C(1 + |x - y|\langle \eta \rangle^{\rho} + |\xi - \eta|\langle \eta \rangle^{-\delta})^{N}
$$

of σ-moderateness of the metric g for $x = y$, but in this case it is equivalent to the estimate of form (3.31) and is valid if and only if $\delta < 1$.

Similarly, $p \in O(C^1, g)$ if $\delta < 1$ and $\delta \leqslant \rho \leqslant 1$.

We note that if $\delta < \rho \leqslant 1$, then g is σ-moderate and $p \in O(C^1, g)$ as a result of Lemmas 1.2, 1.3, 3.2, and 3.3.

Example 3.2. Let a function $a \in C^{\infty}(R^{2n})$ permit an expansion in a sum of terms of degree j, $a_j : a \sim a_m + a_{m-1} + \cdots$,

positively homogeneous with respect to ξ; i.e., for $|\xi| \geqslant 1$ and for any N,

$$
\left| a(x, \xi) - \sum_{0 \leqslant j < N} a_{m-j}(x, \xi) \right| \leqslant C_N |\xi|^{m - N}.
$$

Then we say that a $[\in S_{1,0}^{m}]$ is a classical symbol and that a_{ℓ} is a classical PDO of order m : ord $a_{\ell} = m$. The function a_m is called its principal symbol.

Example 3.3. Let a^{ij} be classical scalar symbols of orders $m_i + m_j$ and let $a = [a^{ij}]_{i,j=1}^{s}$. Then $a \in S(g, \Lambda, \Lambda)$, where

$$
\Lambda = [\delta_{jk} p^{mj}]_{i,j=1}^{s}
$$

and g, p are the same as in Example 3.1.

Example 3.4.   Let

$$g_{x,\xi}(y,\eta) = \varphi(x,\xi)^{-2}|y|^2 + \Phi(x,\xi)^{-2}|\eta|^2.$$

(Different functions could be introduced for different coordinates, but we omit such generalizations for the sake of brevity.) Then

$$g^\sigma_{x,\xi}(y,\eta) = \Phi(x,\xi)^2|y|^2 + \varphi(x,\xi)^2|\eta|^2;$$

hence, the condition $g^\sigma \geqslant g$ is equivalent to the condition $\Phi\varphi \geqslant 1$. The metric g varies slowly if and only if

(3.33)
$$\exists c, C > 0 \,\big|\, |x - y| \leqslant c\varphi(x,\xi), \quad |\xi - \eta| \leqslant c\Phi(x,\xi)$$
$$\Rightarrow \varphi(x,\xi) \leqslant C\varphi(y,\eta), \quad \Phi(x,\xi) \leqslant C\Phi(y,\eta);$$

and the metric g is σ-temperate if there exist C, N such that

$$\Phi(x,\xi)/\Phi(y,\eta) + \varphi(x,\xi)/\varphi(y,\eta) \leqslant C(1 + \varphi(y,\eta)|\xi - \eta| + \Phi(y,\eta)|x - y|)^N.$$

3.8. We conclude this section by citing the asymptotic summation theorem (and its corollaries) and the corollaries of Theorems 3.1 and 3.2.
   Let

$$S^{-\infty}(g, p_1, p_2) = \bigcap_m S(g, h^m p_1, p_2).$$

Theorem 3.3.   *Let* $0 = m_1 < m_2 < \cdots$, *let* $m_j \to +\infty$ *as* $j \to \infty$, *and let* $a_j \in S(g, h^{m_j} p_1, p_2)$.
   Then there exists a function $a \in S(g, p_1, p_2)$ *such that*

(3.35)
$$\operatorname{supp} a \subset \bigcup_j \operatorname{supp} a_j$$

and $a \sim \sum a_j$, *in a sense that for any* $N > 1$,

(3.36)
$$a - \sum_{1 \leqslant j \leqslant N-1} a_j = r_N \in S(g, h^{mN} p_1, p_2);$$

here the seminorms of the symbol a *(r_N) depend only on the seminorms of the symbols* a_j *(and on* N).

Proof.   Let us fix a function $\theta \in C_0^\infty(R^1)$, $0 \leqslant \theta \leqslant 1$, $\theta|_{|t| \leqslant \frac{1}{2}} = 1$, $\theta|_{|t| \geqslant 1} = 0$ and a sequence $\{t_j\}$, $t_j \to \infty$ as $j \to \infty$, and let $\theta_j = \theta(t_j h)$. Lemmas 2.5 and 1.5 imply that $h \in S(g, h)$; therefore,

(3.37)
$$n_k(g, 1, \theta_j) \leqslant c_k \quad \forall k, j.$$

Indeed, $\frac{1}{2} \leqslant h(w)t_j \leqslant 1$ on a support of derivatives of the function θ_j; hence,

$$|\langle d\theta_j(w), w'\rangle| \leqslant t_j \max_{R^1} |\theta'(t)| |\langle dh(w), w'\rangle|$$
$$\leqslant Ct_j h(w)|\langle dh(w), w'\rangle h(w)^{-1}|$$
$$\leqslant Cn_1(g, h, h)g_w(w')^{1/2}.$$

Higher derivatives are estimated similarly, and (3.37) is proved.

Let $b_j - \theta_j a_j$. Then $h(w) \leqslant t_j^{-1}$ on supp b_j; thus, (3.37) and (1.4) imply

$$n_k(g, h^{m_j-1}p_1, p_2, b_j) \leqslant C_k t_j^{m_j-1-m_j} \quad \forall j, k,$$

where C_k does not depend on j. Therefore, t_j can be chosen such that for all j and $k \leqslant j$,

$$n_k(g, h^{m_j-1}p_1, p_2, b_j) \leqslant 2^{-j}.$$

This implies the convergence of the series

$$a(w) = \sum_{j=0}^{\infty} b_j(w)$$

together with all derivatives, which is uniform on each compact set. And furthermore, for all N, k,

$$n_k\left(g, h^{mN}p_1, p_2, \sum_{j=N+1}^{\infty} b_j\right) \leqslant C(k, N) + \sum_{j=N}^{\infty} 2^{-j}.$$

By the same token, we have proved (3.36); (3.35) follows from the way the function a was constructed. ∎

If $W_1 \subset W$ and $a \in S(g, p_1, p_2)$, and if there exists a function $a_1 \in S(g, p_1, p_2)$ such that supp $a_1 \subset W_1$ and $a - a_1 \in S^{-\infty}(g, p_1, p_2)$, then we write $\mathrm{supp}_\infty a \subset W_1$.

Let $\mathrm{supp}_d a = \{w \mid \exists k > 0, (d^k a)(w) \neq 0\}$.

Combining Theorems 3.1–3.3, we obtain the following corollary:

Corollary 3.1. *Let the conditions of Theorem 3.1 be satisfied. Then*

a) *if* $\mathrm{supp}_\infty a_1 \cap \mathrm{supp}_\infty a_2 = \emptyset$, *then* $a_{1,w}a_{2,w} \in \mathcal{L}^{-\infty}(g, p_1, p_3)$;

b) *if* a_1 *is a scalar function, and one of the sets* $\mathrm{supp}_d a_1 \cap \mathrm{supp}_\infty a_2$, $\mathrm{supp}_\infty a_1 \cap \mathrm{supp}_d a_2$, $\mathrm{supp}_d a_1 \cap \mathrm{supp}_d a_2$ *is empty, then*

$$[a_{1,w}, a_{2,w}] = a_{1,w}a_{2,w} - a_{2,w}a_{1,w} \in \mathcal{L}^{-\infty}(g, p_1, p_3).$$

Corollary 3.2. *Let the conditions of Theorem 3.2 be satisfied. Then* $\mathrm{supp}_\infty \sigma^w$ $(a_\ell) \subset \mathrm{supp}\, a$, *and the same holds for all the combinations of indices* w, r, ℓ.

We shall often apply Corollaries 3.1 and 3.2 in Chapter 2 when referring to Theorems 3.1 and 3.2.

§4 Continuity of Pseudodifferential Operators

4.1. In this section, we shall prove that the pseudodifferential operators introduced in §3 are continuous in S and S', and that operators with symmetric symbols from the class $S(g, I_{(1)}, I_{(2)})$ are continuous mappings from $L_2(V, H_2)$ to $L_2(V, H_1)$. The first result implies that when the conditions of the Theorem 3.1 are satisfied, $a_{1,w} a_{2,w}$ is a composition of the operators

$$a_{1,w} : S(V; H_2) \to S(V; H_1), \qquad a_{2,w} : S(V; H_3) \to S(V; H_2);$$

and the same is true for S' in place of S.

As in §2, the proof is based on the expansion of a symbol into a series by use of a partition of unity $\sum \varphi_j = 1$ in $W = R^{2n}$. Some terms will be estimated with the help of the following simple lemma:

Lemma 4.1.   *For an arbitrary* $a \in S(R^{2n}, B_{12})$,

$$(4.1) \quad \|a_w\|_{L_2(R^n, H_2) \to L_2(R^n, H_1)} \leqslant (2\pi)^{-2n} \|\hat{a}\|_{L_1(R^{2n} B_{12})} \stackrel{\text{def.}}{=} \|a\|_{FL_1(R^{2n}, B_{12})}.$$

Proof.   The kernel of an operator a_w is

$$A(x, y) = \int a(\tfrac{1}{2}(x + y), \xi) \exp(i \langle x - y, \xi \rangle) d'\xi$$

$$= (2\pi)^{-2n} \int \exp(i \langle x - y, \xi \rangle / 2) \hat{a}(\theta, y - x) d\theta;$$

therefore,

$$\int \|A(x, y)\|_{12} dx \leqslant \|a\|_{FL_1}, \qquad \int \|A(x, y)\|_{12} dy \leqslant \|a\|_{FL_1}.$$

This implies (4.1).   ∎

In order to apply Lemma 4.1 we need to estimate $\|a\|_{FL_1}$.

Lemma 4.2.   *Let* $a \in S(R^{2n}, B_{12})$, *let* g *be a positive-definite quadratic form on* R^{2n}, *and let* $w \in R^{2n}$. *Then*

$$(4.2) \qquad \|a\|_{FL_1} \leqslant C \sup_{j \leqslant 4n} \sup_{R^{2n}} |a|^{1,1}_{g,j}(w')(1 + g(w - w'))^{2n},$$

where the constant C *does not depend on* a, w.

Proof.   Without loss of generality, we can assume that $g(w) = \sum b_i^2 w_j^2$, where $w = (w_1, \ldots, w_{2n})$. Let $\tilde{w}_j = b_j(w'_j - w_j)$, $\tilde{w} = (\tilde{w}_1, \ldots, \tilde{w}_{2n})$, $\tilde{a}(\tilde{w}) = a(w')$. Then, if g_0 is a square of a Euclidean norm $|\cdot|^2$ in R^{2n}, then for all k,

$$(4.3) \qquad |\tilde{a}|^{1,1}_{g_0,k}(\tilde{w}) = |a|^{1,1}_{g,k}(w');$$

and

(4.4) $g(w - w') = g_0(\widetilde{w}) \quad [= |\widetilde{w}|^2]$.

Obviously, $\|a\|_{FL_1} = \|\tilde{a}\|_{FL_1}$. Furthermore,

$$\langle t \rangle^{-4n} \langle D_w \rangle^{4n} \exp(-i\langle w, t \rangle) = \exp(-i\langle w, t \rangle);$$

hence, integrating by parts gives us

$$\|\tilde{a}\|_{FL_1} = (2\pi)^{-2n} \int \left\| \int \exp(-i\langle \widetilde{w}, \tau \rangle) \langle \tau \rangle^{-4n} \langle D_{\widetilde{w}} \rangle^{4n} \tilde{a}(\widetilde{w}) \, d\widetilde{w} \right\|_{12} d\tau$$

$$\leqslant C \sup_{j \leqslant 4n} \sup_{\widetilde{w}} |\tilde{a}|_{g_0,j}^{1,1}(\widetilde{w})(1 + |\widetilde{w}|^2)^{2n}.$$

Now (4.2) follows from (4.3) and (4.4).   ∎

4.2.

Theorem 4.1. *If g is σ-temperate and p_1 and p_2 are, respectively, left and right σ,g-temperate functions, then the mapping*

$$\mathcal{S}(R^{2n}; B_{12}) \ni a \to a_w \in \mathrm{Hom}(\mathcal{S}(R^n; H_2), \mathcal{S}(R^n; H_1))$$

can be uniquely continued to a weakly continuous mapping from $S(g, p_1, p_2)$ to $\mathrm{Hom}(\mathcal{S}(R^n; H_2), \mathcal{S}(R^n; H_1))$.

The same holds when $\mathcal{S}(R^n, H_i)$ is replaced with $\mathcal{S}'(R^n, H_i)$.

Proof. It is sufficient to prove continuity in $\mathcal{S}$, since continuity of $(a^*)_w$ in $\mathcal{S}$ implies continuity of a_w in $\mathcal{S}'$.

Let $\sum \varphi_j = 1$ be a partition of unity on $W = V \oplus V' = R^{2n}$ introduced in §1. We define the neighborhoods $U_j \subset U_j'$ of sets $\mathrm{supp}\,\varphi_j$, as in §2 prior to Lemma 2.3, and assume that $a = \sum a_j$, $a_j = \varphi_j a$.

If w_j is a center of U_j, then the seminorms of the terms a_j in $S_j(g, p_1, p_2) \overset{\mathrm{def.}}{=} S(g_{w_j}, p_1(w_j), p_2(w_j))$ are bounded uniformly with respect to j and $g_{w_j}(w - w_j) < c$ on $\mathrm{supp}\,a_j$. Therefore, Lemmas 4.1 and 4.2 imply that

(4.5) $\|a_{j,w} u\|_{L_2(V;H_2)} \leqslant C \|p_1(w_j)\|_{11} \|p_2(w_j)\|_{22} \|u\|_{L_2(V;H_1)}$

uniformly with respect to j.

In order to obtain a better estimate, we apply constructions similar to those used in Lemma 2.2. Let $L = L(x, \xi)$ be a linear function on W, let $L(x, D)$ be a corresponding differential operator, and let $a \in \mathcal{S}(W, B_{12})$. By integrating by parts within the integral in (3.8), we have

$$(L(x, D)a_w - (La)_w)u(x) = \iint \exp(i\langle x - y, \xi \rangle)((L(\tfrac{1}{2}(x - y), 0)$$

$$+ L(0, D_x))a(\tfrac{1}{2}(x + y), \xi)u(y) \, dy \, d'\xi.$$

Since

$$L(x - y, 0) \exp(i\langle x - y, \xi\rangle) = L(D_\xi, 0) \exp(i\langle x - y, \xi\rangle),$$

we can integrate by parts to obtain

$$\sigma^w(L(x, D)a_w)(x, \xi) = L(x, \xi)a(x, \xi) + \tfrac{1}{2}L(-D_\xi, D_x)a(x, \xi).$$

By using the Poisson bracket

$$\{f, g\} = \left\langle \frac{\partial f}{\partial \xi}, \frac{\partial g}{\partial x} \right\rangle - \left\langle \frac{\partial g}{\partial \xi}, \frac{\partial f}{\partial x} \right\rangle,$$

we write

$$(4.6) \qquad \sigma^w(L(x, D)a_w) = La + \{L, a\}/2i.$$

Substitution of a^* for a and consequent conjugation yield

$$(4.6') \qquad \sigma^w(a_w \overline{L}(x, D)) = \overline{L}a + \{a, \overline{L}\}/2i.$$

Now let L be a linear function that is positive on U'_j. Lemma 2.1 implies

$$n_k(g_{wj}, 1, L(0)/L, \operatorname{supp} \varphi_j) < \infty, \quad k = 0, 1, \dots$$

uniformly with respect to j; therefore, by applying $(4.6')$ with $a = a_j/L$, we obtain

$$(4.7) \qquad a_{j,w}u = (L^{-1}a_j)_w L(x, D)u + \tfrac{1}{2}i\{L^{-1}a_j, L\}_w u.$$

If $L(w) = \sigma(w, w_0)$, then $\{L^{-1}a_j, L\} = -L\langle w_0, da_j \rangle$; hence,

$$(4.8) \qquad \{L^{-1}a_j, L\} \in Sj(g, g(w_0)^{1/2}p_1, p_2)$$

(uniformly with respect to j, as with all the estimates that follow). In W let us introduce the norm

$$\|\cdot\|'_j = \max(g_0(\cdot)^{1/2}, gw_j(\cdot)^{1/2}),$$

and identify W' with W using the form σ. Then a dual to the norm $\|\cdot\|'_j$ is

$$\|w\|_j = \inf(g^\sigma_{w_j}(w')^{1/2} + g^\sigma_0(w'')^{1/2}), \quad w' + w'' = w;$$

and for $R_j = \min_{vj} \|w\|_j$, the Hahn–Banach theorem yields $R_j \leqslant L(w_j)$ $[= \sigma(w_j, w_o)]$ if $g_0(w_0) \leqslant 1$ and $g_{w_j}(w_0) \leqslant 1$.

Therefore, by integrating (4.7) and using Lemmas 4.1 and 4.2 and estimate (4.8), we obtain the estimate

$$(4.9)\quad \|a_{j,w}u\|_{L_2(R^n;H_1)} \leqslant C_N \|p_1(w_j)\|_{11} \|p_2(w_j)\|_{22} (1+R_j)^{-N} \|u\|_{S,N} \quad \forall N,$$

where $|\cdot|_{S,N}$ are seminorms in $S(R^n, H_2)$ and

$$(4.10)\qquad\qquad C_N = C' \sup_{j \leqslant N'} n_j(g, p_1, p_2, a).$$

Here C', N' depend only on g, p_1, p_2, N, n. Later we shall show that for some N,

$$(4.11)\qquad\qquad \sum_j (1+R_j)^{-N} < \infty,$$

and that there exist points $w_j \in U_j$, $j = 1, 2, \ldots$ such that

$$(4.12)\qquad\qquad g_0(w_j') \leqslant C(1+R_j)^{N_1}.$$

But p_1 is a left σ,g-temperate function; therefore, (4.12) yields

$$(4.13)\quad \begin{aligned} \|p_1(w_j)\| &\leqslant C\|p_1(w_j')\| \leqslant C\|p_1(0)\| \, \|p_1(0)^{-1}p_1(w_k')\| \\ &\leqslant C_1(1+g_0^\sigma(w_j'))^N = C_1(1+g_0(w_l'))^N \leqslant C_2(1+R_j)^{N_2}. \end{aligned}$$

A similar proof leads to the same estimate for $\|p_2(w_j)\|$; and (4.9), (4.11)–(4.13) yield

$$(4.14)\qquad\qquad \|a_w u\|_{L_2} \leqslant \sum_j \|a_{j,w}u\|_{L_2} \leqslant C \, |u|_{S,N} ,$$

where the constant C is given by (4.10).

Let us estimate the derivatives of the function $a_w u$. If M is a linear function on W, then (4.6) and (4.13) imply that for some N,

$$M(x, D)a_{j,w} \in \mathcal{L}(g_{wj}, (1+R_j)^N I_{(1)}, I_{(2)});$$

therefore, estimate (4.14) holds for $M(x, D)a_w$ as well. Iteration leads to the estimate

$$(4.15)\qquad\qquad |a_w u|_{S,N} \leqslant C \sup_{j \leqslant m} n_j(g, p_1, p_2, a) \, |u|_{S,M} ,$$

$N = 0, 1, \ldots$, where C, M depend only on N, n, and on the constants that characterize g, p_1, p_2.

Now let $a \in S(g, p_1, p_2)$ be arbitrary and let $\{a^{[j]}\} \subset S(w, B_{12})$ be a sequence converging weakly to a. Let

$$a_w u = \lim_{k \to \infty} a_w^{[k]} u, \quad u \in S(V; H_2).$$

If $\{\tilde{a}^{[k]}\}$ is another sequence with the same properties, then $\{\tilde{a}^{[k]} - a^{[k]}\}$ converges weakly to zero; and $\tilde{a}_w u - a_w u = 0$ as a result of (4.9) and (4.10). Hence, a_w is defined correctly, and estimate (4.15) holds for an arbitrary $a \in S(g, p_1, p_2)$.

4.3. In order to complete the proof, we have to prove (4.11) and (4.12). Following the pattern of the proof of Lemma 2.4, we write $M_k = \{j \mid R_j^2 \leqslant k\}$. If $j \in M_k$, then we can choose $w_j' \in U_j$ and w_j'' such that

$$g_{w_j}^\sigma(w_j' - w_j'') \leqslant k, \qquad g_0^\sigma(w_k'') \leqslant k.$$

In the first inequality, by replacing the norm $(g_{w_j}^\sigma)^{1/2}$ with the equivalent norm $(g_{w_j}'^\sigma)^{1/2}$ and then applying (3.16′), we obtain

$$g_{w_j''}^\sigma(w_j' - w_j'') \leqslant C g_{w_j'}^\sigma(w_j' - w_j'')(1 + g_{w_j'}^\sigma(w_j' - w_j''))^N \leqslant C_1 k^{N_1};$$

therefore, the second inequality and (3.16′) yield

$$\begin{aligned} g_0^\sigma(w_j' - w_j'') &\leqslant C g_{w_j''}^\sigma(w_j' - w_j'')(1 + g_{w_j''}^\sigma(w_j''))^N \\ &\leqslant C g_{w_j''}^\sigma(w_j' - w_j'')(1 + C g_0^\sigma(w_j'')(1 + g_0^\sigma(w_j'')))^N \leqslant C_1 k^{N_2}. \end{aligned}$$

By applying (3.16′) again, we obtain

$$\begin{aligned} g_{w_j}^\sigma(t) &\leqslant C g_{w_j'}^\sigma(t) \leqslant C_1 g_{w_j''}^\sigma(t)(1 + g_{w_j''}^\sigma(w_j' - w_j'))^N \leqslant C_2 k^{N_1} g_{w_j''}^\sigma(t) \\ &\leqslant C_3 k^{N_1} g_0^\sigma(t)(1 + g_0^\sigma(w_j''))^n \leqslant C_4 k^{N_3} g_0^\sigma(t). \end{aligned}$$

(4.16)

But $g^\sigma \geqslant g$; therefore, the previous inequality yields

$$g_0(w_j') \leqslant 2(g_0(w_j' - w_j'') + g_0(w_j'')) \leqslant C_5 k^{N_4},$$

which proves (4.12).

Furthermore, inequality (4.16) has the same form as inequality (2.10); therefore, the same calculations that led us from (2.10) to (2.9) are used here to prove (4.11).

Theorem 4.1 is proved.   ∎

4.4.

Theorem 4.2. *Let g be a σ-temperate metric and let $a \in S(g, I_{(1)}, I_{(2)})$. Then*

$$(4.17) \qquad \|a_w\|_{L_2(R^n; H_2) \to L_2(R^n; H_1)} \leqslant C \max_{s \leqslant N} n_s(g; I_{(1)}; I_{(2)}; a),$$

where C, N depend only on n and on the constants that characterize the metric g.

Proof. Let $a = \sum a_j$ be the same series as above. Its partial sums $a^{[k]}$ are uniformly bounded in $S(g, I_{(1)}, I_{(2)})$; therefore, if we prove (4.17) for finite symbols, we obtain

$$(4.17') \qquad \|a_w^{[k]} u\| \leqslant C \|u\| \quad \forall k, \ \forall u \in L_2(R^n, H_2),$$

where C has the same form as in the right-hand side of (4.17). But

$$a_w^{[k]} u \to a_w u \quad \forall u \in S(R^n; H_2);$$

and since $S(R^n; H_2) \subset L_2(R^n; H_2)$ densely, then (4.17) follows from (4.17').

Thus, it suffices to prove (4.17) for finite symbols, i.e., symbols for which the sum $\sum a_{j,w}$ is finite. In this case, it is convenient to apply the following lemma by Cotlar, Knapp and Stein on a sum of almost orthogonal operators.

Lemma 4.3. *Let H_1 and H_2 be Hilbert spaces and let $A_1, \ldots, A_N : H_1 \to H_2$ be bounded operators such that*

$$(4.18) \qquad \sum_k \|A_j^* A_k\|^{1/2} \leqslant M, \qquad \sum_k \|A_j A_k^*\|^{1/2} \leqslant M.$$

Then

$$(4.19) \qquad \left\| \sum_k A_k \right\| \leqslant M.$$

Proof. The spectral theorem yields $\|A\|^{2m} = \|(A^* A)^m\|$. We write $\sum A_j$ instead of A and apply the estimate

$$\|A_{j_1}^* A_{j_2} \cdots A_{j_{2m-1}}^* A_{j_{2m}}\| \leqslant \min\{\|A_{j_1}^* A_{j_2}\| \cdots \|A_{j_{2m-1}}^* A_{j_{2m}}\|,$$
$$\|A_{j_1}^*\| \|A_{j_2} A_{j_3}^*\| \cdots \|A_{j_{2m-2}} A_{j_{2,-1}}^*\| \|A_{j_{2m}}\|\}.$$

We take the geometric mean and notice that (4.18) implies $\|A_j\| \leqslant M$. We obtain

$$\|A\|^{2m} \leqslant M \sum \|A_{j_1}^* A_{j_2}\|^{1/2} \|A_{j_2} A_{j_3}^*\|^{1/2} \cdots \|A_{j_{2m-1}}^* A_{j_{2m}}\|^{1/2},$$

where the summation is taken over all $j_1, \ldots, j_{2m}$.

If, by using (4.18), we successively estimate the sums over $j_{2m}, j_{2m-1}, \ldots, j_2$, then only the sum over j_1 will be left:

$$\|A\|^{2m} \leqslant \sum_j M^{2m} = NM^{2m}.$$

By taking the root of the power $2m$ and passing on to the limit as $m \to \infty$, we obtain (4.19).

4.5. We proceed with the proof of Theorem 4.2. Estimate (4.5) implies $\|a_{j,w}\| \leqslant C_j$; and in order to prove (4.18), we have to examine the compositions

$$q_{jk,w} = a_{j,w}^* a_{k,w}, \qquad b_{jk,w} = a_{j,w} a_{k,w}^*.$$

Obviously, it will be sufficient to estimate the symbols a_{jk},

$$a_{jk}(x,\xi) = \exp(\tfrac{1}{2} i\sigma(D_x, D_\xi; D_y, d_\eta)) a_j^*(x,\xi) a_k(y,\eta)\big|_{(y,\eta)=(x,\eta)}.$$

Let the metric G and the form B be as in the proof of Theorem 3.1. The metric G is B-temperate with respect to each point on the diagonal $W \times W$; therefore, we can apply (2.11)—substituting $U_j \times U_k$ for U_j; $I_{(i)}$ for p_i; and

$$d_{jk}(w) = \min_{U_j} g_{w'}^\sigma(w - w') + \min_{U_k} g_{w''}^\sigma(w - w'')$$

for $d_j(x)$. By denoting

$$(4.20) \qquad d_{jk}'(w) = \min_{U_j} g_w^\sigma(w - w') + \min_{U_k} g_w^\sigma(w - w''),$$

we obtain, as a result of (3.16′),

$$1 + d_{jk}'(w) \leqslant C(1 + d_{jk}(w))^N.$$

In (2.11), by also replacing $d_{jk}(w)$ with $d_{jk}'(w)$, we obtain for any N,

$$(4.21) \qquad \|a_{jk}(w)\|_{12} \leqslant C_N (1 + d_{jk}'(w))^{-N}.$$

If we choose w' and w'' such that the minimums in (4.20) are reached, then

$$g_w^\sigma(w' - w'') \leqslant 2(g_w^\sigma(w - w') + g_w^\sigma(w - w'')) \leqslant 2d_{jk}'(w);$$

and (3.16′) yields

$$(4.22) \qquad g_{w_j}^\sigma(w' - w'') \leqslant C g_{w'}^\sigma(w' - w'')$$
$$\leqslant C_1 g_w^\sigma(w' - w'')(1 + g_w^\sigma(w - w'))^N$$
$$\leqslant C_2 (1 + d_{jk}'(w))^{N+1},$$

$$(4.23) \qquad 1 + g_{w_j}(w - w_j) \leqslant 1 + C g_{w'}(w - w_j)$$
$$\leqslant 1 + 2C g_{w'}(w - w') + 2C g_{w'}(w' - w_j)$$
$$\leqslant C_1 + 2C g_{w'}(w - w') \leqslant C_2(1 + g_{w'}^\sigma(w - w'))$$
$$\leqslant C_3 (1 + g_w^\sigma(w - w'))^{N+1} \leqslant C_4 (1 + d_{jk}'(w))^{N+1}.$$

We write

$$(4.24) \qquad f_{jk} = \min_{\substack{w' \in U_j, \\ w'' \in U_k}} g^\sigma_{w_j}(w' - w'').$$

Now (4.22) implies that for some C, N,

$$1 + f_{jk} \leqslant C(1 + d_{jk}(w))^N \quad \forall w \in W;$$

therefore, taking (4.23) into account, we derive from (4.21) the estimate

$$(4.25) \qquad \|a_{jk}(w)\|_{12} \leqslant C_N(1 + f_{jk})^{-N}(1 + g_{w_j}(w - w_j))^{-N},$$

where N is arbitrary and where C_N depends only on N and not on j, k.

The same estimate is also valid for any seminorm of the symbol a_{jk}. Indeed, if $g_{w_j}(t) \leqslant 1$ and if we apply the differential operator $\langle t, D \rangle$ to a_{jk}, then we obtain one term in which a_j is differentiated [this term obviously satisfies an estimate of the type (4.25)], and another term in which a_k is differentiated, the estimate of which requires the added factor

$$(g_{w_k}(t)/g_{w_j}(t))^{1/2}.$$

Since σ is a nondegenerate form, (3.16′) yields the estimate

$$g_{w_k}(t) \leqslant Cg_{w_j}(t)(1 + g^\sigma_{w_j}(w_k - w_j))^N.$$

Hence, if w' and w'' provide the minimum in (4.24), then

$$(4.26) \qquad \begin{aligned} g_{w_k}(t)/g_{w_j}(t) &\leqslant C_1(1 + g^\sigma_{w'}(w_k - w') + g^\sigma_{w'}(w' - w_j))^N \\ &\leqslant C_2(1 + g^\sigma_{w_k}(w_k - w') + g^\sigma_{w_j}(w' - w_j))^{N_2} \\ &\leqslant C_3(1 + g^\sigma_{w_k}(w_k - w'') + g^\sigma_{w_k}(w'' - w'))^{N_3} \\ &\leqslant C_4(1 + g^\sigma_{w''}(w'' - w'))^{N_4} \leqslant C_5(1 + g^\sigma_{w'}(w'' - w'))^{N_5} \\ &\leqslant C_6(1 + g^\sigma_{w_j}(w'' - w'))^{N_5} \leqslant C_7(1 + f_{jk})^{N_5}. \end{aligned}$$

Thus, we have shown that for $s = 0, 1$,

$$(4.27) \quad \|d^s a_{jk}(w; t_1, \ldots, t_s)\|_{12} \prod_{1 \leqslant i \leqslant s} g_{w_j}(t_i)^{-1/2}$$
$$\leqslant C_{sN}(1 + f_{jk})^{-N}(1 + g_{w_j}(w - w_j))^{-N},$$

where N is arbitrary. Similarly, estimate (4.27) could be proved for all s; and Lemmas 4.1 and 4.2 imply that for any N,

$$(4.28) \qquad \|a_{jk,w}\|_{L_2(R^n, H_2) \to L_2(R^n, H_1)} \leqslant C_N(1 + f_{jk})^{-N}.$$

Furthermore, we shall show that for some C, N depending only on the constants that characterize the metric g (and on n),

$$(4.29) \qquad \sum_j (1 + f_{jk})^{-N} \leqslant C, \qquad \sum_k (1 + f_{jk})^{-N} \leqslant C;$$

and since the constants in (4.25), (4.27), and (4.28) are estimated by

$$C_1 \sup_{j,k} \sup_{s,r \leqslant M} n_s(g_{w_j}, I, I, a_j) n_r(g_{w_k}, I, I, a_k) \leqslant C_2 [\sup_{m \leqslant M} n_m(g, I, I, a)]^2,$$

where C_1 and C_2 depend only on the constants that characterize the metric g (and on n), then (4.28), (4.29), and Lemma 4.3 yield (4.17).

4.6. To complete the proof we need to verify estimate (4.29). Since for all w_j, $w' \in U_j$ and w_k, $w'' \in U_k$,

$$1 + g^\sigma_{w_j}(w' - w'') \leqslant C(1 + g^\sigma_{w'}(w' - w'')) \leqslant C_1 (1 + g^\sigma_{w''}(w' - w''))^{N_1}$$
$$\leqslant C_3 (1 + g^\sigma_{w_k}(w' - w''))^{N_1},$$

then

$$1 + f_{jk} \leqslant C'(1 + f_{kj})^{N'};$$

and it suffices to prove $(4.29)_2$. This proof is a variant of the proof of Lemma 2.4. We can choose the coordinate system such that g_{w_j} is the Euclidean metric. Then $g^\sigma_{w_j}$ is not smaller than the Euclidean metric; and if $f_{jk} \leqslant m$, then U_k contains the point w'', whose Euclidean distance from 0 is not larger than $Cm^{1/2}$. Furthermore, similarly to (4.26),

$$g_{w_k}(t) \leqslant C_1 m^{N_1} g_{w_j}(t).$$

Therefore, there exist C, N such that a ball of the radius Cm^{-N} with the center at the point w_k is contained in U'_k; and, as in the proof of Lemma 2.4, it immediately follows that the number of the indices k satisfying the condition $f_{jk} \leqslant m$ is bounded by some power of m.

This proves $(4.29)_2$, and the proof of Theorem 4.2 is completed.

4.7. The following theorem generalizes the famous result about compactness of PDO of negative order.

Theorem 4.3. *For all k, let*

$$(4.30) \qquad |a|^{I,I}_{g,k}(w) \to 0 \quad \text{as} \quad |w| \to \infty,$$

and either $\dim H_1 < \infty$ *or* $\dim H_2 < \infty$.
    Then $a_w : L_2(R^n; H_2) \to L_2(R^n; H_1)$ *is compact.*

Proof. Let $a^{[k]} = \sum_{j<k} a_j$. Then $a^{[k]} \in \mathcal{S}(R^{2n}, B_{12})$ and $a^{[k]}_w$ are integral operators with kernels of the class $\mathcal{S}(R^{2n}, B_{12})$. Hence, all $a^{[k]}_w$ are compact. (4.30) implies that $a^{[k]} \to a$ in $S(g, I_{(1)}, I_{(2)})$; and by Theorem 4.2, $a^{[k]}_w \to a_w$ with respect to the norm.

Therefore, a_w is also compact.

§5 Weight Spaces of Sobolev Type

5.1. We shall assume that the metric g is σ-temperate and that the functions p, p', etc., are σ,g-temperate and positive. In view of the remarks following Definitions 1.3 and 3.1, the latter assumption does not result in loss of generality.

Definition 5.1. Let $p \in O(H; g)$. We write

$$\mathcal{H}(p, H) \; [= \mathcal{H}(p, H, g)] \; = \operatorname{span}\{Au \mid u \in L_2(R^n; H), \; A \in \mathcal{L}(g, p^{-1}, I_H)\}.$$

This is a subspace in $S'(R^n; H)$. We introduce in $\mathcal{H}(p, H)$ the strongest topology in which each of the following operators is continuous:

$$A : L_2(R^n; H) \to \mathcal{H}(p; H), \qquad A \in \mathcal{L}(g, p^{-1}, I).$$

Theorem 5.1.
  a) $\mathcal{H}(I; H) = L_2(R^n; H)$, *as topological spaces.*
  b) *If* $A \in \mathcal{L}(g, p, p')$, *then* $A : \mathcal{H}(p', H) \to \mathcal{H}(p^{-1}, H)$ *is continuous.*
  c) *If* $\|p^{-1}p'\| \leqslant C$, *then* $\mathcal{H}(p; H) \subset \mathcal{H}(p'; H)$ *continuously.*
  d) $S(R^n; H) \subset \mathcal{H}(p; H)$ *and* $\mathcal{H}(p; H) \subset S'(R^n; H)$ *densely and continuously.*

Proof.
  a) By Theorem 4.2, $\mathcal{H}(I, H) \subset L_2(R^n; H)$ continuously, but the identity operator with the symbol I_H belongs to $\mathcal{L}(g, I)$; hence, $\mathcal{H}(I, H) \supset L_2(R^n; H)$ continuously as well.
  b) Let $U \subset \mathcal{H}(p^{-1}; H)$ be an open set. We have to show that $A^{-1}(U)$ is open in $\mathcal{H}(p', H)$, i.e., for any $B \in \mathcal{L}(g, p'^{-1}, I)$, the set $B^{-1}A^{-1}(U) = (AB)^{-1}(U)$ is open in $L_2(R^n; H)$. But $AB \in \mathcal{L}(g, p, I)$ for such B; hence, $(AB)^{-1}(U)$ is indeed open.
  c) It is sufficient to notice that $S(g, p^{-1}, I) \subset S(g, p'^{-1}, I)$.
  d) Let $f \in S(R^n, H)$. Then an integral operator with the kernel $A(x, y) = \|f\|_{L_2}^{-2} f(x) \otimes f(y)^*$ is PDO with the Weyl symbol $a \in S(R^{2n}; \operatorname{End} H) \subset S(g, p^{-1}, I)$. Therefore, $f = a_w f \in \mathcal{H}(p, H)$ and $S(R^n; H) \subset \mathcal{H}(p, H)$. Furthermore, since by definition $\mathcal{H}(p, H) \subset S'(R^n; H)$ and since $A \in \mathcal{L}(g, p^{-1}, I)$ is continuous in $S'(R^n, H)$, it is a continuous operator from $L_2(R^n, H)$ to $S'(R^n, H)$. Therefore, $\mathcal{H}(p, H) \subset S'(R^n, H)$ continuously [and densely, since $S(R^n; H) \subset S'(R^n, H)$ densely]. ∎

5.2. More interesting statements about the spaces $\mathcal{H}(p, H)$ can be obtained if the following condition holds:
  There exist $A_p \in \mathcal{L}(g, p^{-1})$ and $A^p \in \mathcal{L}(g, p)$ such that

$$A_p : L_2(R^n; H) \to \mathcal{H}(p, H) \quad \text{and} \quad A^p : \mathcal{H}(p, H) \to L_2(R^n, H)$$

are topological isomorphisms.
  We denote the class of functions satisfying such conditions by $O'(H, g)$. Next, it will be shown that $O(H, g) = O'(H, g)$ in the cases of interest to us.

Theorem 5.2.   *Let $p \in O'(H,g)$ and $p' \in O'(H',g)$. Then*

a) $\mathcal{H}(p,H)$ *has the topology of a Hilbert space. It can be defined by the norm* $\|u\|_p = \|A^p u\|_{L_2}$.

b) $\mathcal{S}(R^n, H) \subset \mathcal{H}(p,H) \subset \mathcal{S}'(R^n; H)$ *densely and continuously.*

c) *Part b) implies that in a natural way, we can identify $\mathcal{H}(p,H)^*$ with a subspace in $\mathcal{S}'(R^n; H)$. Then $\mathcal{H}(p,H)^* = \mathcal{H}(p^{-1}, H)$ as topological spaces [and $p^{-1} \in O'(H,g)$].*

d) *If $A \in \mathcal{L}(g,p,p')$, then $A : \mathcal{H}(p',H') \to \mathcal{H}(p^{-1},H)$ is bounded, and its norm satisfies the estimate*

$$(5.1) \qquad \qquad \|A\| \leqslant C \max_{j \leqslant M} n_j(g,p,p',\sigma^w(A)).$$

Proof.

a) Obvious.

b) Due to part d) of Theorem 5.1, we must prove the denseness and continuity of the first embedding. But $u_n \xrightarrow{S} u \Rightarrow A^p u_n \xrightarrow{S} A^p u \Rightarrow A^p u_n \xrightarrow{L_2} A^p u \Rightarrow u_n \to u$ [in $\mathcal{H}(p,H)$], since $A^p : \mathcal{H}(p,H) \to L_2(R^n, H)$ is an isomorphism. Furthermore, $\mathcal{S}(R^n, H) \subset L_2(R^n, H)$ densely; therefore, $A_p(\mathcal{S}(R^n; H))$ [$\subset \mathcal{S}(R^n, H)$ by Theorem 4.1] is also dense in $\mathcal{H}(p,h)$.

c) If $A \in \mathcal{L}(g,p,I)$, then $A^* \in \mathcal{L}(g,I,p)$; hence, $A^* : \mathcal{H}(p,H) \to L_2(R^n, H)$ is continuous. Therefore, $A : L_2(R^n; H) \to \mathcal{H}(p,H)^*$ is continuous, and $\mathcal{H}(p^{-1},H) \subset \mathcal{H}(p,H)^*$ continuously. But $(A^p)^*(L_2(R^n, H)) = \mathcal{H}(p,H)^*$ and $(A^p)^* \in \mathcal{L}(g,p,I)$; hence, if $U \subset \mathcal{H}(p^{-1},H)$ is an open set, then $((A^p)^*)^{-1}(U) \subset L_2(R^n; H)$ is an open set, and $U \subset \mathcal{H}(p,H)^*$ is also an open set. Therefore, $\mathcal{H}(p,H)^* \subset \mathcal{H}(p^{-1},H)$ continuously. Finally, $p^{-1} \in O'(H,g)$, since we can assume that $A^{p^{-1}} = (A_p)^*$ and $A_{p^{-1}} = (A^p)^*$.

d) The operator $A : \mathcal{H}(p',H') \to \mathcal{H}(p,H)$ is similar to the operator

$$A' = A^p A A_p : L_2(R^n, H') \to L_2(R^n, H);$$

therefore, (5.1) follows from Theorems 3.1 and 4.2.   ∎

The following lemma shows that the space $\mathcal{H}(p,H) = \mathcal{H}(p,H,g)$ depends mainly on the function p and not on the metric g.

Lemma 5.1.   *Let $G \geqslant g$ be two σ-temperate metrics and let $p \in O'(H,g) \cap O'(H,G)$. Then $\mathcal{H}(p,H,G) = \mathcal{H}(p,H,g)$.*

Proof. $S(G,p^{\pm 1},I) \supset S(g,p^{\pm 1},I)$ and $S(G,I,p^{\pm 1}) \supset S(g,I,p^{\pm 1})$; hence, $\mathcal{H}(p,H,g) \subset \mathcal{H}(p,H,G)$ and $\mathcal{H}(p,H,g)^* = \mathcal{H}(p^{-1},H,g) \subset \mathcal{H}(p^{-1},H,G) = \mathcal{H}(p,H,G)^*$.   ∎

5.3.

Theorem 5.3. *Let*

$$(5.2) \qquad\qquad h(X) \to 0 \quad \text{as} \quad |X| \to \infty.$$

Then $O(H, g) = O'(H, g)$.

The proof is based on the following lemmas.

Lemma 5.2. *Let scalar function $p \in O'(H, g)$, let $q \in O(H, g)$, $q \geqslant cI$, and let there exist operators $A \in \mathcal{L}(g, q)$, $B \in \mathcal{L}(g, q^{-1})$ such that $AB, BA : \mathcal{H}(p, H) \to \mathcal{H}(p, H)$ are invertible. Then $qp \in O'(H, g)$.*

Proof. $A : \mathcal{H}(qp, H) \to \mathcal{H}(p, H)$ is a surjective operator with a continuous right inverse $B(AB)^{-1}$. In addition, A is injective on $\mathcal{H}(p, H) \supset \mathcal{H}(qp, H)$ with the left inverse $(BA)^{-1}B$. Thus, $A : \mathcal{H}(qp, H) \to \mathcal{H}(p, H)$ is a topological isomorphism, and we can assume that $A^{pq} = A^p A$. On the other hand, $B : \mathcal{H}(p, H) \to \mathcal{H}(pq, H)$ has a continuous left inverse $(AB)^{-1}A$. Since $(AB)^{-1}$ and A are topological isomorphisms, then B is also a topological isomorphism, and we can assume that $A_{pq} = BA_p$. $\blacksquare$

Lemma 5.3. *Let $\delta \in [0, \frac{1}{2})$, let $h^\delta \leqslant b \leqslant 1$, and let*

$$(5.3) \qquad\qquad |\langle db(w), w' \rangle| \leqslant C g_w(w')^{1/2} \quad \forall w, w' \in R^{2n}.$$

Then the metric $G = b^{-2}g$ is σ-temperate, and $h_G \leqslant h_g^\varepsilon$ where $\varepsilon = 1 - 2\delta > 0$. If $p_j \in O(H_j, g)$, $j = 1, 2$ and if $a \in S(g, p_1, p_2)$, then $p_j \in O(H_j, G)$ and $a \in S(G, p_1, p_2)$.

Proof. Corollary 1.3 implies that g satisfies the conditions of Lemma 1.3. Then (5.3) implies that G also satisfies the conditions of Lemma 1.3. Hence, G varies slowly. Since $G \geqslant g$, then p_j are G-continuous, and $a \in S(G, p_1, p_2)$.

Both the σ-temperateness of the metric G and the σ, G-temperateness of the functions p_j follow from the next lemma:

Lemma 5.4. *Let g be σ-temperate, let $p \in O(H, g)$, and let a slowly varying metric G satisfy the conditions $g \leqslant G \leqslant Ch_g^{-C}g$ and $h_G \leqslant Ch_g^\varepsilon$, where $\varepsilon > 0$. Then G is σ-temperate and $p \in O(H, G)$.*

Proof. If $G_Y(X - Y) > c$, then [see the proof of Lemma 3.2]

$$G_X^\sigma(\cdot)/G_Y^\sigma(\cdot) \leqslant C g_X^\sigma(\cdot)/g_Y^\sigma(\cdot) h_g(Y)^{-C} \leqslant C' h_G(Y)^{-C'}(1 + g_Y^\sigma(X - Y))^N$$
$$\leqslant C_1 h_G(Y)^{-C_1}(1 + G^\sigma(X - Y))^N \leqslant C_2(1 + G_Y^\sigma(X - Y))^{N_2};$$

and if $G_Y(X - Y) < c$, where $c > 0$ is sufficiently small, then this estimate is valid with $N_2 = 0$, since G varies slowly.

The inclusion $p \in O(H, G)$ is proved similarly.   ∎

Proof of Theorem 5.3.   Let $p \in O(H, g)$ be a positive function. Then $\tilde{p} = p + p^{-1} \in O(H, g)$ [the proof is similar to that of Proposition 1.1]; and since for positive-definite operators $A \leqslant B \Rightarrow \|A\| \leqslant \|B\|$ and since the conditions of g-continuity and σ, g-temperateness are, respectively,

$$p(X)^2 \leqslant C p(Y)^2, \qquad p(X)^2 \leqslant C p(Y)^2 (1 + g_Y^\sigma (X - Y))^N,$$

then $p_1 = \|\tilde{p}\| \in O(H, g)$. In addition, $p_1 \geqslant 1$.

Let us show that the pair $I \, [=p]$, $p_1 \, [=q]$ satisfies the conditions of Lemma 5.2, for which reason we construct, using Lemma 1.5, the function $P \in S(g, p_1)$, such that $P \asymp p_1$, and let $P_\lambda = P + \lambda$. Then $P_\lambda \asymp p_1 + \lambda$ uniformly with respect to $\lambda > 0$, and $P_\lambda \in O(H, g) \cap S(g, P_\lambda)$. Therefore, if $\delta \in (0, 1)$ and $k \geqslant 1$, then condition (5.2) implies that as $\lambda \to \infty$,

$$n_k(h^{-\delta} g, P_\lambda, P_\lambda) = \sup_X \sup_{0 \neq Y_j} \|P_\lambda(X)^{-1} (d^k P_\lambda)(X; Y_1, \ldots, Y_k)\|$$
$$\times \prod_{1 \leqslant j \leqslant k} (h^{-\delta}(X) g_X(Y_j))^{1/2}$$
$$\leqslant c_k \sup_X \|(P(X) + \lambda)^{-1} P(X)\| h(X)^{\delta/2} \to 0.$$

Therefore, Theorem 3.1 implies that if $A_\lambda = P_{\lambda, w}$ and $B_\lambda = (P_\lambda^{-1})_w$, then

(5.4) $$A_\lambda B_\lambda = I + r_{\lambda, w}, \quad \text{and} \quad B_\lambda A_\lambda = I + r'_{\lambda, w},$$

where r_λ, $r'_\lambda \to 0$ in $S(h^{-\delta} g, 1)$ as $\lambda \to \infty$. According to Theorem 4.2, for $\lambda \geqslant \lambda_0$, $A_\lambda B_\lambda$ and $B_\lambda A_\lambda$ are invertible in $L_2(R^n, H)$; and Lemma 5.2 yields $p_1 \in O'(H, g)$.

Part c) of Theorem 5.2 implies $p_1^{-1} \in O'(H, g)$. Letting $q = p p_1$, we construct, as above, operators $A_\lambda \in \mathcal{L}(g, q)$, $B_\lambda \in \mathcal{L}(g, q^{-1})$, for which (5.4) holds. Lemma 5.1 yields $\mathcal{H}(p_1^{-1}, H) = \mathcal{H}(p_1^{-1}, H, h^{-\delta} g)$; therefore, (5.4) and (5.1) imply that $A_\lambda B_\lambda$ and $B_\lambda A_\lambda$ are invertible in $\mathcal{H}(p_1^{-1}, H)$ for $\lambda \geqslant \lambda_0$; and Lemma 5.2 yields $p = q p_1^{-1} \in O'(H, g)$.   ∎

5.4. Definition of an admissible norm in $\mathcal{H}(p, H)$ [i.e., a norm that defines a topology] by the equality $\| \cdot \|_p = \|A^p \cdot \|_{L_2}$ is not always convenient. In order to show a simple method of introducing admissible norms, we need the following definition and lemma.

Definition 5.2.   Let $a \in S(g, p_1, p_2)$ be such that $a = a_1 + a_2$, where $\|p_2 a_1^{-1} p_1\| < c$, $a_2 \in S(g, h^\varepsilon p_1, p_2)$, and $\varepsilon = \varepsilon(a) > 0$. Then we write $a \in SI(g, p_1, p_2)$ and denote the corresponding class of operators by $\mathcal{L}I(g, p_1, p_2)$.

Example 5.1. Let A_{ij}, ord $A_{ij} = \ell_{1j} + \ell_{2j}$ $[i, j = 1, \ldots, m]$ be classical scalar PDOs with principal symbols a_{ij} and let

$$A = [A_{ij}]_{i,j=1,\ldots,m}.$$

Then A is called a classical PDO of order $(\ell_{11}, \ldots, \ell_{1m}; \ell_{21}, \ldots, \ell_{2m})$, and $a = [a_{ij}]$ is its principal symbol. Obviously, $A \in \mathcal{L}(g, p_1, p_2)$, where $g_{x,\xi}(y, \eta) = \langle \xi \rangle^{-2} |\eta|^2 + |y|^2$ and

$$p_s(x, \xi) = [\delta_{ij} \langle \xi \rangle^{\ell_{si}}]_{i,j=1,\ldots,m}.$$

If a is uniformly invertible on $R^n \times S_{n-1}$, then $A \in \mathcal{L}I(g, p_1, p_2)$ and is called a Douglis–Nirenberg elliptic operator [invertibility on $\Omega \times S_{n-1}$, where $\Omega \subset R^{n-1}$ is an open set, is usually required; and it is said to be Douglis–Nirenberg elliptic on Ω].

If $\ell_{ij} = \ell$ $\forall i, j$, then $\mathcal{L}I(g, p_1, p_2) = \mathcal{L}I(g, p) \stackrel{\text{def.}}{=} \mathcal{L}I(g, I, p) \cap \mathcal{L}I(g, p, I)$, where $p(x, \xi) = \langle \xi \rangle^{2\ell}$, and the operator A is called elliptic [a "local" definition is usually used here as well].

Lemma 5.5. *Let $a \in SI(g, p_1, p_2)$. Then there exists $b \in SI(g, p_2^{-1}, p_1^{-1})$, such that $b - a_1^{-1} \in S(g, h^{\varepsilon_0} p_2^{-1}, p_1^{-1})$, where $\varepsilon_0 = \min\{\varepsilon(a), 1\}$, and*

$$(5.5) \qquad a_w b_w - I \in \mathcal{L}^{-\infty}(g, p_1, p_1^{-1}), \qquad b_w a_w - I \in \mathcal{L}^{-\infty}(g, p_2^{-1}, p_2).$$

Operator b_w is called a paramatrix of the operator a_w.

Proof. Theorem 3.1 implies

$$a_w (a_1^{-1})_w - I = K \in \mathcal{L}(g, h^{\varepsilon_0} p_1, p_1^{-1}).$$

Applying Theorem 3.3, we construct the symbol

$$r \sim I - \sigma^w(K) + \sigma^w(K^2) - \sigma^w(K^3) + \cdots$$

and introduce $B_1 = (a_1^{-1})_w r_w \in \mathcal{L}I(g, p_2^{-1}, p_1^{-1})$.

This satisfies $(5.5)_1$. Similarly, to satisfy $(5.5)_2$ we construct B_2; but

$$\begin{aligned}
B_1 - B_2 &= B_1 + (B_2 a_w - I_{(2)}) B_1 a_w B_1 - B_2 + T_1 \\
&= B_1 - B_1 a_w B_1 + B_2 a_w B_1 a_w B_1 - B_2 + T_1 \\
&= (I_{(2)} - B_2 a_w) B_1 (I_{(1)} - a_w B_1) - B_2 (I_{(1)} - a_w B_1) + T_1 = T_2,
\end{aligned}$$

where $T_j \in \mathcal{L}^{-\infty}(g, p_2^{-1} p_1^{-1})$. Therefore, both B_1 and B_2 are paramatrices of the operator a_w. ∎

Remark. One-sided analogies of Definition 5.2 and Lemma 5.5 can be formulated without much difficulty.

Lemma 5.6. *Let $p \in O'(H, g)$, $cI \leqslant p \leqslant Ch^{-C}I$, let $a_1, \ldots, a_s \in \mathcal{L}(g, I, p)$, and let $p^2 \leqslant C \sum\limits_{1 \leqslant j \leqslant s} a_j^* a_j$. Then the norm*

$$\|u\|_p' = \|u\|_{L_2} + \sum_{1 \leqslant j \leqslant s} \|a_{j,w} u\|_{L_2}$$

is admissible in $\mathcal{H}(p, H)$.

Proof. Let $A = \sum\limits_{j} a_{j,w}^* a_{jw} \in \mathcal{L}I(g, p, p)$ and let B be a paramatrix of the operator A. Then for any admissible norm $\| \cdot \|_p$ in $\mathcal{H}(p, H)$, we have

$$\|u\|_p \leqslant C \|A^p u\|_{L_2} \leqslant C_1 \sum_{j} \|A^p B a_{j,w}^*\|_{L_2 \to L_2} \|a_{j,w} u\|_{L_2} + \|A^p T\|_{L_2 \to L_2} \|u\|_{L_2},$$

where $T \in \mathcal{L}^{-\infty}(g, I)$.

Theorem 3.1 implies $A^p B a_{j,w}^*$, $A^p T \in \mathcal{L}(g, I)$; hence, Theorem 4.2 yields $\|u\|_p \leqslant C\|u\|_p'$. The inequality $\|u\|_p' \leqslant C\|u\|_p$ is a corollary of part d) of Theorem 5.2.

§6 Action of Pseudodifferential Operators in Weight Spaces

6.1. In this section all H_j are finite-dimensional and $p_j \in O(H_j; g)$.

Theorem 6.1.
  a) *If $p_j \in O'(H, g)$ for $j = 1, 2$ and if $\|p_1(X) p_2(X)^{-1}\| \to 0$ as $|X| \to \infty$, then $\mathcal{H}(p_2, H) \subset \mathcal{H}(p_1, H)$ is compact.*
  b) *If $p_j \in O'(H_j, g)$ for $j = 1, 2$ and if for all k,*

$$(6.1) \qquad\qquad |a|_{g,k}^{p_1^{-1}, p_2}(X) \to 0 \quad \text{as} \quad |X| \to \infty,$$

then $a_w : \mathcal{H}(p_2, H_2) \to \mathcal{H}(p_1, H_1)$ is a compact operator.

Proof.
  a) Let $a = I$. Then $a_w : \mathcal{H}(p_2, H_2) \to \mathcal{H}(p_1, H_1)$ is an inclusion operator that satisfies condition (6.1). Therefore, it is sufficient to prove part b).
  b) This part reduces to Theorem 4.3 in the same way that estimate (5.1) reduces to Theorem 4.2. ∎

Theorem 6.2. *Let the metric g satisfy condition (5.2) and let $a \in SI(g, p_1^{-1}, p_2)$. Then $a_w : \mathcal{H}(p_2, H_2) \to \mathcal{H}(p_1, H_1)$ is a Fredholm operator; i.e., its image is closed, and $\dim \operatorname{Ker} A < \infty$, $\operatorname{codim} \operatorname{Im} A < \infty$.*

Proof. Let b_w be a paramatrix of the operator a_w. Then

$$(6.2) \qquad\qquad a_w b_w = I + T_1, \qquad b_w a_w = I + T_2,$$

where operators $T_1 \in \mathcal{L}^{-\infty}(g, p_1^{-1}, p_1)$ and $T_2 \in \mathcal{L}^{-\infty}(g, p_2^{-1}, p_2)$ are compact in $\mathcal{H}(p_1, H_1)$ and $\mathcal{H}(p_2, H_2)$, respectively, as a result of condition (5.2) and of part b) of Theorem 6.1. Hence, $b_w : \mathcal{H}(p_1, H_1) \to \mathcal{H}(p_2, H_2)$ is a regularizator of the operator $a_w : \mathcal{H}(p_2, H_2) \to \mathcal{H}(p_1, H_1)$, and the latter is a Fredholm operator.   ∎

6.2.

Lemma 1. *Let there exist $C, \varepsilon > 0$ such that*

$$(6.3) \qquad\qquad h(X) \leqslant C\langle X \rangle^{-\varepsilon} \quad \forall X.$$

Then

$$(6.4) \qquad S^{-\infty}(g, p_1, p_2) = S^{-\infty}(g, I_{(1)}, I_{(2)}) = \mathcal{S}(R^{2n}; B_{12}),$$

$$(6.5) \qquad \mathcal{H}^{-\infty}(H) \overset{\text{def.}}{=} \bigcap_{p \in O(H;g)} \mathcal{H}(p, H) = \bigcap_N \mathcal{H}(h^N; H) = \mathcal{S}(R^n; H),$$

and the operator $a_w \in \mathcal{L}^{-\infty}(g, p_1, p_2)$ is a continuous operator from $\mathcal{S}'(R^n; H_2)$ to $\mathcal{S}(R^n; H_1)$.

Proof.   (6.4) follows from (6.3) and (3.19). (6.4) implies that a_w is an integral operator with the kernel from the class $\mathcal{S}(R^{2n}; B_{12})$, which yields the last statement of the theorem.

Finally, (6.3) and (3.19) imply that any linear function p_j belongs to $S(g, h^m)$ for some $m = m(p_j)$. Therefore, for any $u \in \mathcal{H}^{-\infty}(H)$ and any linear functions p_j,

$$p_{1,w} \cdots p_{k,w} u \in \mathcal{H}^{-\infty}(H) \subset L_2(R^n; H).$$

Hence, $\mathcal{H}^{-\infty}(H) \subset \mathcal{S}(R^n; H)$; but $\mathcal{S}(R^n; H) \subset \mathcal{H}^{-\infty}(H)$, according to Theorem 5.1.   ∎

From this point on, condition (6.3) is assumed to hold.

Corollary 6.1. $\mathcal{S}'(R^n; H) = \bigcup_N \mathcal{H}(ph^N, H) \quad \forall p \in O(H; g).$ ∎

Corollary 6.2.   *Provided that the conditions of Theorem 6.2 hold, we have*

$$(6.6) \qquad \operatorname{Ker} A \subset \mathcal{S}(R^n; H_2), \qquad \operatorname{Ker} A^* \subset \mathcal{S}(R^n; H_1).$$

Proof.   Let B be a paramatrix of the operator A. Then, according to Lemma 6.1,

$$\forall u \in \operatorname{Ker} A, \quad u = (I - BA)u \in \mathcal{S}(R^n; H_2),$$

and $(6.6)_1$ is proved. $(6.6)_2$ is proved exactly the same way, since $A^* \in \mathcal{L}I(g, p_2, p_1^{-1})$.   ∎

Lemma 6.2. *Let* $A \in \mathcal{L}I(g, p_1^{-1}, p_2)$ *and let* $A : \mathcal{H}(p_2 h^s; H_2) \to \mathcal{H}(p_1 h^s, H_1)$ *be invertible for some s.*
  Then the same holds for all s.

Proof. Theorem 6.2 and Corollary 6.2 imply that the operators

$$A : \mathcal{H}(p_2 h^s; H_2) \to \mathcal{H}(p_1 h^s; H_1), \qquad A^* : \mathcal{H}(p_1^{-1} h^{-s}; H_1) \to \mathcal{H}(p_2^{-1} h^{-s}; H_2)$$

are Fredholm operators for any s and that their kernels do not depend on s.
    ∎

  Corollary 6.1 and Lemma 6.2 yield the following corollary:

Corollary 6.3.   *Under the conditions of Lemma 6.2,* $A : \mathcal{S}'(R^n; H_2) \to \mathcal{S}'(R^n; H_1)$ *is invertible.*
    ∎

Lemma 6.3.   *Under the conditions of Lemma 6.2,* $A^{-1} \in \mathcal{L}(g, p_2^{-1}, p_1)$.

Proof. Let B be a paramatrix of the operator A. Then

$$T = A^{-1} - B = (I_{(2)} - BA)A^{-1}(I_{(1)} - AB) - B(I_{(1)} - AB)$$

maps $\mathcal{S}'(R^n, H_1)$ to $\mathcal{S}(R^n; H_2)$, as a result of Lemma 6.1 and Corollary 6.3. Therefore, T is an integral operator with a kernel of the class $\mathcal{S}(R^{2n}; B_{21})$, and (3.10) yields $T \in \mathcal{L}^{-\infty}(g, p_2^{-1}, p_1)$.
  Thus, $A^{-1} = B + T \in \mathcal{L}(g, p_2^{-1} p_1)$.
    ∎

6.3. Let $p_1, p_2 \in O(H; g)$, $p_2 \geqslant cI$, let one of the functions p_1, p_2 be scalar, and let $a \in S(g, p_2)$. Then an operator A_0 in $\mathcal{H}(p_1, H)$ with a domain $\mathcal{H}(p_1 p_2, H)$ can be defined by the equality

$$A_0 u = a_w u \quad \forall u \in D(A_0) = \mathcal{H}(p_1 p_2, H).$$

Theorem 6.3.   *Let* $a \in SI(g, p_2)$ *and let*

$$\|p_2(X)^{-1}\| \to 0 \quad \text{as} \quad |X| \to \infty.$$

Then
  a) *a spectrum of the operator* A_0 *is either* $\mathbb{C}$ *or any discrete set, and each point of its spectrum is an eigenvalue of finite multiplicity;*
  b) *all eigenfunctions and adjoint functions of the operator* A_0 *belong to* $\mathcal{S}(R^n, H)$ *and, therefore, do not depend on* p_1;

c) *for any z from the resolvent set, $(A_0 - zI)^{-1} \in \mathcal{L}(g, p_2^{-1})$.*

Proof. For any $z \in \mathbf{C}$, there exists $C(z) > 0$ such that for $|X| > C(z)$, $\|(a(X) - zI)^{-1}p_2(X)\| \leqslant C(z)$. Therefore, $a - zI \in SI(g, p_2)$; and Theorem 6.2 implies that $a_w - zI : \mathcal{H}(p_1 p_2, H) \to \mathcal{H}(p_1, H)$ is a Fredholm operator. In particular, each point of the spectrum is an eigenvalue of finite multiplicity.

If $A_0 - z_0 I$ is invertible with the inverse $B : \mathcal{H}(p_1, H) \to \mathcal{H}(p_1 p_2, H)$, then Lemma 6.3 implies $B \in \mathcal{L}(g, p_2^{-1})$, and Theorem 6.1 implies that $B : \mathcal{H}(p_1, H) \to H(p_1, H)$ is compact. But $(A_0 - zI)B = I + (z_0 - z)B$; and, hence, $A_0 - zI$ is invertible for all z, except for z from a discrete set.

If u is an eigenfunction or an adjoint function, then $(a_w - zI)^N u = 0$ for some N. By applying a parametrix of the operator $a_w - zI$ to this equality N times, we obtain $u = Tu$, where $T \in \mathcal{L}^{-\infty}(g, 1)$. Lemma 6.1 yields $u \in \mathcal{S}(R^n; H)$. ∎

6.4. Let $p \in O(H, g)$, $p \geqslant cI$, let $a \in SI(g, p)$, and let $A_0 = a_w$ be an operator in $L_2(R^n, H)$ with a domain $\mathcal{H}(p, H)$.

Theorem 6.4.   *The operator A_0 is closed.*

Proof. Let $u_n \in \mathcal{H}(p, H)$, $n = 1, 2, \ldots$ and let $u_n \to u$, $A_0 u_n \to f$ in $L_2(R^n, H)$. Since convergence in $L_2(R^n, H)$ implies convergence in $\mathcal{S}'(R^n, H)$, and, according to Theorem 4.1, a_w is continuous in $\mathcal{S}'(R^n, H)$, we can conclude that $a_w u = f$. By applying the parametrix $b_w \in \mathcal{L}(g, p^{-1})$ of the operator a_w to this equality, we obtain $u = Tu + b_w f$, where $T \in \mathcal{L}^{-\infty}(g, I)$.

Part b) of Theorem 5.1 implies that $u \in \mathcal{H}(p, H)$ and $A_0 u = a_w u = f$. ∎

Since $C_0^\infty(R^n; H) \subset \mathcal{H}(p; H)$ densely [the proof is the same as in part b) of Theorem 5.2], then the following corollary is valid:

Corollary 6.4.   *The operator A_0 is a closure of the operator*

$$A = a_w : C_0^\infty(R^n; H) \to L_2(R^n; H).$$

Theorem 6.5.   *Let $a = a^* \in SI(g, p)$, $p \geqslant cI$ and let A_0, A be the same as in Theorem 6.4 and Corollary 6.4. Then the operator A is essentially self-adjoint; i.e., its closure A_0 is a self-adjoint operator.*

Proof. The calculations used in the proof of Theorem 5.3 show that $A_0 \pm it I : \mathcal{H}(p, H) \to L_2(R^n; H)$ are invertible for large t; hence, A_0 is self-adjoint. ∎

Example. Let $A = -\Delta + |x|^2$. Then $A \in \mathcal{L}I(g, p)$, where $g_X(Y) = \langle X \rangle^{-2}|Y|^2$ and $p(X) = \langle X \rangle^2$; and the closure A_0 of the operator $A : C_0^\infty(R^n) \to L_2(R^n)$ is a self-adjoint operator with the domain

$$D(A_0) = \mathcal{H}(p, C^1) = \{u \in L_2(R^n) \mid \Delta u, \langle x \rangle^2 u \in L_2(R^n)\}$$

and with a discrete spectrum. Its eigenfunctions belong to $\mathcal{S}(R^n)$.

CHAPTER 2

BASIC THEOREMS OF THE METHOD OF APPROXIMATE SPECTRAL PROJECTION FOR SCALAR AND MATRIX OPERATORS

§7 Formulation of the Basic Theorems

7.1. Let $\mathcal{A}$ be a quadratic form semibounded from below and defined on a vector space V. Let

$$\mathcal{N}(\mathcal{A}, V) = \sup_{L \subset V} \{\dim L \mid \mathcal{A}[u] < 0 \quad \forall (0 \neq) u \in L\}.$$

If A is a symmetric operator in a Hilbert space H, if $V \subset D(A)$ (to the domain of A), and if $\mathcal{A}[u] = \langle Au, u \rangle_H$, then $\mathcal{N}(A, V) = \mathcal{N}(\mathcal{A}, V)$ (H will be specified in context). If $V = C_0^\infty(\Omega, H)$ and if H is specified from context, then we define $\mathcal{N}_0(\mathcal{A}, \Omega) = \mathcal{N}(\mathcal{A}, C_0^\infty(\Omega, H))$,

$$\mathcal{N}_0(A, \Omega) = \mathcal{N}(\langle A \cdot, \cdot \rangle_{L_2}, C_0^\infty(\Omega, H)).$$

7.2. As shown in the Introduction, we must obtain the estimates for a number of negative eigenvalues of the operator A_t that depend on t. If $\overset{\circ}{a}_{t,w}$ is a PDO in $S(R^n)^m$ and if A_t is an operator associated with the variational triple $\langle \overset{\circ}{a}_{t,w} \cdot, \cdot \rangle_{L_2}$, $C_0^\infty(\Omega_t)^m$, $L_2(\Omega_t)^m$, then $\mathcal{N}(A_t, D(A_t)) = \mathcal{N}_0(\overset{\circ}{a}_{t,w}, \Omega_t)$; therefore, it is sufficient to obtain estimates for the latter function.

We shall assume that the following conditions hold:

A. There exists a σ-temperate metric g_t [splitting if $\Omega_t \neq R^n$] and a function $\overset{\circ}{q}_t \in O(C^m; g_t)$ such that

(7.1) $$\overset{\circ}{a}_t \in S(g_t, \overset{\circ}{q}_t, \overset{\circ}{q}_t);$$

46

$\exists \varepsilon > 0, C$ such that

$$(7.2) \qquad h_t(X) = h_{g_t}(X) \leqslant C\langle X \rangle^{-\varepsilon} \quad \forall X \in R^{2n};$$

there exist $c > 0, R_t$ such that for $|X| > R_t$,

$$(7.3) \qquad \overset{\circ}{a}_t(X) > c\overset{\circ}{q}_t(X)^2.$$

Condition (7.3) provides finiteness of $\mathcal{N}_0(\overset{\circ}{a}_{t,w}, \Omega_t)$; under condition (7.2), commutators of PDO from an algebra containing $\overset{\circ}{a}_{t,w}$ are relatively small. If $\Omega_t \neq R^n$, then we must construct PDO with cutoff-type symbols in a neighborhood of $\partial\Omega_t$ and must use Theorem 3.2, which is valid for some σ-temperate metrics but not for all. This explains the condition "metric g_t splits."

In the formulation of basic theorems we shall use the following:

1) a function $q_t \in SI(g_t, \overset{\circ}{q}_t^{-1}, I)$ [by Lemma 1.5 such functions exist];

2) the matrix $\overset{\circ}{a}_t = \sigma^w(q_{t,w}^* \overset{\circ}{a}_{t,w} q_{t,w})$ or any matrix $a_t = a_t^* \in S(g_t, I)$ satisfying the condition

$$(7.4) \qquad a_t - \overset{\circ}{a}_t \in S(g_t, d_t),$$

where $d_t \in S(g_t, d_t)$ and

$$(7.5) \qquad d_t(X) \to 0 \quad \text{as} \quad |X| \to \infty;$$

3) sets and functions that depend on the matrix symbol a_t, on a scalar function f_t, on a constant $c > 0$, and also on a set $\mathcal{M} \subset R^{2n}$:

$$J(\mathcal{M}, a_t, f_t) = (2\pi)^{-n} \int_{\mathcal{M}} N(0, a_t(X)) f_t(X) \, dX,$$

$$V(\mathcal{M}, a_t) = J(\mathcal{M}, a_t, 1),$$

$$W_c(\mathcal{M}, a_t, f_t) = V(\mathcal{M}, a_t - cf_t) - V(\mathcal{M}, a_t + cf_t).$$

All constants characterizing the metric g_t and the functions $\overset{\circ}{a}_t$, a_t, $\overset{\circ}{q}_t$, q_t are bounded [and c in (1.6) is separated from zero] uniformly with respect to t; this remark also applies to all the functions introduced later in conditions B and C. Analysis of the theorems in §1, which will be used to prove all results of this section, then shows that the constants characterizing auxiliary metrics and functions to be constructed in the proof process, will also be bounded [or separated from zero] uniformly with respect to t, despite the dependence on t of both the constant R_t in condition (7.3) and the set Ω_t. Therefore, in this section we omit the index t, although, generally speaking, all the sets, functions, metrics, and operators do depend on t. Unless we specify otherwise, all the constants from here on can be chosen to be the same for all t.

Theorem 7.1. *If condition A is satisfied, then for any $\delta \in (0, \frac{1}{3})$, there exist $c, C > 0$ such that*

$$(7.6) \quad |\mathcal{N}_0(\mathring{a}_w, \Omega) - V(\Omega \times R^n, a)| \leqslant C(W_c(\Omega \times R^n, a, d + h^\delta)$$
$$+ V((\partial\Omega \times R^n)(c, h^{-2\delta}g), a - c(d + h^\delta)) + \mathcal{J}_c^\delta(a, d) + 1),$$

where

$$\mathcal{J}_c^\delta(a, d) = \mathcal{J}_c^\delta(\Omega, a, d) = \mathcal{J}((\Omega \times R^n)(c, h^{-2\delta}g), a - c(d, h^\delta), h^\delta).$$

7.3. In regular situations the right-hand side of (7.6) increases [as $t \to \infty$] slower than the left-hand side; therefore, (7.6) yields the classical formula

$$\mathcal{N}_0(\mathring{a}_{t,w}, \Omega_t) \sim V(\Omega_t \times R^n, a_t)$$

with some estimate for the remainder. This estimate can be sharpened by imposing additional constraints, such as the following condition:

B. There exist sets $\Gamma\Omega$ and $U_1, \ldots, U_{n_0} \subset \overline{\Omega} \times R^n$ and functions

$$(7.7) \qquad\qquad e_i \in S(g, I, U_i), \qquad \mathring{\lambda}_{ij} = \overline{\mathring{\lambda}_{ij}} \in S(g, 1)$$

$[i = 1, \ldots, n_0, \; j = 1, \ldots, m]$ such that for $\partial\Omega \times R^n \subset \Gamma\Omega$, for $U_i \cap U_j = \emptyset$ $\forall i \neq j$, and for $\overline{\Omega} \times R^n \backslash \Gamma\Omega = \bigcup U_j$,

$$(7.8) \qquad\qquad e_i^*(X)\, e_i(X) = e_i(X)\, e_i^*(X) = I \quad \forall X \in U_i,$$

$$(7.9) \qquad e_i^*(X)\, a(X)\, e_i(X) = [\delta_{jk}\mathring{\lambda}_{ij}(X)]_{i,j=1}^m \quad \forall X \in U_i,$$

where δ_{ij} is the Kronecker symbol.

Theorem 7.2. *If conditions A and B hold, then for any $\delta \in (0, \frac{1}{2})$, there exist $c, C > 0$ such that*

$$(7.10) \quad |\mathcal{N}_0(\mathring{a}_w, \Omega) - V(\Omega \times R^n, a)| \leqslant C(W_c(\Omega \times R^n, a, h^\delta + d)$$
$$+ V((\Gamma\Omega)(c, h^{-2\delta}g), a - c(d + h^\delta)) + \mathcal{J}_c^\delta(a, d) + 1).$$

If the boundary $\partial\Omega$ is piecewise smooth, then the estimate for the remainder can be improved. For this we need the following condition:

C. Let condition B hold and let $\Gamma\Omega = \Gamma_1\Omega \cup \Gamma_2\Omega$—where $\Gamma_j\Omega = \{X \in \overline{\Omega} \times R^n \mid \varphi_j(X) = 0\}$,

$$(7.11) \qquad\qquad \varphi_j = \overline{\varphi}_j \in S(g, 1),$$

φ_j does not depend on ξ, and $\partial\Omega \times R^n \subset \Gamma_1\Omega$.

Theorem 7.3. *If conditions* A *and* C *hold, then for any* $\delta \in (0, \frac{2}{3})$, *there exist* $c, C > 0$ *such that*

(7.12) $$|\mathcal{N}_0(\overset{\circ}{a}_w, \Omega) - V(\Omega \times R^n, a)| \leqslant C(W_c^\delta(a, d) + \mathcal{J}_c^\delta(a, d) + 1),$$

where

$$W_c^\delta(a, d) = W_c^\delta(a, d, \Omega) = \sum_{1 \leqslant j \leqslant m} \text{mes } \widetilde{W}_{c,j}^\delta(a, d),$$

(7.12′)
$$\widetilde{W}_{c,j}^\delta(a, d) = \{X \in (\Omega \times R^n)(c, h^{-\delta/2}g) \mid \lambda_1(X) < c(h^{\delta/2} + d)(X),$$
$$|(\varphi_1\varphi_2\lambda_j)(X)| \leqslant c(h^\delta(X) + d(X))\},$$

and λ_j *are eigenvalues of the matrix* a.

Remarks.
   a) If $m = 1$, then $\Gamma\Omega = \partial\Omega \times R^n$, and the term $\mathcal{J}_c^\delta(a, d)$ can be omitted.
   b) If $\Omega = R^n$, then the terms

$$V((\partial\Omega \times R^n)(c, h^{-2\delta}g), a - c(d + h^\delta)), \quad V((\Gamma_1\Omega)(c, h^{-2\delta}g), a - c(d + h^\delta))$$

in (7.6) and (7.10) can be omitted, and in (7.12) we can assume that $\varphi_1 = 1$.
   c) If $\Gamma_1\Omega = \emptyset$, then in (7.12) we have $\varphi_2 = 1$.

7.4. Now let $\overset{\circ}{a}_{t,w} = A - tI$, where A is a Douglis–Nirenberg elliptic operator. Theorems 7.1–7.3 hold in this case, too, but they do not take into account the block structure of the operator A, which leads, in general, to a less accurate result.

The following theorem does take into account the block structure of the operator $\overset{\circ}{a}_w$. In order to formulate it, we denote the right-hand sides of estimates (7.6), (7.10), and (7.12) by $(RV)_c^\delta(\Omega, a, d)$—the choice of the formula [and of δ] is determined by conditions on the function a and on the set Ω—and we suppose that the matrix a is blockwise diagonal:

(7.13) $$a = [\delta_{jk}(a_j + a_j')]_{j,k=1}^s,$$

where a_j, a_j' are $m_j \times m_j$ matrices, $a_j' \in S(g, d_j)$, and the function d_j satisfies the same conditions as the function d.

Then the following theorem holds:

Theorem 7.4. *There exist* $c, C > 0$ *such that*

(7.14) $$\left|\mathcal{N}(\overset{\circ}{a}_w, \Omega) - \sum_{1 \leqslant j \leqslant s} V(\Omega \times R^n, a_j)\right| \leqslant C \sum_{1 \leqslant j \leqslant s} (RV)_c^{\delta_j}(\Omega, a_j, d + d_j).$$

7.5. Now, instead of conditions (7.2) and (7.5), let

$$(7.15) \qquad \max_X h_t(X) \to 0, \quad \max_X d_t(X) \to 0 \qquad \text{as} \quad t \to \infty.$$

Theorem 7.5. *Let conditions (7.1), (7.3), (7.4), and (7.15) hold, and let the metric g_t split. Then there exists t_0 such that for $t \geqslant t_0$, estimate (7.6) holds.*

7.6. Proofs of Theorems 7.1–7.5 are given in §§9–12; and §8 contains proofs of some auxiliary propositions. In order not to overshadow the basic ideas of the ASP method with technical details, we shall first (in §9 and §10) prove Theorems 7.2 and 7.3 for scalar operators. In §11 the same theorems are proved for the matrix case, and in §12 we prove Theorems 7.1, 7.4, and 7.5.

§8 Auxiliary Propositions

8.1. Let $a = a^* \in S(g, p^*, p)$ admit the representation $a = a_1 + a_2$, where $a_1 \geqslant cp^*p$, $a_2 \in S(g, h^\varepsilon p^*, p)$, and $\varepsilon = \varepsilon(a) > 0$, $c = c(a) > 0$. Then we write $a \in SI^+(g, p^*, p)$ and we denote $SI^+(g, p) = S(g, p) \cap SI^+(g, p^{1/2}, p^{1/2})$.

Lemma 8.1. *Let $a \in SI^+(g, p^*, p)$. Then there exists a symbol $b \in SI(g, I, p)$ such that*

$$(8.1) \qquad a_w = b_w^* b_w + T, \quad \text{where } T \in \mathcal{L}^{-\infty}(g, p^*, p),$$

and

$$(8.2) \qquad a_1^{1/2} - b \in S(g, h^{\varepsilon_0}, p), \quad \text{where } \varepsilon_0 = \min\{\varepsilon(a), 1\}.$$

Proof. Using Lemma 1.6, we construct the function $b_1 = a_1^{1/2} \in S(g, I, p)$. Part b) of Lemma 1.1 implies that $q_0 = b_1^{-1} \in S(q, p^{-1}, I)$; thus, by Theorem 3.1, we have for $i = 0$,

$$(8.3) \qquad q_{i,w}^* a_w q_{i,w} = I + k_{i,w}, \quad \text{where } k_i \in S(g, h^{(i+1)\varepsilon_0}).$$

Let $q_1 = \sigma^w(q_{0,w}(I - \frac{1}{2}k_{0,w}))$; then (8.3) is valid for $i = 1$. By induction, for all $i \geqslant 0$, we construct the symbols $q_i \in S(g, p^{-1}, I)$, which satisfy condition (8.3) and the condition

$$q_i - q_{i+1} \in S(g, p^{-1}, h^{(i+1)\varepsilon_0}).$$

Let

$$(S(g, p^{-1}, I) \ni)q \sim q_0 + \sum_{i \geqslant 0}(q_{i+1} - q_1)$$

be a symbol defined by Theorem 3.3. Then

$$(8.4) \qquad\qquad q_w^* a_w q_w = I + T, \quad \text{where } T \in \mathcal{L}^{-\infty}(g, I).$$

Since $q_0 \in SI(g, p^{-1}, I)$, then $q \in SI(g, p^{-1}, I)$. Lemma 5.5 implies that there exists a symbol $b \in SI(g, I, p)$ that satisfies condition (8.2) and the condition $q_w b_w - I \in \mathcal{L}^{-\infty}(g, p^{-1}, p)$. Thus, by multiplying equality (8.4) by b_w on the right and by b_w^* on the left, we obtain (8.1). ∎

8.2. In the following three lemmas, we assume that $\dim H_j < \infty$ and that the metric g satisfies condition (7.2).

Lemma 8.2. *Let* $\omega > 0$ *and let* $a \in S(g, h^\omega I_{(2)}, I_{(1)})$. *Then for any* $\varepsilon > 0$, *there exists a subspace* $L \subset L_2(R^n; H_1)$ *and a constant* C *such that*

$$\operatorname{codim} L \leqslant C, \qquad \text{and} \qquad \|a_w u\|_{L_2} \leqslant \varepsilon \|u\|_{L_2} \quad \forall u \in L.$$

The proof will be given at the end of §9.

Lemma 8.3. *Let* $a \in SI(g, p_1, p_2)$ *and let* $A = a_w : \mathcal{H}(p_2, H_2) \to \mathcal{H}(p_1^{-1}, H_1)$. *Then there exists* C *such that*

$$(8.5) \qquad\qquad \dim \operatorname{Ker} A \leqslant C.$$

Proof. According to Theorem 5.3, $\mathcal{H}(p_j, H_j)$ are Hilbert spaces and there exist isomorphisms

$$\mathcal{L}I(g, p_2^{-1}) \ni A_2 : L_2(R^n; H_2) \to \mathcal{H}(p_2; H_2),$$
$$\mathcal{L}I(g, p_1) \ni A_1 : \mathcal{H}(p_1^{-1}, H_1) \to L_2(R^n; H_1).$$

Therefore, it is sufficient to prove (8.5) for the operator $A_0 = A_1 A A_2 : L_2 \to L_2$. But $A_0 \in \mathcal{L}I(g, I)$; therefore, Lemma 5.5 implies that there exists $B \in \mathcal{L}(g, I)$ such that $BA_0 = I + K$, where $K \in \mathcal{L}(g, h)$; and Lemma 8.2 implies $\dim \operatorname{Ker} A_0 \leqslant C$. ∎

Lemma 8.4. *If* $a \in SI^+(g, p^*, p)$, *then there exists a constant* C *and a subspace* $L \subset \mathcal{H}(p, H)$ *such that* $\operatorname{codim} L \leqslant C$ *and*

$$(8.6) \qquad\qquad \langle a_w u, u \rangle_{L_2} > 0 \quad \forall (0 \neq) u \in L.$$

Proof. Let $(\mathcal{L}I(g, p^{-1}) \ni) P : L_2(R^n; H) \to \mathcal{H}(p, H)$ be isomorphisms. Then it suffices to prove the lemma for the operator

$$(\mathcal{L}I^+(g, I) \ni) A' = P^* a_w P : L_2(R^n; H) \to L_2(R^n; H).$$

Lemma 8.1 implies that $A' = B^*B + T$, where $B \in \mathcal{L}I(g, I)$ and $T \in \mathcal{L}^{-\infty}(g, 1)$. Lemmas 8.2 and 8.3 enable us to find a subspace $L \subset L_2(R^n; H)$ of finite codimension such that for all $u \in L$,

$$\|Bu\|_{L_2} \geqslant c\|u\|_{L_2}, \qquad \|Tu\| < \frac{c^2}{2}\|u\|_{L_2}.$$

This yields (8.6). ∎

Remark. If the metric g and the functions a, p_1, p_2, p in Lemmas 8.2–8.4 depend uniformly on some parameter, then constants C can be chosen to be the same for all values of this parameter. This follows from the theorems in Chapter 1 and from the way we constructed the isomorphisms

$$A^p : \mathcal{H}(p, H) \to L_2(R^n; H), \quad A_p : L_2(R^n; H) \to \mathcal{H}(p; H)$$

in Theorem 5.3: if g, p depend uniformly on some parameter, then the same also holds for these isomorphisms. This remark is a basis of the proof of the theorems in §7.

8.3. Now let the metric g_t and the functions a_t, $p_{j,t}$, p_t depend uniformly on t and let condition $(7.15)_1$ be satisfied. Then Theorem 4.2 implies the following analogue of Lemma 8.2:

Lemma 8.2′. *Let $\omega > 0$ and let $a_t \in S(g_t, h_t^\omega I_{(2)}, I_{(1)})$. Then for any $\varepsilon > 0$, there exists t_0 such that $\|a_{t,w}\|_{L_2 \to L_2} < \varepsilon$ for $t \geqslant t_0$.* ∎

In order to obtain analogues of Lemmas 8.3 and 8.4, we notice that if $a_t \in SI(g_t, p_t)$, $\|p_t a_t^{-1}\| \leqslant c$ and $A_t = a_{t,w}$, $B_t = (a_t^{-1})_w$, then

$$(8.7) \qquad A_t B_t = I + T_{1,t}, \qquad B_t A_t = I + T_{2,t},$$

where $\sigma^w(T_{j,t}) \to 0$ in $S(g_t, I)$ by $(7.15)_1$. By applying (8.7) and Lemma 8.2 and by repeating the proof of Theorem 5.3, we obtain $p_t \in O'(H, g_t)$, provided that t is sufficiently small.

Therefore, by using Lemma 8.2′ instead of Lemma 8.2 and by repeating the proofs of Lemmas 8.3 and 8.4, we obtain their analogues:

Lemma 8.3′. *Let $a_t \in SI(g_t, p_{1,t}, p_{2,t})$. Then $a_{t,w} : \mathcal{H}(p_{2,t}, H_2) \to \mathcal{H}(p_{1,t}^{-1}, H_1)$ is invertible for $t \geqslant t_0$.*

Lemma 8.4′. *Let $a_t \in SI^+(g_t, p_t^*, p_t)$. Then for $t \geqslant t_0$,*

$$\langle a_{t,w}u, u \rangle > 0 \quad \forall (0 \neq)u \in \mathcal{H}(p_t; H_t).$$

8.4.

Lemma 8.5. *Let $A = A^*$ be an integral operator with the kernel $K \in \mathcal{S}(R^{2n};$ End $C^m)$. Then A is of trace class and*

$$\operatorname{Tr} A = \int \operatorname{Tr} K(x,x)dx.$$

Proof. Let $u_i(x) = u_{i1}(x), \ldots, u_{im}(x)$ be an eigenfunction on the operator A with eigenvalue ν_i. For a fixed x, a jth row of the matrix $K(x,y)$ belongs to $L_2(R^n)^m$; and by writing it in terms of the basis $\{u_j\}$, we have

$$K_{jj}(x,x) = \sum_i \nu_i |u_{ij}(x)|^2$$

and the following formula for a trace of the matrix $K(x,x)$:

$$\operatorname{Tr} K(x,x) = \sum_{1 \leqslant j \leqslant m} \sum_i \nu_i |u_{ij}(x)|^2.$$

The sequence

$$S_N(x,x) = \sum_j \sum_{i \leqslant N} \nu_i |u_{ij}(x)|^2$$

is fundamental in $l_1(R^n)$; therefore, the Lebesque theorem yields

$$\sum_i \nu_i = \lim_{N \to \infty} \sum_{i \leqslant N} \nu_i = \lim_{N \to \infty} \int S_N(x,x)dx$$

$$= \int \lim_{N \to \infty} S_N(x,x)dx = \lim_{N \to \infty} \int \operatorname{Tr} K(x,x)dx,$$

since the operation of trace and integration commute. ∎

If $a \in \mathcal{S}(R^{2n}; \text{End } C^m)$, then $A = a_w$ satisfies the condition of Lemma 8.5, and

$$K(x,x) = \int a(x,\xi)d'\xi;$$

therefore, Lemma 8.5 implies the next lemma:

Lemma 8.5′. $\operatorname{Tr} a_w = \iint \operatorname{Tr} a(x,\xi)dxd'\xi.$

§9 Proof of Theorem 7.2 for the Scalar Case.

9.1. Lemmas 2.5 and 1.5 imply that there exists $h' \asymp h$ such that $h' \in S(g, h)$. The metric $g' = h'h^{-1}g$ is equivalent to g, and $h_{g'} = h' \in S(g', h')$. Therefore, we can assume that

$$(9.1) \qquad\qquad h \in S(g, h).$$

We fix a function $\overset{\circ}{\chi} \in C^{\infty}(R^1)$, $0 \leqslant \overset{\circ}{\chi} \leqslant 1$, $\overset{\circ}{\chi}\big|_{t<0} = 1$, $\overset{\circ}{\chi}\big|_{t>1} = 0$ and the constants $\omega \in (0, \frac{1}{4})$ and $C_0 > 0$; and we denote $\delta = \frac{1}{2} - \omega$, $\lambda_0 = h^{\delta - \omega}$, $a^- = a + 8C_0 d + 4\lambda_0$, and $\chi = \overset{\circ}{\chi}(a^- h^{-\delta})$.

Lemma 9.1. $\chi \in S(h^{-2\delta}g, 1)$.

Proof. For $k \geqslant 1$, $n_k(h^{-2\delta}g, 1, \chi)$ is estimated by summing the products of the seminorms $n_s(h^{-2\delta}g, 1, a^- h^{-\delta})$ for $s \geqslant 1$. But (7.1), (7.4), (7.5), and (9.1) imply that

$$(9.2) \qquad\qquad a^- \in S(g, 1), \qquad h^{-\delta} \in S(g, h^{-\delta});$$

therefore,

$$n_S(h^{-2\delta}g, 1, a^- h^{-\delta}) \leqslant n_S(g, h^{-\delta}, a^- h^{-\delta}) < \infty \quad \forall S \geqslant 1.$$

∎

If $\Omega = R^n$, then one can construct ASP using the function χ. Otherwise, we must introduce cutoff-type functions in the neighborhood $\partial\Omega \times R^n$. Let $\mathcal{M} = \{X \mid a(X) < 4\lambda_0 + 8C_0 d\}$ and let us construct cutoff functions $\overset{\circ}{\eta}_c$ and ψ_c corresponding to $h^{-2\delta}g$, c and $\partial\Omega \times R^n$, $\mathcal{M}$, respectively. Furthermore, for $c > 0$, let

$$\eta_{-c}(X) = \begin{cases} 0, & X \notin \Omega \times R^n, \\ (1 - \overset{\circ}{\eta}_c(X))\psi_{c-1}(X), & X \in \Omega \times R^n, \end{cases}$$

$$\eta_c(X) = \begin{cases} \overset{\circ}{\eta}_c(X), & X \notin \Omega \times R^n, \\ 1, & X \in \Omega \times R^n. \end{cases}$$

Corollary 1.2 implies

$$(9.3) \qquad\qquad \eta_{\pm c} \in S(h^{-2\delta}g, 1),$$

and

$$(9.4) \qquad\qquad \eta_c \eta_{2c} = \eta_c, \qquad \eta_{-c}\eta_{-2c} = \eta_{-2c}.$$

Furthermore, (7.1)–(7.5) imply that $\operatorname{supp}\eta_{-c} \subset \Omega \times R^n$ is a bounded set; therefore,

$$(9.5) \qquad \eta_{-c,\ell}u \subset C_0^\infty(\Omega)^m \quad \forall u \in \mathcal{S}(R^n)^m.$$

In addition, Corollaries 3.1 and 3.2 imply that we can assume $\psi \equiv 1$, in the definition of η_{-c}, if the operator $\eta_{-c,\ell}$ is a factor in a product also containing a factor χ_w. Then this product becomes PDO of the class $\mathcal{L}^{-\infty}$. Similarly, (9.4) implies that in a product containing factors $\eta_{c,r}$ and $\eta_{2c,r}$ ($\eta_{-c,r}$ and $\eta_{-2c,r}$), the factor $\eta_{2c,r}$ ($\eta_{-2c,r}$) can be dropped. The same holds for PDO with left and with left and right symbols. These remarks will often be used hereafter.

9.2. Let $\mathcal{E} = \eta_{-2,w}\chi_w\eta_{-2,w}$ and $\mathcal{E}_1 = \mathcal{E}^2(3 - 2\mathcal{E})$. Since $\operatorname{supp}\chi \subset \mathcal{M}$ is a bounded set, then $\mathcal{E}$ and $\mathcal{E}_1$ are integral operators with kernels of the class $S(R^{2n})$. Therefore, they are kernel operators. Let ν_i, ν_{1i} be nonzero eigenvalues of the operators $\mathcal{E}$, $\mathcal{E}_1$, respectively, taken with their multiplicity and numbered in an arbitrary order.

Lemma 9.2. $\operatorname{card}\{i \mid \nu_i \notin [-\frac{1}{4}, \frac{5}{4}]\} \leqslant C.$

Proof. Lemma 9.1 and (9.3) imply that for each $\varepsilon > 0$,

$$\varepsilon I + \mathcal{E} \in \mathcal{L}I^+(h^{-2\delta}g, 1), \qquad (1+\varepsilon)I - \mathcal{E} \in \mathcal{L}I^+(h^{-2\delta}g, 1).$$

Lemma 8.1 implies that $\varepsilon I + \mathcal{E} \geqslant T_1$ and $(1 + \varepsilon)I - \mathcal{E} \geqslant T_2$, where $T_j \in \mathcal{L}^{-\infty}(g, 1)$; and, by applying Lemma 8.2 to the operators T_j, we obtain the estimate in question. ∎

Lemma 9.2 implies that

$$(9.6) \qquad \operatorname{card}\{i \mid \nu_{1i} \notin [0, 1]\} \leqslant C.$$

Indeed, $\nu_{1i} = \nu_i^2(3 - 2\nu_i)$. Hence, if $\nu_{1i} > 1$, then $\nu_i < -\frac{1}{4}$; and if $\nu_{1i} < 0$, then $\nu_i > \frac{5}{4}$.

Let us calculate $\operatorname{Tr}\mathcal{E}_1$. To do this (Lemma 8.5′), it is sufficient to calculate $\sigma^w(\mathcal{E}_1)$. By applying Theorems 3.1–3.3 and Lemmas 9.1 and 9.3, we obtain

$$(9.7) \qquad \sigma^w(\mathcal{E}_1) = \eta_{-2}^2\chi(3 - 2\eta_{-2}^2\chi) + f + r,$$

where $r \in S^{-\infty}(g, 1)$ and $f \in S(h^{-2\delta}g, h^{1-2\delta})$ and where $f(X) = 0$, if either $\chi(X) = 0$, or $\eta_{-2}(X) = 0$, or $\chi(X) = 1$ and $\eta_{-2}(X) = 1$.

Hence, for some $c > 0$,

$$(9.8) \qquad \begin{aligned} \operatorname{supp} f &\subset \{X \mid X \in (\partial\Omega \times R^n)(c, h^{-2\delta}g), a(X) < c(d(X) + \lambda_0(X))\} \\ &\cup\{X \in \Omega \times R^n \mid |a(X)| < c(d(X) + \lambda_0(X))\}. \end{aligned}$$

We denote the set in (9.8) by $\mathcal{RV}$; its measure is $(RV)_c^{\delta-\omega}(\Omega, a, \lambda_0 + d)$. The first summand in (9.7) is equal to 1 on the complement to $\mathcal{RV}$ in the set $\{X \in \Omega \times R^n \mid a(X) < 0\}$ and is equal to zero outside the set $\{X \in \Omega \times R^n \mid a(X) < c(\lambda_0(X) + d(X))\}$, and (7.2) implies $\int |r(X)| dX < C$. Therefore, (9.7), (9.8), and Lemma 8.5 imply the estimate

$$(9.9) \qquad |\operatorname{Tr} \mathcal{E}_1 - V(\Omega \times R^n, a)| \leqslant c(RV)_c^{\delta-\omega}(\Omega, a, \lambda_0 + d).$$

Estimate (9.9) with $\mathcal{E}_1^2$ in place of $\mathcal{E}_1$ is proved similarly. Let us prove that the function $\overset{\circ}{N} = \operatorname{card}\{i \mid \nu_{1i} \geqslant \frac{1}{2}\}$ assumes the same estimate

$$(9.10) \qquad |\overset{\circ}{N} - V(\Omega \times R^n, a)| \leqslant C(RV)_c^{\delta-\omega}(\Omega, a, \lambda_0 + d).$$

Since $\mathcal{E}_1 \in \mathcal{L}(h^{-2\delta}g, 1)$, Theorem 4.2 implies that $\|\mathcal{E}_1\|_{L_2 \to L_2} \leqslant C$. Hence, $|\nu_{1i}| \leqslant C$ for all i, and (9.6) and (9.9) yield

$$\overset{\circ}{N} \leqslant \sum_{\nu_{1i} \geqslant 1/2} \nu_{1i} + \sum_{1/2 \leqslant \nu_{1i} \leqslant 1} (1 - \nu_{1i}) + C_1 \leqslant \sum_i \nu_{1i} + 2 \sum_i \nu_{1i}(1 - \nu_{1i}) + c_2$$

$$= 3 \operatorname{Tr} \mathcal{E}_1 - 2 \operatorname{Tr} \mathcal{E}_1^2 + C_2 \leqslant V(\Omega \times R^n, a) + C(RV)_C^{\delta-\omega}(\Omega, a, \lambda_0 + d)$$

and

$$\overset{\circ}{N} = \sum_{\nu_{1i} \geqslant 1/2} \nu_{1i} + \sum_{\nu_{1i} \geqslant 1/2} (1 - \nu_{1i})$$

$$\geqslant \sum_i \nu_{1i} - \sum_{0 \leqslant \nu_{1i} \leqslant 1/2} \nu_{1i} - c_2$$

$$\geqslant \sum_i \nu_{1i} - 2 \sum_{0 \leqslant \nu_{1i} \leqslant 1/2} \nu_{1i}(1 - \nu_{1i}) - C_2$$

$$\geqslant \sum_i \nu_{1i} - 2 \sum_i \nu_{1i}(1 - \nu_{1i}) - C_1$$

$$\geqslant 2 \operatorname{Tr} \mathcal{E}_1^2 - \operatorname{Tr} \mathcal{E}_1 - C_1$$

$$\geqslant V(\Omega \times R^n, a) - C(RV)_c^{\delta-\omega}(\Omega, a, \lambda_0 + d).$$

9.3. Now we can obtain the estimate from below for the function $\mathcal{N}_0(\overset{\circ}{a}_w, \Omega)$ by means of a subspace $L \subset C_0^\infty(\Omega)$, constructed such that

$$(9.11) \qquad \langle \overset{\circ}{a}_w u, u \rangle < 0 \quad \forall (0 \neq) u \in L.$$

Let $\pi = (\lambda_0 + C_0 d)^{-1/2}$ and $Q = q_w \pi_w$, where C_0 is the same constant as in the definition of the function a^- at the beginning of the section and where q, d are the functions from condition A in §7. Obviously, $\pi \in SI(g, \pi)$; hence,

(Lemma 5.5) there exists a paramatrix R_π of the operator π_w. Condition A of §7 implies that there also exists a paramatrix $R_q \in \mathcal{L}(g, I, \overset{\circ}{q})$ of the operator q_w. Then $R_Q = R_\pi R_q$ is a paramatrix of the operator Q.

Let $L = \mathrm{span}\{v \mid \mathcal{E}_1 v = \nu_{1i} v, \ \nu_{1i} \geqslant \frac{1}{2}\}$ and let $L^- = \eta_{-1,\ell} QL$. (9.5) implies that $L^- \subset C_0^\infty(\Omega)$. Since $\mathrm{supp}_\infty \sigma^w(\mathcal{E}_1) \subset \mathrm{supp}\, \eta_{-2}$ and $\eta_{-2} \eta_{-1} = \eta_{-2}$, we have

$$(9.12) \qquad R_Q \eta_{-1,\ell} Q \mathcal{E}_1 = \mathcal{E}_1 + T,$$

where $T \in \mathcal{L}^{-\infty}(g, I)$. Obviously,

$$(9.13) \qquad \|\mathcal{E}_1 v\| \geqslant \tfrac{1}{2}\|v\| \quad \forall v \in L,$$

and Lemma 8.2 implies that $\|Tv\| \leqslant \frac{1}{8}\|v\|$ for all v from a subspace of $L_2(R^n)$ of finite codimension. Thus, (9.12) and (9.13) yield

$$(9.14) \qquad \dim L^- \geqslant \dim L - C = \overset{\circ}{N} - C.$$

We shall show that some subspace of L^- of finite codimension satisfies (9.11). Let $v \in L$, $u = \mathcal{E}_1 v \in L$, and $v' = \eta_{-1,\ell} Qu \in L^-$. Then $(9.4)_2$ implies that

$$(9.15) \qquad v' = Q\mathcal{E}_1 v + Tv = Qu + Tv,$$

where $T \in \mathcal{L}^{-\infty}(g, \overset{\circ}{q}{}^{-1}, I)$, and that

$$(9.16) \qquad \begin{aligned} \langle \overset{\circ}{a}_w v', v'\rangle &= \langle Q^*[\overset{\circ}{a}_w + 4R_q^*(\lambda_{0,w} + C_0 d_w)R_q]Qu, u\rangle \\ &\quad - 4\langle Q^* R_q^*(\lambda_{0,w} + C_0 d_w)R_q Qu, u\rangle + \langle Tv, v\rangle \\ &\overset{\mathrm{def.}}{=} \langle D_1 u, u\rangle - 4\langle D_2 u, u\rangle + \langle Tv, v\rangle, \end{aligned}$$

where $T \in \mathcal{L}^{-\infty}(g, 1)$. But $R_q Q - \pi_w \in \mathcal{L}^{-\infty}(g, 1)$; therefore,

$$(9.17) \qquad D_2 = I + K, \qquad D_1 = \pi_w D_1' \pi_w + K_1,$$

where $K \in \mathcal{L}(g, h)$ and $K_1 \in \mathcal{L}(g, h)$; and

$$D_1' = q_w^* \overset{\circ}{a}_w q_w + 4(\lambda_0 + C_0 d)_w.$$

If C_0 is sufficiently large, then (7.4) implies that

$$a_w - q_w^* \overset{\circ}{a}_w q_w + 4C_0 d_w \in \mathcal{L}I^+(g, d),$$

and Lemma 8.1 implies that

$$q_w^* \overset{\circ}{a}_w q_w \leqslant a_w + 4C_0 d_w + T, \quad \text{where } T \in \mathcal{L}^{-\infty}(g, 1).$$

Therefore, (9.16), (9.17), and (9.13) yield the estimate

$$(9.18) \qquad \langle \overset{\circ}{a}_w v', v' \rangle \leqslant \langle \pi_w a_w^- \pi_w u, u \rangle - \|v\|^2 + \langle Kv, v \rangle,$$

where $K \in \mathcal{L}(g, h^\omega)$ and $a^- = a + 8C_0 d + 4\lambda_0$. Furthermore,

$$(9.19) \qquad \begin{aligned} \langle \pi_w a_w^- \pi_w u, u \rangle &= \langle \mathcal{E}_1 \pi_w a_w^- \pi_w \mathcal{E}_1 v, v \rangle \\ &= \langle \mathcal{F}^* \chi_w a_w^- \chi_w \mathcal{F} v, v \rangle + \langle \mathcal{F}_1^* K \mathcal{F}_1 v, v \rangle, \end{aligned}$$

where

$$\mathcal{F} = \pi_w \eta_{-2,w} \mathcal{F}_1, \quad \mathcal{F}_1 = \eta_{-2,w} \mathcal{E}(3 - 2\mathcal{E}),$$
$$K = \chi_w \eta_{-2,w} \pi_w a_w^- [\pi_w \eta_{-2,w}, \chi_w] + [\pi_w \eta_{-2,w}, \chi_w] a_w^- \chi_w \pi_w \eta_{-2,w}.$$

By applying Theorems 3.1–3.3, we obtain

$$\mathrm{supp}_\infty \, \sigma^w([\chi_w, \eta_{-2,w} \pi_w]) \subset \mathrm{supp}_d \chi \subset \{X \mid |a^-(X)| \leqslant (\lambda_0 h^\omega)(X)\};$$

and, therefore, by calculating $b^w(K)$, we can replace a^- with

$$a_1 = a \theta_1, \quad \theta_1 = \theta(2^{-1} a^- \lambda_0^{-1} h^{-\omega}),$$

where $\theta \in C_0^\infty(R^1)$, $0 \leqslant \theta \leqslant 1$, $\theta|_{|t|<1} = 1$, $\theta|_{|t|>2} = 0$ [this will add $\sigma^w(K)$ to a function of the class $S^{-\infty}(g, 1)$]. The calculations in Lemma 9.1 show that $\theta_1 \in S(h^{-2\delta}g, 1)$. But $|a^-| \leqslant 4\lambda_0 h^\omega$ on the support θ_1, and $n_k(h^{-2\delta}g, \lambda_0 h^\omega, a^-) < \infty$ for $k \geqslant 1$; therefore,

$$(9.20) \qquad a_1 \in S(h^{-2\delta}g, \lambda_0 h^\omega),$$

and 3.1 and 3.3 yield

$$(9.21) \qquad \mathcal{F}_1^* K \mathcal{F}_1 \in \mathcal{L}(h^{-2\delta}g, h^\omega \pi^2 \lambda_0) \subset \mathcal{L}(h^{-2\delta}g, h^\omega).$$

Also, (9.20) implies that

$$(9.22) \qquad \chi_w a_w^- \chi_w - (\chi a^- \chi)_w \in \mathcal{L}(h^{-2\delta}g, \lambda_0 h^\omega)$$

and that for $k \geqslant 1$,

$$(9.23) \qquad n_k(h^{-2\delta}g, \lambda_0 h^\omega, \chi a^- \chi) < \infty.$$

Substituting (9.21) and (9.22) into (9.19) and then into (9.18), we obtain

$$(9.24) \qquad \langle \overset{\circ}{a}_w v', v' \rangle \leqslant \langle \mathcal{F}^* (\chi a^- \chi)_w \mathcal{F} v, v \rangle - \|v\|^2 + \langle Kv, v \rangle,$$

where $K \in \mathcal{L}(h^{-2\delta}g, h^\omega)$.

Let $f = -\chi a^- \chi + C_1 \lambda_0 h^\omega$. By construction, $-\chi a^- \chi \geqslant -\lambda_0 h^\omega$ and $\lambda_0 h^\omega \in SI^+(g, \lambda_0 h^\omega)$; hence, (9.23) implies that $f \in SI^+(h^{-2\delta}g, f^{1/2}f^{1/2})$ for C_1 sufficiently large. Lemmas 1.2 and 3.2 imply that $f^{1/2} \in O(C^1, h^{-2\delta}g)$; and, therefore, Lemma 8.1 yields $f_w \geqslant T \in \mathcal{L}^{-\infty}(g, 1)$ and $(\chi a^- \chi)_w \leqslant K_1 \in \mathcal{L}(h^{-2\delta}g, \lambda_0 h^\omega)$.

Now (9.24) yields the estimate $\langle \overset{\circ}{a}_w v', v' \rangle \leqslant -\|v\|^2 + \langle Kv, v \rangle$, where $K \in \mathcal{L}(h^{-2\delta}g, h^\omega)$. According to Lemma 8.2, the operator $-I + K$ is negative definite on a subspace of $L_2(R^n)$ of finite codimension; therefore, (9.14) implies that (9.11) holds for a subspace of $C_0^\infty(\Omega)$ of a dimension not smaller than $\overset{\circ}{N} - C$. By recalling (9.10) and taking into account that $\delta = \frac{1}{2} - \omega$, where $\omega > 0$ is arbitrarily small, we obtain the estimate from below for $\mathcal{N}_0(\overset{\circ}{a}_w, \Omega)$.

9.4. Let us verify an estimate from above. Let q, R_q, π, Q, R_Q, $u \in C_0^\infty$, $v = R_Q u$, and the function η_c be the same as in the beginning of the section. Then $\eta_{c,r} u = u$; and, therefore,

$$\eta_{c,r} v = v + Tu, \quad \text{where } T \in \mathcal{L}^{-\infty}(g, \pi^{-1}, \overset{\circ}{q}).$$

Hence,

$$
\begin{aligned}
\langle \overset{\circ}{a}_w u, u \rangle &= \langle R_Q^* Q^* \overset{\circ}{a}_w Q R_Q u, u \rangle + \langle T_1 u, u \rangle \\
&= \langle \eta_{1,\ell} \pi_w [q_w^* \overset{\circ}{a}_w q_w - 4(\lambda_{0,w} + C_0 d_w)] \pi_w \eta_{1,r} v, v \rangle \\
&\qquad + 4\langle \pi_w(\lambda_{0,w} + C_0 d_w) \pi_w v, v \rangle + \langle T_2 u, u \rangle \\
&\overset{\text{def.}}{=} \langle \eta_{1,\ell} \pi_w D_1 \pi_w \eta_{1,r} v, v \rangle + 4\langle D_2 v, v \rangle + \langle T_2 u, u \rangle,
\end{aligned}
$$

where $T_j \in \mathcal{L}^{-\infty}(g, \overset{\circ}{q}, \overset{\circ}{q})$. (7.4) and Lemma 8.1 imply that $D_1 \geqslant a_w^+ - T$, where $a^+ = a - 4\lambda_0 - 8C_0 d$ and $T \in \mathcal{L}^{-\infty}(g, 1)$; and since $D_2 - I \in \mathcal{L}(g, h)$, we obtain

$$(9.25) \qquad \langle \overset{\circ}{a}_w u, u \rangle \geqslant \langle (\eta_{1,\ell} \pi_w a_w^+ \pi_w \eta_{1,r} + I)v, v \rangle + \|v\|^2 + \langle T'u, u \rangle,$$

where $T' \in \mathcal{L}(g, h\pi^{-1}\overset{\circ}{q}, h\pi^{-1}\overset{\circ}{q})$.

Since $R_Q \in LI(g, \pi^{-1}, \overset{\circ}{q})$, we have $R_Q^* R_Q + T' \in LI^+(g, \pi^{-1}\overset{\circ}{q}, \pi^{-1}\overset{\circ}{q})$. Therefore, Lemma 8.4 yields

$$\|v\|^2 + \langle T'u, u \rangle = \langle (R_Q^* R_Q + T')u, u \rangle > 0$$

for nonzero u from a subspace of $C_0^\infty(\Omega)$ of finite codimension. For such u, (9.25) yields

$$(9.26) \qquad \langle \overset{\circ}{a}_w u, u \rangle \geqslant \langle A'v, v \rangle, \quad \text{where } A' = \eta_{1,\ell} \pi_w a_w^+ \pi_w \eta_{1,r} + I.$$

We shall show that (9.26) implies

(9.27) $$\mathcal{N}_0(\overset{\circ}{a}_w, \Omega) \leqslant \mathcal{N}_0(A', R^n) + C.$$

Theorem 5.3 implies that for any t, $\mathcal{H}(\overset{\circ}{q}_t, C^1)$ and $\mathcal{H}(\pi_t, C^1)$ are Hilbert spaces; part b) of Theorem 5.1 implies that the operators

$$\overset{\circ}{a}_{t,w} : \mathcal{H}(\overset{\circ}{q}_t, C^1) \to \mathcal{H}(\overset{\circ}{q}_t^{-1}, C^1), \qquad A'_t : \mathcal{H}(\pi_t, C^1) \to \mathcal{H}(\pi_t^{-1}, C^1)$$

are bounded; and part c) of Theorem 5.2 implies that $\mathcal{H}(\overset{\circ}{q}_t^{-1}, C^1) = \mathcal{H}(\overset{\circ}{q}_t, C^1)^*$ and $\mathcal{H}(\pi_t^{-1}, C^1) = \mathcal{H}(\pi_t, C^1)^*$. Therefore, $|\langle \overset{\circ}{a}_{t,w} u, u \rangle| \leqslant C_t \|u\|^2_{\overset{\circ}{q}_t}$ and $|\langle A'_t v, v \rangle| \leqslant C_t \|v\|^2_{\pi_t}$; and the two forms in (9.26) are closable in $\mathcal{H}(\overset{\circ}{q}_t, C^1)$ and $\mathcal{H}(\pi_t, C^1)$, respectively. Lemma 8.3 implies that dim Ker $R_Q \leqslant C$; thus, (9.26) and Lemma E.5 yield the estimate

$$\mathcal{N}_0(\overset{\circ}{a}_w, \Omega) \leqslant \mathcal{N}(\mathcal{A}, \mathcal{H}(\pi, C^1)) + C,$$

where $\mathcal{A}[u] = \langle A'u, u \rangle$; and since $C_0^\infty(R^n) \subset \mathcal{H}(\pi, C^1)$ is dense, we obtain (9.27).

9.5. In order to estimate $\mathcal{N}_0(A', R^n)$, we adapt the following notations:

$$\chi = \overset{\circ}{\chi}(a^+ h^{-\delta}), \qquad \mathcal{E} = (\eta_2^2 \chi)_w, \qquad \mathcal{E}_1 = \mathcal{E}^2(3 - 2\mathcal{E}).$$

The proofs, similar to those in Sections 9.1–9.3, show that $\mathcal{E}_1$ is an operator of a trace class whose nonzero eigenvalues satisfy the estimate

(9.28) $$\sum_{\nu_{1i} \geqslant 1/2} 1 = V(\Omega \times R^n, a) + O((RV)_c^{\delta - \omega}(\Omega, a, d)).$$

Let $L^+ = \text{span}\{v \mid \mathcal{E}_1 v = \nu_{1i} v, \nu_{1i} \geqslant \frac{1}{2}\}$, $v \in {}^\perp L^+ \cap \mathcal{S}(R^n)$, $u = (I - \mathcal{E}_1)v$, $\mathcal{E}'_1 = \chi_w^2(3 - 2\chi_w)$. Since $\eta_1 \eta_2 = \eta_1$, we have $(\eta_{1,r}\mathcal{E}_1 - \eta_{1,r}\mathcal{E}'_1) \in \mathcal{L}^{-\infty}(g, 1)$; hence,

(9.29) $$\begin{aligned} \langle A'u, u \rangle &\geqslant \langle (I - \mathcal{E}'_1)\eta_{1,\ell}\pi_w a_w^+ \pi_w \eta_{1,r}(I - \mathcal{E}'_1)v, v \rangle + \tfrac{1}{4}\|v\|^2 + \langle Tv, v \rangle \\ &\overset{\text{def}}{=} \langle Dv, v \rangle + \tfrac{1}{4}\|v\|^2 + \langle Tv, v \rangle, \end{aligned}$$

where $T \in \mathcal{L}^{-\infty}(g, 1)$. The proofs, similar to those in Section 9.3 [see (9.19) and the following], show that

$$\begin{aligned} \langle Dv, v \rangle &= \langle \eta_{1,\ell}\pi_w(I - \mathcal{E}'_1)a_w^+(I - \mathcal{E}'_1)\pi_w \eta_{1,r}v, v \rangle + \langle K_1 v, v \rangle \\ &= \langle (\eta_{1,\ell}\pi_w(\chi' a^+ \chi')_w \pi_w \eta_{1,r} + K_2)v, v \rangle, \end{aligned}$$

where $K_j \in \mathcal{L}(h^{-2\delta}g, h^\omega)$, $\chi' = 1 = \chi^2(3 - 2\chi)$, and that

$$f = \chi' a^+ \chi' + C_1 \lambda_0 h^\omega \in SI^+(g, f)$$

for C_1 sufficiently large. Applying Lemma 8.1 to f_w, we obtain

$$(\chi' a^+ \chi')_w \geqslant K \in \mathcal{L}(h^{-2\delta}g, \lambda_0 h^\omega);$$

therefore, $D \geqslant K_1 \in \mathcal{L}(h^{-2\delta}g, h^\omega)$, and (9.29) yields the estimate

$$\langle A'u, u \rangle \geqslant \tfrac{1}{4}\|v\|^2 + \langle K'v, v \rangle,$$

where $K' \in \mathcal{L}(h^{-2\delta}g, h^\omega)$. Lemma 8.2 yields $\langle A'u, u \rangle \geqslant 0$ $\quad \forall u \in L'$, where $L' \subset {}^\perp L^+ \cap S(R^n)$ is a subspace of finite codimension.

Thus, if $L \subset S(R^n)$ is a subspace such that

$$\langle A'u, u \rangle < 0 \quad \forall (0 \neq) u \in L',$$

then $L \cap L' = \{0\}$ and $\dim L \leqslant \dim L^+ + C$.

Therefore, $\mathcal{N}_0(A', R^n) \leqslant \dim L^+ + C$; and (9.27) and (9.28) yield an estimate from above for $\mathcal{N}_0(\overset{\circ}{a}_w, \Omega)$.   ∎

9.6.

Proof of Lemma 8.2.   Let $b_w = a_w^* a_w$. Then $b_w = b_w^* \in \mathcal{L}(g, h^{2\omega})$, and

$$(9.30) \qquad \|a_w u\| \leqslant \varepsilon \|u\| \iff \langle b_w u, u \rangle \leqslant \varepsilon^2 \|u\|^2.$$

We fix $\theta \in C_0^\infty(R^1)$, $0 \leqslant \theta \leqslant 1$, $\theta|_{|t|<1/2} = 1$, $\theta|_{|t|>1} = 0$ and denote $e_\tau = \theta(h^{-\omega}\tau)$. Obviously, $e_\tau \in S(g, 1)$ uniformly with respect to $\tau \in (0, 1)$; and since $c\tau \leqslant h^\omega \leqslant C\tau$ on $\mathrm{supp}_d\, e_\tau$, then for all $\chi \in (0, \tfrac{1}{2})$ and all k,

$$(9.31) \qquad n_k(h^{-2\chi}g, 1, e_\tau) \leqslant C_k \tau^{k\chi\omega^{-1}}.$$

(9.31) and Lemma 8.1 imply that for any $S > 0$,

$$(9.32) \qquad sI + e_{\tau,w} \geqslant T_{1,\tau,w}, \qquad (1+s)I - e_{\tau,w} \geqslant T_{2,\tau,w},$$

where $T_{j,\tau} \in S(h^{-2\chi}g, 1)$. Analysis of the proof of Lemma 8.1 shows that seminorms of the symbols $T_{j,\tau}$ are estimated through a sum of products that include the factors $n_k(h^{-2\chi}g, 1, e_\tau)$ with $k \geqslant 1$. Hence, Theorem 4.2 implies that $\|T_{j,\tau,w}\| \to 0$ as $\tau \to 0$ and that, as follows from (9.32), for small τ, all eigenvalues of the kernel operators $e_{\tau,w}$ belong to $[-\tfrac{1}{4}, \tfrac{5}{4}]$. Thus, all the eigenvalues $\nu_{1\tau i}$ of the operator $\mathcal{E}_{1\tau} = e_{\tau,w}^2(3 - 2e_{\tau,w})$ belong to $[0, 1]$; and, similar to (9.10),

$$(9.33) \qquad \mathrm{card}\{i \mid \nu_{1\tau i} \in [\tfrac{1}{2}, 1]\} \leqslant C\tau^{-C}.$$

Let $L_\tau^+ = \mathrm{span}\{v \mid \mathcal{E}_{1\tau}v = \nu_{1\tau i}v, \ \nu_{1\tau i} \geqslant \frac{1}{2}\}$, $v \in {}^\perp L_\tau^+$, $u = (I - \mathcal{E}_{1\tau})v$. Then $\|u\| \geqslant \|v\|/2$, and (9.31) yields

$$K = (I - \mathcal{E}_{1\tau})b_w(I - \mathcal{E}_{1\tau}) \in \mathcal{L}(h^{-2\chi}g, \tau^2).$$

Thus, Theorem 4.2 yields

$$\langle(\varepsilon^2 - b_w)u, u\rangle = \varepsilon^2\|(I - \mathcal{E}_{1\tau})v\|^2 - \langle Kv, v\rangle \geqslant \frac{\varepsilon^2}{4}\|v\|^2 - C_1^2\tau^2\|v\|^2,$$

where C^1 does not depend on τ. If $\tau < \varepsilon/2C_1$, then

$$\langle b_w u, u\rangle \leqslant \varepsilon^2\|u\|^2 \quad \forall u \in {}^\perp L_\tau^+;$$

hence, the statement of Lemma 8.2 follows from (9.33) and (9.30).

§10 Proof of Theorem 7.3 for the Scalar Case

10.1. Let $\Omega = R^n$. Then ASP $\mathcal{E}_1 = \mathcal{E}^2(3 - 2\mathcal{E})$ is constructed with the functions $\chi^\pm = \overset{\circ}{\chi}(a^\pm h^{-\delta}) : \mathcal{E} = \chi_w^\pm$. In §9 we have shown that $\chi^\pm \in S(h^{-2\delta}g, 1)$, and since the metric $h^{-2\delta}g$ is σ-temperate for $\delta < \frac{1}{2}$ (Lemma 5.3), we were able to apply fundamental theorems of PDO calculus. One could consider $\delta \in (0, \frac{2}{3})$; however, in this case, another metric must be constructed, since the metric $h^{-2\delta}g$ is not σ-temperate for $\delta > \frac{1}{2}$.

Lemma 10.1. *Let g be a σ-temperate metric and let $f \in S(g, 1)$ and $\delta \in (0, \frac{2}{3})$. Then the metric $G = h^{-2\delta}|df|^2 + h^{-\delta}g$ is σ-temperate, and*

$$(10.1) \qquad\qquad h_G \leqslant Ch^\varepsilon, \quad \varepsilon = 2 - 3\delta > 0.$$

If $p_j \in O(H_j, g)$, $j = 1, 2$ and $a \in S(g, p_1, p_2)$, then $p_j \in O(H_j; g)$ and $a \in S(G, p_1, p_2)$.

Proof. Lemmas 2.5 and 1.5 imply that we can assume that $h \in S(g, h)$. Furthermore, for all $k \geqslant 1$ and $X, Z_i \in R^{2n}$,

$$h^{-\delta}(X)|d^k\langle df(X), Z_0\rangle(Z_1, \ldots, Z_k)| \leqslant C_k \prod_{1 \leqslant j \leqslant k} (h^{-\delta}(X)g_X(Z_i))^{1/2};$$

thus, the metric G satisfies the conditions of Lemma 1.3 and, therefore, varies slowly. Since $G \geqslant g$, the functions p_j are G-continuous, and $a \in S(G, p_1, p_2)$. Finally, due to Lemma 5.4, our proof will be completed upon proving (10.1). For this we need the following lemma:

Lemma 10.2. *Let g be a positive-definite quadratic form in a simplectic space W, let*

$$(10.2) \qquad G(t) = \sigma(t,p)^2 + g(t), \quad t \in W,$$

where $p \in W$ is fixed, and let $g \leqslant h^2 g^\sigma$.
  Then $G \leqslant 2(g(p)+h^2)G^\sigma$.

Proof. We have to show that if $G^\sigma(w) \leqslant 1$, then

$$G(w) \leqslant 2(g(p)+h^2).$$

But by definition, the condition $G^\sigma(w) \leqslant 1$ means that

$$\sigma(t,w)^2 \leqslant \sigma(t,p)^2 + g(t), \quad t \in W;$$

hence, the form $(\sigma(t,p),t) \to \sigma(t,w)$ can be continued on $R \oplus W$ to a form with unit norm, i.e., $\sigma(t,w) = a\sigma(t,p) + \sigma(t,b)$, where $a \in R$, $b \in W$, and $a^2 + g^\sigma(b) \leqslant 1$. Therefore, $w = ap + b$ and

$$\begin{aligned}
G(w) = \sigma(b,p)^2 + g(ap+b) &\leqslant g^\sigma(b)g(p) + 2(a^2 g(p)+g(b)) \\
&\leqslant g^\sigma(b)g(p) + 2h^2 g^\sigma(b) + 2a^2 g(p) \leqslant g^\sigma(b)(g(p)+2h^2) + 2a^2 g(p) \\
&\leqslant 2(g(p)+h^2).
\end{aligned}$$ ∎

We return to the metric G from Lemma 10.1. At a fixed point X, it has the form (10.2) with $h^{-\delta}g$ substituted for g and with $p = h^{-\delta}H_f$, where H_f is a Hamilton vector field of the function f; therefore,

$$(10.3) \qquad G \leqslant 2(h^{-3\delta}g(H_f) + h^{2-2\delta})G^\sigma.$$

Since $f \in S(g,1)$, we have

$$g_X^\sigma(H_f(X)) = \sup_Z \frac{\sigma(H_f(X),Z)}{g_X(Z)} = \sup_Z \frac{\langle df(X),Z\rangle^2}{g_X(Z)} \leqslant C.$$

Thus, $g(H_f) \leqslant h^2 g^\sigma(H_f) \leqslant Ch^2$, and (10.3) implies (10.1). ∎

Let $\overset{\circ}{\chi}$ be the same as in the beginning of §9. We will show that

$$(10.4) \qquad \chi_f = \overset{\circ}{\chi}(h^{-\delta}f) \in S(G,1).$$

Obviously, (10.4) follows from the inclusion

$$(10.5) \qquad h^{-\delta}f \in S(G,1,\mathcal{N}),$$

where $\mathcal{N} = \{X \mid |f(X)| \leqslant h(X)^\delta\}$. But $h^{-\delta} \in S(g, h^{-\delta})$ and $h^{-\delta}f \in S(g, h^{-\delta})$; hence, for $k \geqslant 2$,

$$
\begin{aligned}
|d^k(h^{-\delta}f)(X; Y_1, \ldots, Y_k)| &\leqslant C_k h^{-\delta}(X) \prod_{1 \leqslant j \leqslant k} g_X(Y_j)^{1/2} \\
&\leqslant C_k \prod_{1 \leqslant j \leqslant k} (h^{-\delta}(X) g_X(Y_j))^{1/2} \\
&\leqslant C_k \prod_{1 \leqslant j \leqslant k} G_X(Y_j)^{1/2};
\end{aligned}
$$

(10.6)

and if $|f(X)| \leqslant h(X)^\delta$, then
(10.7)
$$
\begin{aligned}
|\langle d(h^{-\delta}f)(X), Y\rangle| &\leqslant |\langle f(X)dh^{-\delta}(X), Y\rangle| + |\langle h^{-\delta}(X)df(X), Y\rangle| \\
&\leqslant Ch^{-\delta}(X)|f(X)|g_X(Y)^{1/2} + G_X(Y)^{1/2} \leqslant C_1 G_X(Y)^{1/2}.
\end{aligned}
$$

Then (10.6) and (10.7) yield (10.5) and (10.4).

10.2. Using estimate (10.4), one can prove Theorem 7.3 for the case $\Omega = R^n$, $m = 1$ by making only minor changes in the proof in §9. We shall not cite this proof here, since all its new points are contained in the proof of Theorem 7.3 for the case $\Omega \neq R^n$, $m = 1$, to which we now proceed.

We succeeded in improving the estimate for the remainder in the case $\Omega = R^n$, due to the more accurate approximation of the characteristic functions for the sets $\{X \mid a^{\pm}(X) < 0\}$, but here the symbol ASP is described by a metric that depends on the functions $a^{\pm}$. Similarly, if we more closely approximate the characteristic functions for the set $\Omega \times R^n$, then we must use a metric that depends on the function φ_1 vanishing on the boundary. Metrics constructed with respect to $a^{\pm}$ and φ_1 will differ; and, therefore, we will not be able to apply the calculus of PDO. Thus, we are compelled to compromise by approximating characteristic functions in a less accurate (and more complicated) way.

We fix the function $\theta \in C_0^\infty(R^1)$, $\theta|_{|t|<1} = 1$, $\theta|_{|t|>2} = 0$, $0 \leqslant \theta \leqslant 1$ and the constants $C_0 > 0$, $\omega \in (0, 1/g)$, $\delta = \frac{2}{3} - 3\omega$, $\delta_1 = \delta + 2\omega$, and denote

$$
\begin{aligned}
b_3 &= (\varphi_1^2 + h^\delta)^{1/2}, \qquad \lambda_0 = h^{\delta-\omega}b_3^{-1}, \qquad a^- = a + 8C_0 d + 4\lambda_0, \\
\chi &= \overset{\circ}{\chi}(h^{-\delta}b_3 a^-), \qquad b_1 = ((a + 8C_0 d)^2 + h^\delta)^{1/2}, \\
\overset{\circ}{\eta}_c &= \theta(c^{-1}\varphi_1 h^{-\delta}b_1), \qquad G^{(1)} = h^{-2\delta_1}|d\varphi_1|^2 + h^{-\delta_1}g, \\
G &= h^{-2\delta_1}|d(\varphi_1(a + 8C_0 d))|^2 + h^{-\delta_1}g, \\
\mathcal{M}_{c,b}^\delta &= \{X \mid |b(X)| \leqslant Ch(X)^{\delta/2}\}.
\end{aligned}
$$

Lemma 10.3.

a) $b_3^{-2}g$, G, and $G^{(1)}$ *are σ-temperate metrics that satisfy condition* (10.1) *with $\varepsilon = \omega$.*

b) *The following estimates and inclusions hold:*

$$(10.8) \qquad ch^{\delta/2} \leqslant b_i \leqslant C, \qquad ch^{\delta-\omega} \leqslant \lambda_0 \leqslant Ch^{\delta/2-\omega};$$

$$(10.9) \qquad b_i \in S(b_i^{-2}g, b_i), \qquad \lambda_0 \in S(b_3^{-2}g, \lambda_0);$$

$$(10.10) \qquad \chi \in S(h^{-\delta_1}g, 1, \mathcal{M}_{c,\varphi_1}^{\delta});$$

$$(10.11) \qquad \overset{\circ}{\eta}_c \in S(h^{-\delta}g, 1, \mathcal{M}_{c,a+8C_0d}^{\delta});$$

$$(10.12) \qquad \chi \in S(G, 1), \qquad \overset{\circ}{\eta}_c \in S(G, 1) \cap S(G^{(1)}, 1).$$

Proof.

a) This part follows from (10.9) and Lemmas 5.4 and 5.1.

b) (10.8) follows from the way in which the functions b_i are constructed; and (10.9) follows from (10.8) and the inclusions

$$a, d, \varphi_1 \in S(g, 1), \qquad h \in S(g, h).$$

In order to prove (10.10) we have to show that if $X \in \mathcal{N} \cap \mathcal{M}_{c,\varphi_1}^{\delta}$, where $\mathcal{N} = \{X \mid |a^-(X)| \leqslant (h^{\delta}b_3^{-1})(X)\}$, then for any k,

$$(10.13) \qquad |d^k(a^-h^{-\delta}b_3)(X; Y_1, \ldots, Y_k)| \leqslant C_k \prod_{1 \leqslant j \leqslant k} (h^{-\delta_1}(X)g_X(Y_j))^{1/2}.$$

Leibnitz's rule implies that it suffices to prove (10.13) for the products

$$(d^s a^-)(X; Y_1, \ldots, Y_s)d^{k-s}(h^{-\delta}b_3)(X; Y_{s+1}, \ldots, Y_k).$$

For $s = 0$, the first factor is estimated by $(h^{\delta}b_3)(X)$; and for $s > 0$, by

$$|(d^s(a + 8C_0d) + d^s\lambda_0)(X; Y_1, \ldots, Y_s)|$$
$$\leqslant C_s(1 + (h^{\delta-\omega}b_3)(X)b_3^{-s}(X)) \prod_{1 \leqslant j \leqslant s} g_X(Y_j)^{1/2}$$
$$\leqslant C_s' h(X)^{-\omega-(s-1)/2} \prod_{1 \leqslant j \leqslant s} g_X(Y_j)^{1/2},$$

due to $(10.9)_2$ and (10.8). If $r = k - s > 0$, then the second factor is estimated by $(10.9)_1$ and $(10.8)_1$:

$$|d^r(h^{-\delta}b_3)(X; Y_1, \ldots, Y_r)| \leqslant C_r h(X)^{-\delta-(r-1)\delta/2} \prod_{1 \leqslant j \leqslant r} g_X(Y_j)^{1/2};$$

and $(h^{-\delta}b_3)(X) \leqslant Ch^{-\delta/2}(X)$, since $b_3(X) \leqslant Ch(X)^{\delta/2}$.

These estimates and the equality $\delta_1 = \delta + 2\omega$ yield (10.13).

Thus, (10.10) is proved; (10.11) can be proved similarly (and more simply).

In order to prove $(10.12)_1$, it suffices to verify estimate (10.13) with $X \in \mathcal{N}$ and with $G_X(Y_r)$ substituted for $h^{-\delta_1}(X)g_X(Y_r)$. If $X \in \mathcal{M}_{c,\varphi_1}^\delta$, then (10.13) is already proved; and if $X \in \mathcal{M}_c' = R^{2n}\backslash\mathcal{M}_{c,\varphi_1}^\delta$, then $b_3(X)^{\pm 1} < C|\varphi_1(X)|^{\pm 1}$ and

$$(10.14) \qquad (a^-h^{-\delta}b_3)(X) = (a^-\varphi_1)(X)(h^{-\delta}\varphi_1^{-1}b_3)(X) \overset{\text{def.}}{=} \pi_1(X)\pi_2(X).$$

Leibnitz's rule implies that it suffices to consider only products of the form $d^s\pi_1(X;\cdot,\ldots,\cdot)d^{k-s}\pi_2(X;\cdot,\ldots,\cdot)$. Similarly to proving $(10.9)_1$, we obtain

$$(10.15) \qquad \varphi_1^{\pm 1} \in S(h^{-\delta}g, b_3^{\pm 1}, \mathcal{M}_c');$$

then (10.8), (10.9), and (10.15) yield

$$(10.16) \qquad \varphi^{-1}b_3 \in S(h^{-\delta}g, 1, \mathcal{M}_c'), \qquad \lambda_0\varphi_1 \in S(h^{-\delta}g, h^{\delta-\omega}, \mathcal{M}_c').$$

Inclusion $(10.16)_1$ implies (10.13) for $s = 0$. As for $s > 0$, we notice that $\pi_1 = (a + 8C_0d)\varphi_1 + 4\lambda_0\varphi_1 = \pi_3 + 4\lambda_0\varphi_1$, that

$$|\langle d\pi_3(X), Y\rangle| \leqslant Ch^\delta(X)G_X(Y)^{1/2},$$

and if $r > 1$, that

$$|d^r\pi_3(X; Y_1,\ldots,Y_r)| \leqslant c_r h^{r\delta/2} \prod_{1\leqslant j\leqslant r} (h^{-\delta}(X)g_X(Y_j))^{1/2}$$
$$\leqslant c_r'h^\delta(X) \prod_{1\leqslant j\leqslant r} G_X(Y_j)^{1/2}.$$

Now (10.13) follows from (10.14), $(10.16)_2$, and the equality $\delta_1 = \delta + 2\omega$. Thus, $(10.12)_1$ is proved.

Instead of proving $(10.12)_2$, we shall prove a more general statement: if $s, r \geqslant 0$ and $s + r = k > 0$, then

$$(10.12') \quad |d^k\overset{\circ}{\eta}_c(X; Y_1,\ldots,Y_k)| \leqslant C_k \prod_{1\leqslant j\leqslant s} G_X(Y_j)^{1/2} \prod_{s+1\leqslant j\leqslant k} G_X^{(1)}(Y_j)^{1/2}.$$

Obviously, it suffices to prove (10.12') with $\varphi_1 h^{-\delta}b_1$ substituted for $\overset{\circ}{\eta}_c$ and with X such that $|\varphi_1(X)| \leqslant (h^\delta b_1^{-1})(X)$.

First, let $k \geqslant 2$. Then it suffices to consider the product $d^p(\varphi_1 h^{-\delta})(X; Y_1,\ldots,Y_p)d^{k-p}b_1(X; Y_{p+1},\ldots,Y_k)$. $(10.9)_1$ and $(10.8)_1$ imply that for $p = 0$, this product is estimated by

$$b_1^{-1}(X) \cdot b_1(X) \prod_{1\leqslant j\leqslant k} (h^{-\delta}(X)g_X(Y_j))^{1/2};$$

and for $p > 0$, by

$$h^{-\delta}(X)b_1(X)^{1-k+p} \prod_{1 \leqslant j \leqslant k} g_X(Y_j)^{1/2} \leqslant C_k \prod_{1 \leqslant j \leqslant k} (h^{-\delta}(X)g_X(Y_j))^{1/2}.$$

Since $h^{-\delta}(X)g_X(Y) \leqslant \min\{G_X(Y), G_X^{(1)}(Y)\}$, then (10.12′) is proved for the case $k \geqslant 2$. (10.12′) is verified for $k = 1$ similarly to how $(10.12)_1$ was proved. ∎

10.3. Letting $\mathcal{M} = \{X \mid a(X) < 4\lambda_0 + 8C_0d + C_0h^{\delta/2}\}$, we construct cutoff functions ψ_c associated with $h^{-\delta}g$, c, $\mathcal{M}$. As in §9, we define functions η_{-c} through the functions $\overset{\circ}{\eta}_c$, ψ_c. Then, we still have (9.4) and (9.5); but, unlike (9.3),

$$(10.17) \qquad \eta_{-c} \in S(G,1) \cap S(G^{(1)},1) \cap S(h^{-\delta_1}g,1,\mathcal{M}_{c,a+8C_0d}),$$

which follows from (10.11) and $(10.12)_2$. Since the metric g splits and since φ_1 does not depend on ξ, then the metric $G^{(1)}$ also splits. Therefore, Theorem 3.2 implies that $\eta_{-c,r}$, $\eta_{-c,\ell} \in \mathcal{L}(G^{(1)},1)$. But we shall have to consider compositions of these operators with the operator $\chi_w \in \mathcal{L}(G,1)$; hence, we need to show that

$$(10.18) \qquad \eta_{-c,r}, \; \eta_{-c,\ell} \in \mathcal{L}(G,1).$$

Theorem 3.2 implies that $\forall N$,

$$\sigma^w(\eta_{-c,\ell})(X) = \sum_{j < N} C_j \langle D_x, D_\xi \rangle^j \eta_{-c}(X) + r_N(X),$$

where $r_N \in S(G^{(1)}, h^{N\omega})$. Since $G_X(Y)^{-1} \leqslant Ch^{-\delta}(X)G_X^{(1)}(Y)^{-1}$, then $n_k(G,1,r_N) < \infty$ for $k < 2N\omega/3$. Hence, it suffices to prove that for $k > 0$ and $Y_j \neq 0$,

$$\left| d^k(\langle D_x, D_\xi \rangle^j \eta_{-c})(X; Y_1, \ldots, Y_k) \prod_{1 \leqslant j \leqslant k} G_Z(Y_j)^{-1/2} \big|_{Z=X} \right| \leqslant c_k h^{j\omega}(X).$$

By rearranging d^k and $\langle D_x, D_\xi \rangle^j$ and taking into account the fact that for $b \in S(G^{(1)}, 1)$,

$$\langle D_x, D_\xi \rangle^j b \in S(G^{(1)}, h^{j\omega}),$$

we see that the required estimate follows from the inclusion

$$d^k \eta_{-c}(X; Y_1, \ldots, Y_k) \prod_{1 \leqslant j \leqslant k} G_Z(Y_j)^{-1/2} \in S(G^{(1)}, 1, U_Z)$$

uniformly with respect to $Y_j \neq 0$, where U_Z is an arbitrarily small neighborhood of the point Z. But this inclusion is implied by (10.12).

10.4. Thus, we have proved (10.18). Now let $\mathcal{E} = \eta_{-2,w}\chi_w\eta_{-2,w}$ and $\mathcal{E}_1 = \mathcal{E}^2(3-2\mathcal{E})$. (10.18) and (10.12)$_1$ enable us to repeat the arguments from §9 to prove Lemma 9.2 and estimate (9.6), and to obtain (9.7) for $r \in S^{-\infty}(G,1)$ and $f \in S(G,h^\omega)$. As in (9.7), $f(X) = 0$, if either $\chi(X) = 0$, or $\eta_{-2}(X) = 0$, or $\chi(X) = 1$ and $\eta_{-2}(X) = 1$; therefore, there exists $c > 0$ such that

$$\operatorname{supp} f \subset \{X \in (\Omega \times R^n)(1, h^{-\delta}g) \mid a(X) \leq c(\lambda_0(X) + d(X) + h^{\delta/2}(X),$$
$$|(\varphi_1 a)(X)| \leq c(h^{\delta-\omega}(X) + d(X))\},$$

and, similarly to (9.9) and (9.10), we obtain estimate (9.10) with $(RV)_c^{\delta-\omega}$ (Ω, a, d), the right-hand side of estimate (7.12). Furthermore, as in §9, we let $\pi = (\lambda_0 + C_0 d)^{-1/2}$, $Q = q_w \pi_w$. Then (10.9)$_2$ and the inclusion $d \in S(g, d) \subset S(g, 1)$ imply $\pi \in SI(b_3^{-2}g, \pi)$; therefore, there exists a paramatrix R_Q of the operator Q; and if, as in §9, we define a subspace $L^- \subset C_0^\infty(\Omega)$ through the operator $\mathcal{E}_1$, we obtain estimate (9.14) as well. Equalities (9.16) and (9.17) will also be valid, but the operators K and K_1 in (9.17) will now belong to $\mathcal{L}(b_3^{-2}g, b_3^{-2}h)$. (10.8) implies that both operators belong to $\mathcal{L}(b_3^{-2}g, h^\omega)$; and, therefore, estimate (9.18) is valid.

Defining a_1 as in §9, instead of (9.20) we now have

$$(10.19) \qquad\qquad a_1 \in S(G, \lambda_0 h^\omega),$$

the proof of which is an obvious synthesis of proofs of formulas (9.20) and (10.12)$_1$. Using (10.19), we obtain (19.21)–(19.24), where the metric $h^{-2\delta}g$ is replaced by the metric G; and the function $\lambda_0 = h^{\delta-\omega}$, by the function $\lambda_0 = h^{\delta-\omega}b_3^{-1}$. Then the proof of the estimate from below for $N_0(\overset{\circ}{a}_w, \Omega)$ is completed exactly as in §9.

We omit the proof of the estimate from above, since the changes are exactly the same as in the proof of the estimate from below. In all our constructions, the function a^- must be replaced by the function $a^+ = a - 8C_0 d - 4\lambda_0$. The functions χ, $\overset{\circ}{\eta}_c$, η_c and the metrics G, $G^{(1)}$ are substituted accordingly.

§11 Proofs of Theorems 7.2 and 7.3 for the Matrix Case

11.1. First, in condition B in §7, let $n_0 = 1$ and $U_1 = \Omega \times R^n$ and let the function $e = e_1$ exist globally. Then, by substituting the operator $q_w e_w$ for q_w in condition A in §7, we obtain the diagonal matrix

$$a = [\delta_{jk}\lambda_j]_{j,k=1}^m;$$

and since Lemma 8.1 implies

$$-C_0 d_w - T_1 \leq q_w^* \overset{\circ}{a}_w q_w - a_w \leq C_0 d_w + T_2,$$

where $T_j \in \mathcal{L}^{-\infty}(g,1)$, then the analysis of the matrix operator is reduced to the analysis of a direct sum of scalar operators. Thus, in this case, our proof differs very little from the one cited above.

On the other hand, if $n_0 > 1$, then we can "almost-diagonalize" the operator a_w microlocally only; therefore, in addition to the cutoff functions in the neighborhood $\partial\Omega \times R^n$, we must introduce cutoff functions in a neighborhood of the set $\Gamma\Omega$, which divides domains with different diagonalizations. Like the functions $\overset{\circ}{\eta}_c$, these functions are constructed in Theorem 7.2 differently from in Theorem 7.3. First, we shall prove the more difficult Theorem 7.3 and, at the end of the section, indicate the changes necessary for the proof of Theorem 7.2.

11.2. In the matrix case, the cutoff functions must be described by all metrics constructed with the functions $\overset{\circ}{\lambda}_{ij}$ (see condition B in §7); and, therefore, their construction becomes substantially more complicated. Let the functions $\theta \in C_0^\infty(R^1)$, $\chi \in C^\infty(R^1)$ and the constants $C_0, \omega > 0$, $\delta = \frac{2}{3}-3\omega$, $\delta_1 = \delta+2\omega$ be the same as in §10; let λ_j $[j = 1,\ldots,m]$ be eigenvalues of the matrix a; and let functions $\overset{\circ}{\lambda}_{ij}$ $[i = 1,\ldots,n_0,\ j = 1,\ldots,m]$ and $\varphi_1, \varphi_2 \in S(g,1)$ be as introduced in conditions B and C in §7. Let

$$b_3 = (\varphi_1^2\varphi_2^2 + h^\delta)^{1/2}, \qquad \lambda_0 = h^{\delta-\omega}b_3^{-1}, \qquad \lambda_{ij} = \overset{\circ}{\lambda}_{ij} + 8C_0d + 4\lambda_0,$$

$$\chi_{ij} = \overset{\circ}{\chi}(\lambda_{ij}h^{-\delta}b_3), \qquad \overset{\circ}{b}_1 = \sum_{1 \leqslant i \leqslant m} (\lambda_i + 8C_0d)^{-2}\varphi_2^{-2},$$

$$\overset{\circ}{b}_2 = \sum_{1 \leqslant i \leqslant m} (\lambda_i + 8C_0d)^{-2},$$

$$b_k = h^{\delta/2}(1 - \theta(h^\delta\overset{\circ}{b}_k)) + \overset{\circ}{b}_k^{-1/2}\theta(h^\delta\overset{\circ}{b}_k) \quad [k = 1,2],$$

$$\overset{\circ}{\eta}_{1,c} = \theta(c^{-1}\varphi_1h^{-\delta}b_1), \qquad \overset{\circ}{\eta}_{2,c} = \theta(c^{-1}\varphi_1\varphi_2h^{-\delta}b_2),$$

$$G^{(1)} = h^{-2\delta_1}|d\varphi_1|^2 + h^{-\delta_1}g, \qquad G^{(2)} = h^{-2\delta_1}|d(\varphi_1\varphi_2)|^2 + h^{-\delta_1}g,$$

$$G_{ij} = h^{-2\delta_1}|d(\varphi_1\varphi_2(\overset{\circ}{\lambda}_{ij} + 8C_0d))|^2 + h^{-\delta_1}g.$$

Obviously,

$$b_3 \asymp \max\{|\varphi_1\varphi_2|, h^{\delta/2}\},$$

$$b_1 \asymp \max\{\min_{1 \leqslant i \leqslant m} |(\lambda_i + 8C_0d)\varphi_2|, h^{-\delta/2}\},$$

$$b_2 \asymp \max\{\min_{1 \leqslant i \leqslant m} |\lambda_i + 8C_0d|, h^{\delta/2}\};$$

therefore, estimates (10.8) and (10.9) hold; and similarly to (10.12),

(11.1) $\chi_{ij} \in S(G_{ij}, 1)$, $\overset{\circ}{\eta}_{1,c}, \overset{\circ}{\eta}_{2,c} \in S(G_{ij}, 1, U_i)$, $\forall i,j$;

(11.2) $\overset{\circ}{\eta}_{2,c} \in S(G^{(2)}, 1)$, $\overset{\circ}{\eta}_{1,c} \in S(G^{(1)}, 1) \cap S(G^{(2)}, 1)$.

For example, the proof of inclusion $(11.1)_1$ follows from substituting $\varphi_1\varphi_2$ for φ_1 in the proof of inclusion (10.12). As in §10,

(11.3) $b_3^{-2}g$, G_{ij}, and $G^{(k)}$ are σ-moderate metrics

satisfying condition (10.1) with $\varepsilon = \omega$; and $G^{(1)}$ splits.

We shall construct the cutoff functions ψ_c associated with $h^{-\delta}g$, c, $\{X \mid \lambda_1(X) < 4\lambda_0 + 8C_0 d + C_0 h^{\delta/2}\}$; and we denote

$$\eta_{-c}(X) = \begin{cases} 0, & X \notin \Omega \times R^n, \\ (1 - \overset{\circ}{\eta}_{1,c}(X))\psi_{c-1}(X), & X \in \Omega \times R^n, \end{cases}$$

(11.4)

$$\eta_{j,c}(X) = \begin{cases} 0, & X \notin U_j, \\ (1 - \overset{\circ}{\eta}_{2,c}(X))\psi_{c-1}(X), & X \in U_j. \end{cases}$$

We remark that

(11.5) $\eta_{i,c}\eta_{j,c} = 0 \quad \forall i \neq j,$

(11.6) $\eta_{i,c}\eta_{i,2c} = \eta_{i,2c}, \qquad \eta_{-c}\eta_{i,2c} = \eta_{i,2c}.$

Then (11.1)–(11.4) yield

(11.7) $\eta_{j,c} \in S(G_{ji}, 1) \cap S(G^{(2)}, 1), \qquad \eta_{-c} \in S(G^{(1)}, 1) \cap S(G^{(2)}, 1);$

and, similarly to (10.18), we can prove the inclusion

(11.7') $\eta_{-c,\ell} \in \mathcal{L}(G^{(1)}, 1) \cap \mathcal{L}(G^{(2)}, 1).$

11.3. Let $e_j (j = 1, \ldots, n_0)$ be the functions introduced in condition B of §7. Let

$$\overset{1}{e}_j = \eta_{j,1}e_j, \qquad \overset{1}{\chi}_j = [\delta_{ik}\chi_{ji}]_{i,k=1}^m, \qquad \overset{1}{\mathcal{E}}_j = \eta_{j,2,w}\overset{1}{\chi}_{j,w}\eta_{j,2,w},$$

$$\mathcal{E}_j = \overset{1}{e}_{j,w}\overset{1}{\mathcal{E}}_j\overset{1}{e}_{j,w}^*, \qquad \mathcal{E} = \sum_{1 \leqslant j \leqslant n_0} \mathcal{E}_j, \qquad \mathcal{E}_1 = \mathcal{E}^2(3 - 2\mathcal{E}).$$

Since $e_j \in S(g, 1, U_j)$ and $\operatorname{supp}\eta_{j,1} \subset U_j$, we have

(11.8) $\overset{1}{e}_j \in S(G_{ji}, 1) \cap S(G^{(2)}, 1).$

Conditions (7.1)–(7.4) imply that $\operatorname{supp}\eta_{j,c}$ are bounded sets; therefore, $\sigma^w(\mathcal{E}) \in S(R^{2n}; \operatorname{End} C^m)$. Hence, $\mathcal{E}$ and $\mathcal{E}_1$ are operators of a trace class. Let ν_i and ν_{1i} be their nonzero eigenvalues, counted with their multiplicity and numbered in arbitrary order.

As in §9 and §10, Lemma 9.2 holds, but its proof here is not as simple. This is due to the fact that now the operator $\mathcal{E}$ is a sum of products of PDO from algebras not imbedded in each other; therefore, we cannot apply PDO calculus directly. The following two lemmas will help us to find a way around this obstacle.

Lemma 11.1. *Let the metric G satisfy the condition*

(11.9) $$h_G \leqslant Ch_g^\varepsilon, \quad g \leqslant G \leqslant Ch_g^{-C}g.$$

Then $S^{-\infty}(g, p_1, p_2) \cong S^{-\infty}(G, p_1, p_2)$ as countably-normalized spaces.

Proof. Obvious.   ∎

Lemma 11.2. *Let $N > 0$ be an integer, let $p_j \in O(C^1, g)$, let the σ-temperate metrics G_j satisfy condition (11.9), and let $A_j \in \mathcal{L}(G_j, p_j)$, $j = 1, \ldots, N$. Then there exists $K \in \mathcal{L}(g, p)$, where $p = \prod p_j$, such that*

$$|\langle A_1 \cdots A_N u, u \rangle| \leqslant \langle Ku, u \rangle \quad \forall u \in \mathcal{S}(R^n)^m.$$

Proof. We use Lemma 1.5 to construct the functions $\tilde{p}_i \in SI(g, p_i)$, denote

$$q_0^2 = q_N^2 = \sqcap \tilde{p}_i, \quad q_j = \tilde{p}_j^{-1} q_{j-1}, \quad j = 1, \ldots, N-1,$$

and use Lemma 5.5 to construct the paramatrices $q'_{j,w}$ for the operators $q_{j,w}$. Then $A'_j = q'_{j-1,w} A_j q_{j,w} \in \mathcal{L}(G_j, 1)$ and $\|A'_j\| \leqslant C$, due to Theorem 4.2. Therefore,

$$|\langle A_1 \cdots A_N u, u \rangle| \leqslant C_1 \|q_{0,w} u\|^2 + \langle Tu, u \rangle,$$

where $T \in \mathcal{L}^{-\infty}(g, p)$; and we can assume that $K = q^*_{0,w} q_{0,w} + T$.   ∎

11.4. Now we can prove Lemma 9.2. As in §9, it suffices to show that

(11.10) $$\tfrac{1}{8}I + \mathcal{E} \geqslant T_1, \quad \tfrac{9}{8}I - \mathcal{E} \geqslant T_2,$$

where $T_j \in \mathcal{L}^{-\infty}(g, 1)$. But $\overset{1}{\mathcal{E}}_j = [\delta_{ik}\mu_{ji,w}]^m_{i,k=1}$, where the functions $\mu_{ji} \in S(G_{ji}, 1)$ admit the representation

$$\mu_{ji} = \mu^1_{ji} + \mu^2_{ji}, \quad 0 \leqslant \mu^1_{ji} \leqslant 1, \quad \mu^2_{ji} \in S(G_{ji}, h^\omega);$$

hence, $\varepsilon + \mu_{ji} \in SI^+(G_{ji}; 1)$ $\forall \varepsilon > 0$; and, applying Lemmas 8.1 and 11.1, we obtain

$$\varepsilon I + \overset{1}{\mathcal{E}}_j \geqslant T_j \in \mathcal{L}^{-\infty}(g, 1).$$

Now Lemma 11.2 yields

$$\tfrac{1}{8}I + \mathcal{E} \geqslant \tfrac{1}{8} + \sum_{1 \leqslant j \leqslant n_0} (\varepsilon K_j + T_{\varepsilon, j}),$$

where $K_j \in \mathcal{L}(g,1)$ does not depend on ε and $T_{\varepsilon,j} \in \mathcal{L}^{-\infty}(g,1)$. Theorem 4.2 implies $\|K_j\| < C$; and, by choosing $\varepsilon < (8Cn_0)^{-1}$, we obtain $(11.10)_1$. Similarly, $(11.10)_2$ is proved.

Thus, Lemma 9.2 is proved. As in §9, it yields estimate (9.6). The next step—the calculation of $\operatorname{Tr}\mathcal{E}_1$—is again more difficult than in §9 and §10. Direct calculation of $\sigma^w(\mathcal{E}_1)$ using the theorems in Chapter 1 is not possible; therefore, using Lemmas 11.1 and 11.2, we shall construct the operators $\mathcal{F}$, $\mathcal{F}'$ such that $\mathcal{F} \leqslant \mathcal{E}_1 \leqslant \mathcal{F}'$ and shall use the obvious corollary of the minimax principle:

$$(11.11) \qquad \operatorname{Tr}\mathcal{F}' \geqslant \operatorname{Tr}\mathcal{E}_1 \geqslant \operatorname{Tr}\mathcal{F}.$$

Now (11.5) and Corollaries 3.1 and 3.2 yield

$$\mathcal{E}_1 = \sum_{1 \leqslant j \leqslant n_0} \mathcal{E}'_{1,j} + T,$$

where $T \in \mathcal{L}^{-\infty}(g,1)$ and $\mathcal{E}'_{1,j} = \mathcal{E}_j^2(3 - 2\mathcal{E}_j)$. Let $\overset{1}{\mathcal{E}}_{1,j} = \overset{1}{\mathcal{E}_j^2}(3 - 2\overset{1}{\mathcal{E}}_j)$. Condition (11.6) implies that $e_j^* = e_j^{-1}$ on $\operatorname{supp}_\infty \sigma^w(\mathcal{E}'_{1,j})$ $\overset{1}{e}_j = e_j$; hence, by applying Theorem 3.1 to $\overset{1}{e}_{j,w}\overset{1}{e}_{j,w}^*$ and then using Lemma 11.2, we obtain

$$\mathcal{E}_{1,j}^1 \geqslant \overset{1}{e}_{j,w}\overset{1}{\mathcal{E}}_{1j}\overset{1}{e}_{j,w}^* + K_j + T_j \overset{\text{def.}}{=} \mathcal{F}_j + K_j + T_j,$$

where $T_j \in \mathcal{L}^{-\infty}(g,1)$, $K_j \in \mathcal{L}(g,h)$, and

$$(11.12) \qquad \operatorname{supp}\sigma^w(K_j) \subset \operatorname{supp}_\infty \sigma^w(\mathcal{E}'_{1,j}).$$

Let $\mathcal{F} = \sum(\mathcal{F}_j + K_j + T_j) + T$. Then (11.12) and condition (7.2) yield the estimate

$$\int \|\sigma^w(K_j + T_j + T)(X)\|dX \leqslant C + C \int_{\mathcal{M}} h(X)dX,$$

where $\mathcal{M} = \{X \in (\Omega \times R^n)(c, h^{-\delta}g) \mid \lambda_1(X) < c(h^{\delta/2} + d)(X)\}$; therefore,

$$(11.13) \qquad \operatorname{Tr}\mathcal{F} \geqslant \operatorname{Tr}\sum_j \mathcal{F}_j - J_c^\delta(a,d) - C.$$

In order to calculate $\sigma^w(\mathcal{F}_j)$ and $\operatorname{Tr}\sum \mathcal{F}_j$, we make the following remarks:
1) $\eta_{j,2}(X) = 0 \quad \forall X \notin \Omega \times R^n$;

2) If $X \in \operatorname{supp}_\infty \sigma^w(\overset{1}{\mathcal{E}}_{1,j})\backslash(\bigcup_{1 \leqslant k \leqslant m} \widetilde{W}_{c,k}^{\delta-w}(a,d))$, where the sets $\widetilde{W}_{c,k}^{\delta-w}(a,d)$ are introduced by equality (7.12), then $\eta_{j,2}(X) = 1$, and $\chi_{ji}(X)$ are equal to either zero or one;

3) For the same X, we have $\overset{1}{e}_j(X) = e_j(X) = e_j^*(X)^{-1}$ and, therefore,

$$\sigma^w(\overset{1}{e}_{j,w}\overset{1}{\mathcal{E}}_{1,j}\overset{1*}{e}_{j,w})(X) = (e_j\overset{12}{\chi}_j^2(3-2\overset{1}{\chi}_j)e_j^*)(X) + O(h(X));$$

4) If $ee^* = e^*e = I$, then $\operatorname{Tr} eAe^* = \operatorname{Tr} A$;

5) $\sum_j (2\pi)^{-n} \int\limits_{U_j} \operatorname{Tr} \overset{12}{\chi}_j^2(3-2\overset{1}{\chi}_j)(X)dX = V(\Omega \times R^n; a) + O(W_c^{\delta-\omega}(a,d)+1),$

where the function $W_c^{\delta-\omega}(a,d)$ is the same as in (7.12). Hence, by applying Lemma 8.5' to $\operatorname{Tr} \sum \mathcal{F}_j$ and then using (11.13) and (11.11), we obtain

(11.14) $\operatorname{Tr} \mathcal{E}_1 \geqslant V(\Omega \times R^n, a) + O((RV)_c^{\delta-\omega}(\Omega, a, d)),$

where $(RV)_c^{\delta-\omega}(\Omega, a, d)$ is the right-hand side of estimate (7.12) with $\delta - \omega$ instead of δ.

In the same way we obtain the estimate from above for $\operatorname{Tr} \mathcal{E}_1$ and similar estimates for $\operatorname{Tr} \mathcal{E}_1^2$; hence, as in §9, (11.14) implies estimate (9.10).

11.5. Proof of the estimate from below for the function $\mathcal{N}_0(\overset{o}{a}_w, \Omega)$ up to the formula (9.18) is almost the same as in §9 and §10. The proof of equality (9.12) becomes somewhat more difficult:

$$R\eta_{-1,\ell}Q\mathcal{E}_1 = R\eta_{-1,\ell}Q \sum_{1 \leqslant j \leqslant n_0} \overset{1}{e}_{j,w}\eta_{j,2,w}\overset{1}{\chi}_{j,w}\eta_{j,2,w}\overset{1*}{e}_{j,w}\mathcal{E}(3-2\mathcal{E})$$

$$= RQ\mathcal{E}_1 + T_1 = \mathcal{E}_1 + T_2,$$

where $T_j \in \mathcal{L}^{-\infty}(G^{(2)}, 1)$ [we use the equality $\eta_{-1}\eta_{j,2} = \eta_{j,2}$ and the inclusions $\eta_{-1,\ell}$, $\eta_{j,2,w}$, $\overset{1}{e}_{j,w} \in \mathcal{L}(G^{(2)}, 1)$].

Formulas (9.13)–(9.18) are proved in exactly the same way. We point out that in (9.17) $K, K_1 \in \mathcal{L}(b_3^{-2}g, h)$, and estimation of the term $\langle \pi_w \overset{-}{a}_w \pi_w u, u \rangle$ in (9.18) becomes more complicated: (11.6) implies

(11.15) $\mathcal{E}_1 \pi_w \overset{-}{a}_w \pi_w \mathcal{E}_1 = \sum_{1 \leqslant j \leqslant n_0} \Phi_j \overset{1}{\chi}_{j,w}\eta_{j,2,w}\overset{1*}{e}_{j,w}\pi_w \overset{-}{a}_w \pi_w \overset{1}{e}_{j,w}\eta_{j,2,w}\overset{1}{\chi}_{j,w}\Phi_j^* + T,$

where $T \in \mathcal{L}^{-\infty}(g, 1)$ and $\Phi_j = \mathcal{E}(3-2\mathcal{E})\overset{1}{e}_{j,w}\eta_{j,2,w}$. By applying Theorem 3.1 to the operator $K_j = \eta_{j,2,w}[\overset{1*}{e}_{j,w}, \pi_w]$ and taking into account that $\pi \in S(b_3^{-2}g, \pi)$ and

(11.16) $\overset{1}{e}_j\big|_{\operatorname{supp} \eta_{j,2}} \in S(g, 1, \operatorname{supp} \eta_{j,2}),$

we obtain $K_j \in \mathcal{L}(G_{ji}, \pi b_3^{-1}h)$. But $\pi b_3^{-1}h = \lambda_0 h^{1-\delta+\omega}\pi \leqslant \lambda_0 h^\omega$; hence, $K_j \in \mathcal{L}(G_{ji}, \lambda_0 h^\omega)$ for any i. Therefore, in (11.15) by commuting $\overset{1*}{e}_{j,w}$ with π_w and then π_w with $\overset{1}{e}_{j,w}$ and by applying Lemmas 11.2 and 11.1, we obtain

(11.17) $\mathcal{E}_1 \pi_w \overset{-}{a}_w \pi_w \mathcal{E}_1 \leqslant \sum_{1 \leqslant j \leqslant n_0} \Phi_j \overset{1}{\chi}_{j,w}\eta_{j,2,w}\pi_w \overset{1*}{e}_{j,w}\overset{-}{a}_w \overset{1}{e}_{j,w}\pi_w \eta_{j,2,w}\overset{1}{\chi}_{j,w}\Phi_j^* + T,$

where $T \in \mathcal{L}(g, h^\omega)$. Using (11.16) and the equality

$$(e_j^* a^- e_j)(X) = [\delta_{ik} \lambda_{ji}]_{i,k=1}^m (X) \quad \forall X \in U_j,$$

together with Theorems 3.1 and 3.3 and Lemmas 11.2 and 11.1, one can easily show that estimate (11.17) remains valid [with different T] if we replace $e_{j,w}^{1*} a_w^- e_{j,w}^1$ with $[\delta_{ik} \lambda_{ji,w}]_{i,k=1}^m$. Furthermore, we let $\lambda_{ji}^1 = \lambda_{ji} \theta(\lambda_{ji} \lambda_0^{-1} h^{-\omega})$. Similar to (10.19), $\lambda_{ji}^1 \in \mathcal{L}(G_{ji}, \lambda_0 h^\omega)$. But $\lambda_{ji} = \lambda_{ji}^1$ on $\operatorname{supp}_d \chi_{ji}$; hence, by commuting $\chi_{ji,w}$ with $\eta_{j,2,w} \pi_w$, we can replace $\lambda_{ji,w}$ with λ_{ji}^1 to obtain

$$\chi_{ji,w} \eta_{j,2,w} \pi_w \lambda_{ji,w} \pi_w \eta_{j,2,w} \chi_{ji,w} = \eta_{j,2,w} \pi_w (\chi_{ji} \lambda_{ji} \chi_{ji})_w \pi_w \eta_{j,2,w} + K_{ji},$$

where $K_{ji} \in \mathcal{L}(G_{ji}, h^\omega)$.

Using the function λ_{ji}^1, one can easily see, again, (see the end of Section 9.3) that if C is large, then

$$f_{ji} \overset{\text{def.}}{=} -\chi_{ji} \lambda_{ji} \chi_{ji} + C \lambda_0 h^\omega \in SI^+(G_{ji}, f_{ji});$$

thus, Lemma 8.1 implies that $(\chi_{ji} \lambda_{ji} \chi_{ji})_w \leqslant C(\lambda_0 h^\omega)_w + T$, where $T \in \mathcal{L}^{-\infty}(g, 1)$, and estimate (11.17) yields the estimate

$$\mathcal{E}_1 \pi_w a_w^- \pi_w \mathcal{E}_1 \leqslant K \in \mathcal{L}(g, h^\omega).$$

Returning to (11.15) and then to (9.18), we obtain

$$\langle \overset{\circ}{a}_w v', v' \rangle \leqslant -\|v\|^2 + \langle K' v, v \rangle,$$

where $K' \in \mathcal{L}(g, h^\omega)$. Therefore, Lemma 8.2 and (11.14) imply the estimate from below for the function $\mathcal{N}_0(\overset{\circ}{a}_w, \Omega)$.

11.6. In order to prove the estimate from above, we replace $\lambda_{ji} + 8C_0 d + 4\lambda_0$ with $\lambda_{ji} - 8C_0 d - 4\lambda_0$, $\lambda_{ji} + 8C_0 d$ with $\lambda_{ji} - 8C_0 d$, and $\lambda_j + 8C_0 d$ with $\lambda_j - 8C_0 d$ in all the constructions at the beginning of the section; let $\psi_{c-1} = 1$ in (11.4); and, instead of the functions η_{-c}, construct the functions

$$\eta_c(X) = \begin{cases} \overset{\circ}{\eta}_{1,c}(X), & X \notin \Omega \times R^n, \\ 1, & X \in \Omega \times R^n. \end{cases}$$

The functions and metrics so constructed will satisfy all conditions formulated above [in (11.6), (11.7), and (11.7′), η_{-c} must be replaced with η_c]. By construction, $\eta_{c,r} u = u \quad \forall u \in C_0^\infty(\Omega)^m$; therefore, using (11.7′), we obtain (9.27) with $a^+ = a - (8C_0 d + 4\lambda_0)I$ [the proofs are the same as in §9].

In order to estimate $\mathcal{N}_0(A', R^n)$, we let

(11.18) $\qquad \overset{\circ}{\mu} = \eta_{j,8}(j = 1,\ldots,n_0), \qquad \overset{\circ}{\mu}_0 = 1 - \sum_{1 \leqslant j \leqslant n_0} \mu_j,$

(11.19) $\qquad \mu = (\sum_{0 \leqslant j \leqslant n_0} \overset{\circ}{\mu}_j^2)^{-1/2}, \qquad \mu_j = \overset{\circ}{\mu}_j \mu.$

Then

(11.20) $\qquad \mu_j \in S(G^{(2)}, 1), \quad j = 0,\ldots,n_0;$

(11.21) $\qquad \mu_j \in S(G_{ji}, 1), \quad j = 1,\ldots,n_0, \quad i = 1,\ldots,m;$

(11.22) $\qquad \operatorname{supp}_d \mu_j \subset \mathcal{M}_{c,\varphi_1\varphi_2}^\delta \cap U_j, \quad j = 1,\ldots,n_0;$

(11.23) $\qquad \operatorname{supp}_d \mu_0 \subset \bigcup_{1 \leqslant j \leqslant n_0} \operatorname{supp}_d \mu_j.$

Now (11.18)–(11.23) and (11.5) yield

(11.24) $\qquad \sum_{0 \leqslant j \leqslant n_0} \mu_{j,w}^* \mu_{j,w} = I + \sum_{1 \leqslant j \leqslant n_0} k_{j,w} + T,$

where $t \in \mathcal{L}^{-\infty}(g,1)$, $k_j \in S(G^{(2)}, h^\omega) \cap S(G_{ji}, h^\omega)$ $\forall i$, and

(11.25) $\qquad \operatorname{supp} k_j \subset \mathcal{M}_{c,\varphi_1\varphi_2}^\delta \cap U_j.$

We apply Theorems 3.1 and 3.3 to represent the operators $\mu_{j,w}(I - \frac{1}{2}\sum_j k_{j,w})$ in the form $\overset{1}{\mu}_{j,w} + \overset{2}{\mu}_{j,w}$, where $\overset{2}{\mu}_{j,w} \in \mathcal{L}^{-\infty}(g,1)$ and $\overset{1}{\mu}_j$ satisfy conditions (11.22) and (11.23); and we replace the operators $\mu_{j,w}$ with the operators $\overset{1}{\mu}_{j,w}$ in (11.24). Then we shall obtain (11.24) and (11.25) with $k_j \in S(G^{(2)}, h^{2\omega})$. By repeating this procedure a sufficient number of times and denoting the transformed operators again by $\mu_{j,w}$, we obtain (11.20)–(11.25) with $k_j \in S(G^{(2)}, h) \cap S(G_{ji}, h)$ $\forall i$.

Since

$$C\lambda_0(X) \geqslant h^{\delta/2-\omega}(X) \geqslant h^{\delta_1/2-\omega}(X) \quad \forall X \in \mathcal{M}_{c,\varphi_1\varphi_2}^\delta$$

and since

$$n_k(G^{(2)}, h^{\delta_1/2}, a - 8C_0d) < \infty \quad \forall k \geqslant 1,$$

then (11.22), (11.23), and (11.20) imply that $[\mu_{j,w}^*, a_w^\pm] \in \mathcal{L}(G^{(2)}, \lambda_0 h^\omega)$.

Thus,

(11.26) $\qquad A' = \sum_{1 \leqslant j \leqslant n_0} A_j + c_0 I + K,$

where $K \in \mathcal{L}(G^{(2)}, h^\omega)$, $c_0 = (n_0 + 2)^{-1}$, and $A_j = \eta_{1,\ell} \pi_w \mu_{j,w}^* a_w^+ \mu_{j,w} \pi_w \eta_{1,r} + c_0 I.$

Now (11.26) and Lemma 8.4 imply that on a subspace of $\mathcal{S}(R^n)^m$ of finite codimension, we have

$$\langle A'u, u \rangle \geqslant \sum_{0 \leqslant j \leqslant n_0} \langle A_j u, u \rangle;$$

therefore, the same arguments used in §9 to derive (9.27) from (9.26) lead us to the estimate

$$(11.27) \qquad \mathcal{N}(A', \mathcal{S}(R^n)^m) \leqslant \sum_{0 \leqslant j \leqslant n_0} \mathcal{N}_0(A_j, R^n) + C.$$

11.7. In order to estimate $\mathcal{N}_0(A_j, R^n)$ for $j \geqslant 1$, we shall construct $\mathcal{E}_j$ as in the beginning of the section; but, using new functions χ_{ji} and $\eta_{j,2}$, we let $\mathcal{E}_{1j} = \mathcal{E}_j^2(3 - 2\mathcal{E}_j)$ and let L_j be a subspace spanned over eigenfunctions of the operator $\mathcal{E}_{1j}$, which correspond to the eigenvalues $\nu_{1ji} \geqslant \frac{1}{2}$. Arguments similar to those in Section 11.3 show that

$$(11.28) \qquad \dim L_j \leqslant V(\Omega \times R^n \cap U_j, a) + C((RV)_c^{\delta - \omega}(\Omega, a, d)).$$

Let $v \in {}^\perp L_j$ and $u = (I - \mathcal{E}_{1j})v \in {}^\perp L_j$. We consider $\langle A_j u, u \rangle$. By using (11.6), (11.7), and the equality $\eta_1 \mu_j = \mu_j$, we obtain

$$\eta_{1,\ell} \pi_w \mu_{j,w}^* - \pi_w \mu_{j,w}^* \in \mathcal{L}^{-\infty}(G^{(2)}, 1) \cong \mathcal{L}^{-\infty}(g, 1);$$

hence, in the definition of the operator A_j, we can omit the factors $\eta_{1,\ell}$ and $\eta_{1,r}$ by adding a term of the class $\mathcal{L}^{-\infty}(g, 1)$. Furthermore, $\|u\| \geqslant \|v\|/2$; therefore,

(11.29)
$$\langle A_j u, u \rangle \geqslant \langle (I - \mathcal{E}_{1j})\pi_w \mu_{j,w}^* a_w^+ \mu_{j,w} \pi_w (I - \mathcal{E}_{1j})v, v \rangle + \tfrac{c_0}{4}\|v\|^2 + \langle Tv, v \rangle$$
$$\overset{\text{def.}}{=} \langle (D + \tfrac{c_0}{4} + T)v, v \rangle.$$

Since $\eta_{j,2}\mu_j = \mu_j$, we can replace $\mathcal{E}_{1j}$ with $\widehat{\chi}_{j,w}$ in the definition of the operator D, where $\widehat{\chi}_{j,w} = I - \widehat{\chi}_{j,w}^{12}(3 - 2\widehat{\chi}_{j,w}^1)$, adding a term of the class $\mathcal{L}^{-\infty}(g, 1)$. After this, an elementary synthesis of the proofs in §9 of the estimates from above and below give us $D \geqslant K$, where $K \in \mathcal{L}(G^{(2)}, h^\omega)$. By applying Lemma 8.2 thereafter, we see that $A_j \geqslant 0$ on the subspace $L_j' \subset L_j$ of finite codimension; therefore, (11.28) yields the estimate

$$(11.30) \qquad \mathcal{N}_0(A_j, R^n) \leqslant V(\Omega \times R^n \cap U_j, a) + C(RV)_c^{\delta - \omega}(\Omega, a, d),$$

where $j \geqslant 1$.

11.8. In order to estimate $\mathcal{N}_0(A_0, R^n)$, we construct the cutoff functions ψ_c associated with

$$h^{-1/3}g, \quad c, \quad \{X \mid \lambda_1(X) < h^{\delta/2}(X) + 8C_0 d\}$$

and let

$$\tilde{\eta} = \sum_{1 \leqslant j \leqslant n_0} \eta_{j,16}, \qquad \chi_0 \eta_2(1 - \tilde{\eta})\psi_2, \qquad \mathcal{E}_0 = (\chi_0 I)_w,$$

$\mathcal{E}_{10} = \mathcal{E}_0^2(3 - 2\mathcal{E}_0)$. Then $\mathcal{E}_0, \mathcal{E}_{10} \in \mathcal{L}(G^{(2)}, 1)$ and

$$\text{mes supp}_\infty \sigma^w(\mathcal{E}_{10}) \leqslant C(RV)_c^{\delta-w}(\Omega, a, d);$$

and, since $\|\sigma^w(\mathcal{E}_{10})\| \leqslant C$, Lemma 8.5 implies that the same estimate holds for $\text{Tr}\,\mathcal{E}_{10}$ and $\text{Tr}\,\mathcal{E}_{10}^2$. Now the same arguments as in §9 prove that

(11.31) $$\dim L_0 \leqslant C(RV)_c^{\delta-w}(\Omega, a, d).$$

Let $v \in {}^\perp L_0$ and $u = (I - \mathcal{E}_{10})v \in {}^\perp L_0$. Since $\eta_1 \eta_2 = \eta_1$ and $(1 - \tilde{\eta})\mu_0 = \mu_0$, we have

$$\mu_{0,w} \pi_w \eta_{1,r}(I - \mathcal{E}_{10}) = \mu_{0,w} \pi_w \eta_{1,r} \tilde{\psi}_{2,w} + T,$$

where $T \in \mathcal{L}^{-\infty}(g, 1)$ and $\tilde{\psi}_c = 1 - \psi_c^2(3 - 2\psi_c)$. Thus

(11.32) $\langle A_0 u, u \rangle \geqslant \langle \tilde{\psi}_{2,w} \eta_{1,r} \pi_w \mu_{0,w}^* a_w^+ \mu_{0,w} \pi_w \eta_{1,\ell} \tilde{\psi}_{2,w} v, v \rangle + \langle Tv, v \rangle + \dfrac{c_0}{4} \|v\|^2,$

where $T \in \mathcal{L}^{-\infty}(g, 1)$. Since $\psi_1 \psi_2 = \psi_1$, then also $\tilde{\psi}_1 \tilde{\psi}_2 = \tilde{\psi}_2$; therefore, we can replace a^+ with $\tilde{a} = \tilde{\psi}_1 a^+ + (1 - \tilde{\psi}_1) h^{\delta/2}$ in (11.32).

By construction, $c_1 h^{\delta/2} \leqslant \tilde{a} \leqslant C$; and if $\delta/2 + \frac{1}{6} < \delta_1 < \frac{1}{2}$, then $n_k(h^{-2\delta_1}g, h^{\delta/2}, \tilde{a}) < \infty$ for $k \geqslant 1$. Hence, $\tilde{a} \in SI^+(h^{-\delta_1}g, \tilde{a}^{1/2}, \tilde{a}^{1/2})$, and Lemma 8.1 and (11.32) yield estimate (11.29) with $j = 0$, $D \in \mathcal{L}(G^{(2)}, h^w)$. By applying Lemma 8.2, we see that $\mathcal{N}_0(A_0, R^n)$ admits estimate (11.31).

Since $w > 0$ is arbitrary and $\delta = \frac{2}{3} - 3w$, then (9.27), (11.27), (11.30), and (11.31) imply the estimate from above in Theorem 7.3.   ∎

11.9.

Proof of Theorem 7.2.   As in §9, we now let $\delta = \frac{1}{2} - w$ and $\lambda_0 = h^{\delta-w}$, and then, in all the constructions from this section, replace the cutoff functions $\overset{\circ}{\eta}_{1,c}$ and $\overset{\circ}{\eta}_{2,c}$ with cutoff functions associated, respectively, with $h^{-2\delta}g$, c and with the sets $\partial\Omega \times R^n$, $\Gamma\Omega$ ($\Gamma\Omega$ is introduced in condition B in §7). All the symbols are described by the metric $h^{-2\delta}g$, and the proof becomes much simpler.

§12 Proofs of Theorems 7.1, 7.4, and 7.5

12.1.

Proof of Theorem 7.1. Let $\omega \in (0, 1/g)$, $\delta = \frac{1}{3} - 2\omega$, $\nu_0 > 0$, $\nu \in (0, \nu_0)$, and let $\sum \varphi_j = 1$, $\mathrm{supp}\,\varphi_j = U_j$, be a partition of unity associated with $h^{-2\delta}g$, ν^2, and $(\nu/2)^2$.

Lemma 12.1. *If $\nu_0 > 0$ is sufficiently small, then there exists $C > 0$ such that*

$$(12.1) \qquad\qquad |\lambda_j(X) - \lambda_j(Y)| \leqslant C\nu h(X)^\delta$$

for all $\nu \in (0, \nu_0)$, $j = 1, 2, \ldots, m$, $k = 1, 2, \ldots$, and $X, Y \in U_k$.

Proof. If $C_1 > 0$ is sufficiently large, then $a_1 = a + C_1 I \geqslant C_1/2I$, $a_1 \in S(g, 1)$; therefore, estimate (1.11) holds with a_1 substituted for $\tilde{p}$. Hence, for small ν,

$$a_1(Y) \leqslant a_1(X) + Cg_X(X - Y)^{1/2}a_1(Y)$$

and

$$(1 - C_1\nu h(X)^\delta)a_1(Y) \leqslant a_1(X) \leqslant (1 + C_1\nu h(X)^\delta)a_1(Y).$$

If ν is sufficiently small, then the last estimate and the minimax principle yield (12.1).   ∎

Let $\lambda_0 = h^{\delta - \omega}$, $\lambda_j^- = \lambda_j + 8C_0 d + 4\lambda_0$, $R_{jk} = \{\lambda_j^-(X) \mid X \in U_k\}$ and let K_{jk} be a union with respect to μ of circles with the radius $h(X)^\delta$ and with centers at the points $\mu \in R_{jk}$. We unite K_{jk} $[j = 1, \ldots, m]$ into nonintersecting connected sets $L_{\alpha, k}$ $[\alpha = 1, \ldots, \alpha(k)]$. Then we select $L_{\alpha, k}$ containing the points with $\mathrm{Re}\,\mu < 0$, unite them with respect to α into the sets L_k, and let $\gamma_k = \partial L_k$. If $\nu_0 > 0$ is sufficiently small, then (12.1) implies that for all $X \in U_k$,

$$(12.2) \qquad\qquad \mathrm{dist}(\lambda_j^-(X), \gamma_k) \geqslant c_1 h(X)^\delta;$$

$$(12.3) \qquad\qquad \text{if } \lambda_j^-(X) \in L_k, \quad \text{then } \lambda_j(X) < c_2 h(X)^\delta;$$

$$(12.4) \qquad\qquad \text{if } \lambda_j^-(X) \notin L_k, \quad \text{then } \lambda_j(X) > c_1 h(X)^\delta,$$

where c_1, c_2 do not depend on j, k.

Let $a^- = a + 8C_0 d + 4\lambda_0$ and $e = \sum \varphi_k p_k$, where

$$p_k(X) = -\frac{1}{2\pi i}\int_{\gamma_k}(a^-(X) - \lambda I)^{-1}d\lambda, \quad X \in U_k.$$

Lemma 12.2. $e \in S(h^{-2\delta}g, 1)$.

Proof. Since $\varphi_k \in S(h^{-2\delta}g, 1)$ and since the multiplicity of the covering $\{U_k\}$ is finite, it suffices to prove that $p_k \in S(h^{-2\delta}g, 1, U_k)$. But $\|p_k\| \leqslant 1$, and for $j \geqslant 1$ and $X \in U_k$, (12.2) yields

$$\|d^j p_k(X; Y_1, \ldots, Y_j)\|$$

$$\leqslant c \sum_{\Sigma r_i = j, \ r_i > 0} \left\| \int_{\gamma_k} \prod_i (a^-(X) - \lambda I)^{-1} (d^{r_i} a^-)(X; Y_{j_1}, \ldots, Y_{j_{r_i}})(a^-(X) - \lambda I)^{-1} \right\|$$

$$\leqslant c_1 \operatorname{mes}_1 \gamma_k \max_{\gamma_k} \|(a^-(X) - \lambda I)^{-1}\|^{j+1} \sum_{\Sigma r_i = j, \ r_i > 0} \prod_i \|d^{r_i} a(X; Y_{j_1}, \ldots, Y_{j_i})\|$$

$$\leqslant c_2 h(X)^{\delta} h(X)^{-\delta(j+1)} \prod_{1 \leqslant i \leqslant j} g_X(Y_i)^{1/2} = c_2 \prod_{1 \leqslant i \leqslant j} (h(X)^{-2\delta} g_X(Y_i))^{1/2}.$$

∎

Then (12.3) and (12.4) imply that for $X \in U_k$, $p_k(X)$ is a projection onto a subspace generated by eigenvectors of the matrix $a^-(X)$ corresponding to the eigenvalues $\lambda_j^-(X) \leqslant ch(X)^{\delta}$, where $c = c(X) \leqslant C$; therefore,

(12.5) $$0 \leqslant e(X) \leqslant I,$$
(12.6) $$e(X)a^-(X)e(X) \leqslant Ch(X)^{\delta}I.$$

12.2. We define the functions $\overset{\circ}{\eta}_c$ and $\eta_{\pm c}$ with respect to $\delta = \frac{1}{3} - 2\omega$ as was done in §9. Then (9.3) and (9.4) hold. Let $\mathcal{E} = \eta_{-2,w} e_w \eta_{-2,w}$ and $\mathcal{E}_1 = \mathcal{E}^2(3 - 2\mathcal{E})$. Lemma 12.2, (9.3), and (12.5) yield

$$\mathcal{E} + \varepsilon I \in \mathcal{L}I^+(h^{-2\delta}g, 1), \quad (1+\varepsilon)I - \mathcal{E} \in \mathcal{L}I^+(h^{-2\delta}g, 1), \quad \forall \varepsilon > 0;$$

therefore, as in §9, the eigenvalues ν_{1i} of the operator $\mathcal{E}_1$ satisfy (9.6).

By calculating $\sigma^w(\mathcal{E}_1)$ using Theorems 3.1 and 3.3, we find that

(12.7) $$\sigma^w(\mathcal{E}_1) = \eta_{-2}^4 e^2(3 - 2\eta_{-2}^2) + f + r,$$

where $r \in S^{-\infty}(g, 1)$, $f \in S(h^{-2\delta}g, h^{1-2\delta})$, and

$$\operatorname{supp} f \subset \operatorname{supp} \eta_{-2} e \subset \{X \in \Omega \times R^n \mid \lambda_1(X) < c(\lambda_0 + d)(X)\}.$$

Since $1 - 2\delta < \delta$ and $\lambda_0 = h^{\delta-\omega}$, we have

(12.8) $$\int \|f(X) + r(X)\| dX \leqslant c + C\mathcal{J}_c^{\delta-\omega}(a, d).$$

Furthermore, if $X \in \Omega \times R^n \backslash (\partial\Omega \times R^n)(4, h^{-2\delta}g)$, then

$$(\eta_{-2}^4 e^2(3 - 2\eta_{-2}^2 e))(X) = (e^2(3 - 2e))(X);$$

and (12.3) and (12.4) yield

$$\begin{aligned}
\operatorname{Tr}(e^2(3 - 2e))(X) \\
= \operatorname{card}\{i \mid \lambda_i(X) < 0\} + O(\operatorname{card}\{i \mid |\lambda_i(X)| \leqslant C(h(X)^{\delta-\omega} + d(X))\}).
\end{aligned}$$

Therefore, for some c,

$$(2\pi)^{-n} \int \operatorname{Tr}(\eta_{-2}^4 e^2(3 - 3\eta_{-2}^2 e))(X) dX = V(\Omega \times R^n, a) + O((RV)_c^{\delta-\omega}(\Omega, a, d));$$

and (12.8) and Lemma 8.5′ imply that

$$(12.9) \qquad \operatorname{Tr} \mathcal{E}_1 = V(\Omega \times R^n, a) + O((RV)_c^{\delta-\omega}(\Omega, a, d)).$$

12.3. By using (12.9) and the similar estimate for $\operatorname{Tr} \mathcal{E}_1^2$, we obtain estimate (9.10) as in §9; and by defining the operators π_w, Q_w and a space L as in §9, we obtain estimates (9.14) and (9.18). Direct application of Theorem 3.1 yields

$$(12.10) \qquad \langle \pi_w a_w^- \pi_w u, u \rangle = \langle \mathcal{F}^*(ea^- e)_w \mathcal{F}v, v \rangle + \langle Kv, v \rangle,$$

where

$$K \in \mathcal{L}(h^{-2\delta}g, h^\omega), \qquad \mathcal{F} = \pi_w(\eta_{-2}^2)_w \mathcal{E}(3 - 2\mathcal{E}).$$

Let $f = -ea^- e + C_1 h^\delta I$. Furthermore, we shall show that if C_1 is sufficiently large and if $\delta_1 = \frac{1}{2} - \omega$, then

$$(12.11) \qquad f \in SI^+(h^{-2\delta_1}g, f^{1/2}, f^{1/2}).$$

Now (12.11) and Lemmas 1.2 and 3.2 imply that $f^{1/2} \in O(C^m, h^{-2\delta_1}g)$, and Lemma 8.1 yields $f_w \geqslant T \in \mathcal{L}^{-\infty}(g, 1)$. Therefore, $(ea^- e)_w \leqslant K \leqslant \mathcal{L}(h^{-2\delta_1}g, h^\omega)$. Returning to (12.10) and then to (9.18), we obtain

$$\langle \overset{\circ}{a}_w v', v' \rangle \leqslant -\|v\|^2 + \langle Kv, v \rangle, \quad K \in \mathcal{L}(h^{-2\delta_1}, h^\omega).$$

With (12.9), Lemma 8.2 now yields the estimate from below for $\mathcal{N}_0(\overset{\circ}{a}_w, \Omega)$ in exactly the same way as in §9.

12.4. (12.5) implies that if C_1 is large, then $f \geqslant h^\delta$; therefore, (12.11) is proved, provided that

$$ea^- e \in S(h^{-2\delta_1}g, f^{1/2}, f^{1/2}).$$

But for $k \geqslant 1$,

$$(12.12) \qquad n_k(h^{-2\delta_1}g, h^{k(\delta_1-\delta)}, e) \leqslant n_k(h^{-2\delta}g, 1, e) < \infty,$$
$$n_k(h^{-2\delta_1}g, h^{\delta_1 k}, a^-) \leqslant n_k(g, 1, a^-) < \infty;$$

and since $\delta_1 - \delta > \delta/2$, then for $k \geqslant 2$,

$$(12.13) \qquad n_k(h^{-2\delta_1}g, h^\delta, ea^-e) < \infty.$$

Then (12.6) yields $f^{1/2} \geqslant h^{\delta/2}$, provided that C_1 is sufficiently large; therefore, (12.13) yields the estimates

$$(12.14) \qquad n_k(h^{-2\delta_1}g, f^{1/2}, f^{1/2}, ea^-e) < \infty$$

for $k \geqslant 2$. In order to obtain (12.14) with $k = 1$, we remark that

$$\langle d(ea^-e)(X), Y \rangle$$
$$= \langle de(X), Y \rangle(a^-e)(X) + e(X)\langle da^-(X), Y \rangle e(X) + (ea^-)(X)\langle de(X), Y \rangle$$
$$= M_1 + M_2 + M_3.$$

Since $f^{-1/2} \leqslant h^{-\delta/2}$ and $\|\langle da^-(X), Y \rangle\| \leqslant Cg_X(Y)^{1/2}$, then for $j = 2$,

$$(12.15) \qquad \|f^{-1/2}M_j f^{-1/2}\| \leqslant Ch(X)^{-\delta_1}g_X(Y)^{1/2}.$$

Now (12.12) implies that (12.15) is proved for $j = 1,3$—provided that $\|a^-(X)e(X)f^{-1/2}(X)\| \leqslant C$. Recalling that $e(X)$ and, therefore, that $f(X)$ are the functions of the matrix $a^-(X)$, we see that it suffices to prove the estimate

$$(12.16) \qquad \|a^-(X)e(X)f^{-1/2}(X)v_j(X)\| \leqslant C,$$

where $v_j(X)$ is an arbitrary normalized eigenvector of the matrix $a^-(X)$ with the eigenvalue $\lambda_j^-(X)$.

Then (12.3) and (12.4) imply that

$$\text{if} \quad \lambda_j^-(X) \geqslant C_1 h(X)^\delta, \quad \text{then} \quad a^-(X)e(X)v_j(X) = 0;$$
$$\text{if} \quad \lambda_j^-(X) \leqslant -C_1 h(X)^\delta, \quad \text{then}$$
$$\|a^-(X)e(X)f(X)^{-1/2}v_j(X)\|$$
$$\leqslant |\lambda_j^-(X)(-\lambda_j^-(X) + C_1 h(X)^\delta)^{-1/2}|$$
$$\leqslant C_2|\lambda_j^-(X)|^{1/2} \leqslant C_3;$$
$$\text{if} \quad |\lambda_j^-(X)| \leqslant C_1 h(X)^\delta, \quad \text{then}$$
$$\|a^-(X)e(X)f(X)^{-1/2}v_j(X)\|$$
$$\leqslant C_2 h(X)^\delta \|f(X)^{-1/2}\| \leqslant C_3.$$

This proves (12.16), (12.15), (12.14), and, therefore, (12.11), as well as the estimate from below for the function $\mathcal{N}_0(\overset{\circ}{a}_w, \Omega)$.

12.5. The estimate from above is proved similarly. We must construct an operator $\mathcal{E}_1$—starting with the cutoff function η_C associated with $h^{-2\delta}g$, c, $\Omega \times R^n$—and the functions $\lambda_j^+ = \lambda_j - 8C_0 d - 4\lambda_0$, $a^+ = a - (8C_0 d + 4\lambda f_0)I_m$; and then we must make an obvious synthesis of the proofs in this section and the proof of the estimate from above in §9. The role of the function f in the last section is now played by the function

$$f = (I - \tilde{e})a^+(I - \tilde{e}) + C_1 h^\delta, \quad \text{where } \tilde{e} = e^2(3 - 2e);$$

its estimate (12.11) is proved in exactly the same way as above.

12.6. The proof of Theorem 7.5 basically repeats the proof of Theorem 7.1. The changes are connected, first, with the fact that instead of Lemmas 8.2– 8.4, which are valid under condition (7.2), we use Lemmas 8.2'–8.4' [valid under condition (7.15)$_1$]. Of course, these changes are not essential. More essential is the impossibility of obtaining the estimate

$$\int \|r(X)\| dX < C$$

in (12.7). To get around it, we have to make construction of ASP more complicated.

Let $\theta \in C_0^\infty(R^1)$, $0 \leqslant \theta \leqslant 1$, $\theta|_{|t|<1/2} = 1$, $\theta|_{|t|>1} = 0$ and let $\hat{e} \in S(h^{-2\delta}g, 1)$ be a symbol for one of the ASP $\mathcal{E}$, which was used to prove the estimates from above and below in Theorem 7.1. We recall that under the suppositions of Theorem 7.5 the metric g splits:

$$g_{x,\xi}(y, \eta) = g'_{x,\xi}(y) + g''_{x,\xi}(\eta).$$

We regularize it by using Corollary 1.3 and let
(12.17)
$$\hat{e}_\theta(x, y, \xi) = \hat{e}(\frac{x+y}{2}, \xi)\theta(h(x,\xi)^{-2\delta}g'_{x,\xi}(x-y) + h(y,\xi)^{-2\delta}g'_{x,\xi}(x-y)),$$

$$\mathcal{E} = Op\hat{e}_\theta, \quad \text{where}$$

$$((Opa)u)(x) = \int d'\xi \int dy e^{i\langle x-y,\xi\rangle} a(x, y, \xi)u(y).$$

By writing u in the form $u(y) = F_{\eta \to y}^{-1}\hat{u}(\eta)$, we see that $\mathcal{E} = \tilde{e}_\ell$, where

(12.18)
$$\tilde{e}(x, \eta) = \iint d'\xi \, dy \exp(i\langle x - y, \xi - \eta\rangle)\hat{e}_\theta(x, y, \xi)$$
$$= \exp(-i\langle D_\eta, D_y\rangle)\hat{e}_\theta(x, y, \xi)\big|_{y=x, \, \xi=\eta}.$$

If t is large, then condition $(7.15)_1$ implies that $h_t(x,\xi)^{2\delta}$ is smaller than the constant in $(1.6')$; thus, if $(x,y,\xi) \in \operatorname{supp} \hat{e}_\theta$, then $cg_{x,\xi} \leqslant g_{y,\xi} \leqslant Cg_{x,\xi}$; and since $h \in S(g,h)$, $\hat{e} \in S(h^{-2\delta}g, 1)$, and $g'(\tau) \in S(g, g'(\tau))$ uniformly with respect to $\tau \neq 0$, then for any k, we have

$$(12.19) \qquad d_x^k \hat{e}_\theta(x, \cdot, \cdot; y_1, \ldots, y_k) \in S(h^{-2\delta}g, \prod_{1 \leqslant j \leqslant k} (h^{-2\delta}g'(y_j))^{1/2}$$

uniformly with respect to $x, y_1, \ldots, y_k \in R^n$.

Now (12.18), (12.19), and Theorem 2.2 yield

$$(12.20) \qquad \tilde{e} - e \in S(h^{-2\delta}, h^{1-2\delta});$$

therefore, the proofs of Theorem 7.1 hold without change, except for the calculation of $\operatorname{Tr} \mathcal{E}_1$. Lemma 8.5 and (12.17) yield

$$(12.21) \qquad \operatorname{Tr} \mathcal{E}_1 = \int d'\xi \int dx \, \operatorname{Tr} \hat{e}_{1,\theta}(x,x,\xi),$$

where
(12.22)
$$\hat{e}_{1,\theta}(x,y,\xi)$$
$$= \int d'\eta \int dz \int d'\zeta \int dv \exp(i[\langle x-z,\eta\rangle + \langle z-v,\zeta\rangle + \langle v-x,\xi\rangle]) \times$$
$$\hat{e}_\theta(x,z,\eta)(3 - 2\hat{e}_\theta(z,v,\zeta))\hat{e}_\theta(v,t,\xi)$$
$$= \int d'\eta \int dz \exp(i\langle x-z, \eta-\xi\rangle)\hat{e}_\theta(x,z,\eta) \times$$
$$\left\{ \int \int d'\zeta \, dv \exp(i\langle z-v, \zeta-\xi\rangle)(3 - 2\hat{e}_\theta(z,v,\zeta))\hat{e}_\theta(v,y,\xi) \right\}$$
$$= \exp(-i\langle D_z, D_\eta\rangle)\{\hat{e}_\theta(x,z,\eta) \times$$
$$[\exp(-i\langle D_v, D_\zeta\rangle)(3 - 2\hat{e}_\theta(z,v,\zeta))\hat{e}_\theta(v,y,\xi)]|_{v=z,\zeta=\xi}\}|_{z=x,\eta=\xi}.$$

By applying Theorem 2.2 twice, we obtain

$$(12.23) \qquad \hat{e}_{1,\theta}(x,y,\xi) = \hat{e}_{2,\theta}(x,y,\xi) + \hat{e}_{3,\theta}(x,y,\xi),$$

where

$$(12.24) \qquad \begin{aligned} \hat{e}_{2,\theta}(x,x,\xi) &= (\hat{e}^2(3 - 2\hat{e}))(x,\xi), \\ \|\hat{e}_{3,\theta}(x,x,\xi)\| &\leqslant Ch(x,\xi)^{1-2\delta}. \end{aligned}$$

Here we took into account the fact that $\hat{e}_\theta(x,x,\xi) = \hat{e}(x,\xi)$.

Now (12.21), (12.23), and (12.24) imply that for $\mathrm{Tr}\,\mathcal{E}_1$ we can obtain exactly the same estimate as in the proof of Theorem 7.1, provided that

(12.25) $\mathrm{supp}\,\hat{e}_{1,\theta}(x,\cdot,\cdot) \subset \{X \in (\Omega \times R^n)(c, h^{-2\delta}g) \mid \lambda_1(X) < c(h^\delta + d)(X)\}.$

Indeed, (12.23)–(12.25) and (12.21) yield

$$\mathrm{Tr}\,\mathcal{E}_1 = \int d'\xi \int dx\,(\hat{e}^2(3 - 2\hat{e}))(x,\xi) \; + \; O\left(\int_{\mathcal{M}} h(X)^{1-2\delta}dX\right),$$

where $\mathcal{M}$ is a set in (12.25); therefore, unlike as in (12.7), we do not need to estimate

$$\int_{R^{2n}} h(X)^N \, dX.$$

As implied by (12.22), it is sufficient to prove (12.25) for $\hat{e}_\theta$.

By construction, if either

(12.26) $\lambda_1((v + y)/2, \xi) > c(h((v + y)/2, \xi)^\delta + d((v + y)/2, \xi)),$

or, $\forall X \in \Omega,$

(12.27) $h^{-2\delta}\left(\dfrac{v + y}{2}, \xi\right) g'_{\frac{v+y}{2},\xi}\left(\dfrac{v + y}{2} - x\right) > C,$

then $\hat{e}_\theta(v, y, \xi) = 0$; and if $(x, y, \xi) \in \mathrm{supp}\,\hat{e}_\theta$, then

(12.28) $h^{-2\delta}(y, \xi)g'_{y,\xi}\left(\dfrac{v + y}{2} - y\right) \leqslant h^{-2\delta}(y, \xi)g'_{y,\xi}(y - v) \leqslant 1.$

Since the metric g varies slowly, then (12.27) implies the same estimate, except with y instead of $(v + y)/2$ [and with different C]. By using (12.28) and the arguments of Lemma 12.1, we can replace $(v + y)/2$ with y in (12.26) as well.

But changed in this manner, estimates (12.26) and (12.27) mean exactly that $\hat{e}_\theta$, along with $\hat{e}_{1,\theta}$, satisfies condition (12.25).

Thus, we have listed all the changes that must be made in the proof of Theorem 7.1 in order to obtain the proof of Theorem 7.5.

12.7.

Proof of Theorem 7.4.   For $j = 1, \ldots s$, we choose a constant δ_j in accordance with the conditions satisfied by the function a_j and by $\partial\Omega$, and, starting with a_j and δ_j, construct functions $\lambda_0^{(j)}$, $\pi^{(j)}$, $\eta_{\pm c}^{(j)}$, $a_j^\pm$, by using the function $d + d_j$ instead of the function d. Furthermore, let

$$\lambda_0 = [\delta_{jk}\lambda_0^{(j)}I_{m_j}]_{j,k=1}^s, \qquad \pi = [\delta_{jk}\pi^{(j)}I_{m_j}]_{j,k=1}^s,$$
$$\eta_{\pm c} = [\delta_{jk}\eta_{\pm c}^{(j)}I_{m_j}]_{j,k=1}^s, \qquad a^\pm = [\delta_{jk}a_j^\pm]_{j,k=1}^s.$$

As in §9, Lemmas 8.1, 11.1, and 11.2 yield estimate (9.25) and estimate (9.27) with

$$A' = [\delta_{jk} A'_j]^s_{j,k=1}, \quad A'_j = \eta^{(j)}_{1,\ell} \pi^{(j)}_w a^+_{j,w} \pi^{(j)}_w \eta^{(j)}_{1,r} + I_{m_j}.$$

Thus,

(12.29)
$$\mathcal{N}_0(\overset{\circ}{a}_w, \Omega) \leqslant \sum_{1 \leqslant j \leqslant s} \mathcal{N}(A'_j, C_0^\infty(R^n)^{m_j}) + C.$$

Each term in this sum is estimated as in the proofs of Theorems 7.1–7.3, and (12.29) yields the estimate in Theorem 7.4.

In order to prove the estimate from below, we construct operators $\mathcal{E}^{(j)}_1$— starting with the functions a_j in accordance with the conditions that they satisfy, denoting by ν_{1ji} their nonzero eigenvalues counted with their multiplicity and numbered in arbitrary order, and denoting:

$$L_j = \text{span}\{v \mid \mathcal{E}^{(j)}_1 v = \nu_{1ji} v, \ \nu_{1ji} \geqslant \tfrac{1}{2}\},$$
$$L^- = \{v' = (v'_1, \ldots, v'_s) \mid v'_i = \sum_j \eta^{(j)}_{-1,\ell} Q_{ij} u_j, \ u_j = \mathcal{E}^{(j)}_1 v_j \in L_j\},$$

where Q_{ij}, $i,j = 1, \ldots, s$ are blocks of the matrix $Q = q_w \pi_w \in S(h^{-2\delta} g, \overset{\circ}{q}{}^{-1}, \pi)$, $\hat{\delta} \in (0, \tfrac{1}{3})$.

By construction, $\eta^{(j)}_{-1} = 1$ on $\text{supp}_\infty \sigma^w(\mathcal{E}^{(j)}_1)$; therefore,

(12.30)
$$v' = Qu + Tv, \quad T \in \mathcal{L}^{-\infty}(g, 1).$$

Using (12.30), we obtain, as in (9.14) and (9.18),

(12.31)
$$\dim L^- \geqslant \sum_{1 \leqslant j \leqslant s} \dim L_j - C,$$

(12.32) $\langle \overset{\circ}{a}_w v', v' \rangle \leqslant \langle [\delta_{jk} \pi^{(j)}_w a^-_{j,w} \pi^{(j)}_w]^s_{j,k=1} u, u \rangle - \|v\|^2 + \langle Kv, v \rangle,$

where $K \in \mathcal{L}(g, h^\omega)$. Since the matrix in (12.32) is blockwise diagonal, the arguments of Theorems 7.1–7.3 show that

$$\langle \overset{\circ}{a}_w v', v' \rangle < 0 \quad \forall (0 \neq) v' \in L' \subset L^-,$$

where $\text{codim } L' \leqslant C$. Now (12.21) and the formulas for $\dim L_j$, proved above, yield the estimate from below for the function $\mathcal{N}_0(\overset{\circ}{a}_w, \Omega)$.

CHAPTER 3

OPERATORS IN A BOUNDED DOMAIN

§13 Douglis–Nirenberg Elliptic Operators. Dirichlet-Type Problems

13.1. Let $\Omega \subset R^n$ be a bounded domain, satisfying the condition

$$(13.1) \qquad \qquad \mathrm{mes}(\partial\Omega)_\varepsilon \leqslant C\varepsilon,$$

where $(\partial\Omega)_\varepsilon = \{x \mid \mathrm{dist}(x, \partial\Omega) < \varepsilon\}$, and let A and B be classical matrix PDOs, symmetric in $L_2(R^n)^m$, of orders $2\ell_1$ and $2\ell_2$, respectively. In addition, let

$$(13.2) \qquad \qquad \ell = 2\ell_0 = 2\ell_1 - 2\ell_2 > 0$$

and

$$(13.3) \qquad \qquad \langle Au, u \rangle \geqslant c\|u\|_{\ell_1}^2 \quad \forall u \in C_0^\infty(\Omega)^m,$$

where $c > 0$ does not depend on u, and $\|\cdot\|_s$ is a norm in the Sobolov space $H^s(R^n)^m$, $\langle \cdot, \cdot \rangle = \langle \cdot, \cdot \rangle_{L_2}$.

Let us consider the problem

$$(13.4) \qquad \qquad Au = tBu, \quad u \in C_0^\infty(\Omega)^m.$$

Obviously,[1]

$$(13.5) \qquad \qquad |\langle Bu, v \rangle| \leqslant C\|u\|_{\ell_2}\|v\|_{\ell_2} \quad \forall u, v \in C_0^\infty(\Omega)^m;$$

therefore, the spectrum of problem (13.4) can be defined as follows.

---

[1] For example, (13.5) follows from parts d) and c) of Theorem 5.2.

Let $H(A, \Omega)$ be a completion of $C_0^\infty(\Omega)^m$ with respect to the norm $\| \cdot \|_A = \langle A \cdot, \cdot \rangle^{1/2}$. Then (13.2), (13.3), (13.5), and the Riesz theorem on general representation of linear continuous functionals in Hilbert space imply that a closure of the hermitian form $\langle B \cdot, \cdot \rangle$ in $H(A, \Omega)$ determines an operator B_1 bounded in $H(A, \Omega)$. Obviously, B_1 is symmetric; and, therefore, the spectrum of the problem

$$(13.6) \qquad\qquad u = t B_1 u, \quad u \in H(A, \Omega),$$

is real. We shall define the spectrum of problem (13.4) as the spectrum of problem (13.6). One can prove that if $\partial\Omega$ is sufficiently smooth and if A and B are DO or PDO of positive order with transmission property (see Boutet de Monvel [1]), then the spectrum just defined is the spectrum of the problem

$$\widehat{A} u = t \widehat{B} u, \quad u \in \overset{\circ}{H}{}^{2\ell_1}(\Omega)^m = D(\widehat{A}),$$

where $\widehat{A} = r_+ A e_+$, $\widehat{B} = r_+ B e_+$, (and r_+, e_+ are usual operators of contraction of generalized functions to the set Ω and extension by zero of functions from $L_2(\Omega)$ to R^n, respectively), i.e., the set of numbers $t \in \mathbf{C}$ such that the operator $\widehat{A} - t\widehat{B} : D(\widehat{A}) \to L_2(\Omega)^m$ does not have a bounded inverse defined on all $L_2(\Omega)^m$.

Applying the Glazman lemma to the operators $\pm B_1$, we have

$$(13.7) \qquad N_\pm(t) = \mathcal{N}(I \mp t B_1, H(a, \Omega)) = \mathcal{N}_0(A \mp tB, \Omega).$$

Using this equality and the theorems in §7, we shall prove that the spectrum of pencil (13.4) is discrete, and we shall find asymptotics of the distribution functions of its spectrum. For the sake of greater clarity, we shall first consider the case where B is elliptic. At the end of the section, we shall analyze the case where $B = I$ and A is a Douglis–Nirenberg elliptic operator.

But first we shall prove the following lemma:

Lemma 13.1. *If condition (13.3) holds, then the principal symbol a^0 of the operator A satisfies the following condition:* $\forall (x, \xi) \in \Omega \times S_{n-1}, \forall v \in \mathbf{C}^m,$

$$(13.8) \qquad\qquad \langle a^0(x, \xi) v, v \rangle_{C^\ell} \geqslant c \| v \|_{C^\ell}^2.$$

Proof. For the sake of simplicity, we limit ourself to the case of a DO A; if A is a PDO, then we have to apply the theorem on PDO action on rapidly oscillating exponents (see, for example, Shubin [1]).

We fix $(x_0, \xi_0) \in \Omega \times S_{n-1}$, $v \in \mathbf{C}^m$, $\theta \in C_0^\infty(R^1)$ such that $0 \leqslant \theta \leqslant 1$, $\theta\big|_{|t|<1/2} = 1$, $\theta\big|_{|t|>1} = 0$, and denote

$$\varepsilon = (\ln t)^{-1}, \qquad u_t(x) = \theta(\varepsilon^{-1} |x - x_0|) \exp(it \langle x, \xi_0 \rangle) v.$$

Then as $t \to \infty$ and for some N,

$$\langle Au_t, u_t \rangle = t^{2\ell_1} \|\theta(\varepsilon^{-1}| \cdot - x_0|)\|_{L_2}^2 \langle a^0(x_0, \xi_0)v, v \rangle_{C^m}$$
$$+ t^{2\ell_1} \int \theta(\varepsilon^{-1}|x - x_0|)^2 \langle (a^0(x, \xi_0) - a^0(x_0, \xi_0))v, v \rangle_{C^m} dx$$
$$+ O(t^{2\ell_1 - 1} \varepsilon^{-N}) \|v\|_{C^m}^2$$
$$= c_\theta t^{2\ell_1} \varepsilon^n (\langle a^0(x_0, \xi_0)v, v \rangle + o(\|v\|^2));$$

and, similarly,

$$\|u_t\|_{\ell_1}^2 = c_\theta t^{2\ell_1} \varepsilon^n \|v\|_{C^m}^2 (1 + o(1)).$$

When we substitute the last two equalities into (13.3), then divide by $t^{2\ell_1} \varepsilon^n$ and take the limit, we arrive at (13.8). ∎

If $B = I$ and the operator A satisfies condition (13.8) instead of condition (13.3), then the proof of Theorem 13.1 below shows that the spectrum of the Friedrichs extension of the operator

$$r_+ A e_+ : C_0^\infty(\Omega)^m \to L_2(\Omega)^m$$

is discrete and gives asymptotics for the distribution function of this spectrum.

13.2.

Theorem 13.1.   a) *Let conditions* (13.1)–(13.3) *hold and let*

(13.9) $\det b^0(x, \xi) \neq 0 \quad \forall (x, \xi) \in \overline{\Omega} \times (R^n \backslash O).$

Then for all $\delta \in (0, \frac{1}{3})$,

(13.10) $N_\pm(t) = c_\pm t^{n/\ell} + O(t^{(n-\delta)/\ell}),$

where

(13.11) $c_\pm = n^{-1}(2\pi)^{-n} \int\limits_{\Omega} dx \int\limits_{S_{n-1}} \mathrm{Tr}(a_0)_\pm^{n/\ell}(x, \xi) dS(\xi),$

and $(a_0)_\pm$ *are positive* $(+)$ *and negative* $(-)$ *parts of the hermitian matrix* $a_0 = (a^0)^{-1/2} b^0 (a^0)^{-1/2}$.

b) *If conditions* (13.2), (13.3), *and* (13.9) *hold, if* Ω *is a Lipschitz domain, and if the function* a_0 *is microlocally smoothly diagonalizable on* $\overline{\Omega} \times (R^n \backslash O)$, *i.e., for any* $(x, \xi) \in \overline{\Omega} \times (R^n \backslash O)$ *there exists a neighborhood* $V_{x\xi}$ *of the point* (x, ξ) *that is conic, with respect to* ξ, *and a function* $e = (e^*)^{-1} \in$

$C^\infty(V_{x\xi}, \mathrm{End}\, C^m)$ *that is positively homogeneous of degree* 0, *with respect to* ξ, *such that* $\forall X \in V_{x\xi}$,

$$(13.12) \qquad (e^* a_0 e)(X) = [\delta_{jk} \lambda_j^0(X)]_{j,k=1}^m,$$

then estimate (13.10) *is valid for all* $\delta \in (0, \frac{1}{2})$.

   c) *In addition, if* $\partial\Omega \subset \bigcup_{1 \leqslant j \leqslant N_0} \Gamma_j'$, *where* $\Gamma_j' = \{x \mid \varphi_j'(x) = 0\}$, $\varphi_j' \in C_0^\infty(R^n)$, *and*

$$(13.13) \qquad \begin{array}{l} \textit{if } x \in \Gamma_{j_1}' \cap \cdots \cap \Gamma_{j_s}', \quad 1 \leqslant s \leqslant N_0, \quad \textit{then the forms} \\ d\varphi_{j_1}'(x), \ldots, d\varphi_{j_s}'(x) \quad \textit{are linearly independent,} \end{array}$$

then estimate (13.10) *holds for all* $\delta \in (0, \frac{2}{3})$.

Proof. This will be demonstrated for $N_+(t)$. Let

$$(13.14)$$
$$h(x,\xi) = (1 + |x|^2 + |\xi|^2)^{-1/2}, \qquad g_{x,\xi}(y,\eta) = h(x,\xi)^2 |\eta|^2 + |y|^2,$$
$$q_t(X)^2 = \langle X \rangle^{-2\ell_1} + t \langle X \rangle^{-2\ell_2}$$

[the metric (13.14) will be used in all sections of this chapter]. Lemmas 1.2, 1.3, 3.2, and 3.3 imply that the metric g is σ-temperate and $\overset{\circ}{q}_t \in O(C^1; g)$, uniformly with respect to $t \geqslant 1$. Furthermore, for any compact set K, $\sigma^\ell(A - tB) \in S(g, \overset{\circ}{q}_t^2, K \times R^n)$ uniformly with respect to $t \geqslant 1$; but for $u \in C_0^\infty(\Omega)^m$, $\langle (A - tB)u, u \rangle$ depends only on values of the symbol $\sigma^\ell(A - tB)$ on $\Omega \times R^n$. Therefore, we can change it outside $\Omega \times R^n$ in such a way that $\sigma^\ell(A - tB) \in S(g, \overset{\circ}{q}_t^2)$ uniformly with respect to $t \geqslant 1$. Theorem 3.2 also yields $\sigma^w(A - tB) \overset{\text{def}}{=} \overset{\circ}{a}_t \in S(g, \overset{\circ}{q}_t^2)$. Moreover, (13.8) allows us to make $A \in \mathcal{LI}^+(g, h^{-2\ell_1})$; and, therefore, $\overset{\circ}{a}_t \in SI^+(g, \overset{\circ}{q}_t^2) \quad \forall t$.

   Thus, condition A in §7 holds.

   In order to choose q_t, a_t, d_t, we note that by definition of principal symbol, for some $\varepsilon > 0$,

$$\overset{\circ}{a}_t - (a^0 - tb^0) \in S(g, h\overset{\circ}{q}_t^2, \mathcal{M}^{4\varepsilon}),$$

where $\mathcal{M}^\varepsilon = \Omega_\varepsilon \times \{\xi \mid |\xi| > 1\}$; thus, we can construct q_t, a_t such that in $\mathcal{M}^\varepsilon$,

$$q_t(x,\xi) = (1 + t|\xi|^{-\ell})^{-1/2} a^0(x,\xi)^{-1/2},$$
$$d_t(x,\xi) = |\xi|^{-1},$$
$$a_t(x,\xi) = (1 + t|\xi|^{-\ell})^{-1}(I_m - t a_0(x,\xi)),$$
$$\lambda_{t,j}(x,\xi) = (1 + t|\xi|^{-\ell})^{-1}(1 - t\lambda_j^0(x,\xi)),$$

where λ_j^0 denote eigenvalues of the matrix a_0.

Positive homogeneity of the functions λ_j^0 with respect to ξ yields

$$V(\Omega \times R^n, a_t)$$

$$= (2\pi)^{-n} \sum_{1 \leqslant j \leqslant m} \text{mes}\{X \in \Omega \times R^n \mid a_t(X) < 0\}$$

$$= O(\text{mes}\{(x, \xi) \mid x \in \Omega, |\xi| \leqslant 1\})$$

$$(13.15) \qquad + (2\pi)^{-n} \sum_{1 \leqslant j \leqslant m} \text{mes}\{(x, \xi) \in \Omega \times R^n \mid \lambda_j^0(x, \xi/|\xi|)|\xi|^{-\ell} > t^{-1}\}$$

$$= O(1) + n^{-1}(2\pi)^{-n} \int_\Omega dx \int_{S_{n-1}} dS(\xi) \sum_{\lambda_j^0(x,\xi)>0} (t\lambda_j^0(x, \xi))^{n/\ell}$$

$$= c_+ t^{n/\ell} + O(1),$$

where c_+ is defined by equality (13.11).

13.3. Thus, we have estimated the principal term of the asymptotics of the function $N_+(t)$. In order to estimate the remainder in Theorem 7.1, we note that

$$(13.16) \qquad t\lambda_j^0(x, \xi) < \tfrac{1}{2}, \quad \text{if } |\xi| > Ct^{1/\ell};$$

therefore,

$$(13.17) \qquad J_c^\delta(a_t, d_t) \leqslant C \int_{1 \leqslant |\xi| \leqslant Ct^{1/\ell}} |\xi|^{-\delta} d\xi \leqslant C_1 t^{(n-\delta)/\ell},$$

and (13.1) implies
(13.18)

$$V((\partial\Omega \times R^n)(c, h^{-2\delta}g), a_t - c(d_t + h_t^\delta)) \leqslant c \int_{1 \leqslant |\xi| \leqslant Ct^{1/\ell}} \text{mes}(\partial\Omega)_{|\xi|^{-\delta}} d\xi$$

$$\leqslant C_1 \int_{1 \leqslant |\xi| \leqslant Ct^{1/\ell}} |\xi|^{-\delta} d\xi \leqslant C_2 t^{(n-\delta)/\ell}.$$

Finally,

$$(13.19) \quad W_c(\Omega \times R^n, a_t, h^\delta + d_t) \leqslant C \iint_{\Omega \times S_{n-1}} dx\, d\varphi \sum_{1 \leqslant j \leqslant m} \int_{M_j(x,\varphi)} dr \cdot r^{n-1},$$

where

$$M_j(x, \varphi) = \{r \mid |1 - tr^{-\ell}\lambda_j^0(x, \varphi)| \leqslant Ctr^{-\ell-\delta}\}.$$

Now (13.9) yields $c \leqslant |\lambda_j^0(x,\varphi)| \leqslant C$; therefore, there exist $c_1, c_2 > 0$ such that if $r \in M_j(x,\varphi)$, then $c_1 t \leqslant r^\ell \leqslant c_2 t$. Therefore, by substituting $\rho = tr^{-\ell}\lambda_j^0(x,\varphi)$, we obtain estimate (13.18) for $W_c(\Omega \times R^n, a_t, h^\delta + d_t)$ as well.

Estimates (13.15), (13.17) and (13.18) and Theorem 7.1 prove part a) of Theorem 13.1.

If (13.12) holds, then we can satisfy the conditions of Theorem 7.2, letting

$$\Gamma\Omega = \{(x,\xi) \mid (x, |\xi|^{-1}\xi) \in \widetilde{\Gamma\Omega} \subset \overline{\Omega} \times S_{n-1}, |\xi| \geqslant 1\} \cup \overline{\Omega} \times \{\xi \mid |\xi| \leqslant 1\},$$

where $\widetilde{\Gamma\Omega}$ satisfies condition (13.1) as a set of $R^n \times S_{n-1}$ and breaks $\Omega \times S_{n-1}$ into subsets of sufficiently small diameter. Then the term $V((\Gamma\Omega)(c, h^{-2\delta}g), a_t - c(d_t + h^\delta))$ in Theorem 7.2 also admits estimate (13.18), and we obtain estimate (13.10).

13.4. Given the conditions of part c), we apply Theorem 7.3. We denote $\overset{\circ}{\varphi}{}_i^1(x,\xi) = \varphi_i'(x)$ and construct for any $\Delta > 0$ the functions

$$\overset{\circ}{\varphi}{}_i^2 = \overset{\circ}{\varphi}{}_{i,\Delta}^2 \in C^\infty(R^n \times (R^n \backslash O))$$

that are homogeneous of degree 0 with respect to ξ and finite with respect to x, such that the following conditions hold:

1) If $\overset{\circ}{\Gamma}{}_i^j = \{X \mid \overset{\circ}{\varphi}{}_i^j(X) = 0\}$, $\overset{\circ}{\Gamma}_j = \bigcup_i \overset{\circ}{\Gamma}{}_i^j$, $j = 1,2$, and $\overset{\circ}{U}_i$, $i = 1, \ldots, N_2$ are connected components of the set

$$(\overline{\mathcal{M}^\varepsilon} \backslash (\overset{\circ}{\Gamma}_1 \cup \overset{\circ}{\Gamma}_2)) \cap \overline{\Omega}_\varepsilon \times S_{n-1},$$

then

(13.20) $$\operatorname{diam} \overset{\circ}{U}_i < \Delta;$$

2) If $X \in \overset{\circ}{\Gamma}{}_{i_1}^1 \cap \cdots \cap \overset{\circ}{\Gamma}{}_{i_s}^1 \cap \overset{\circ}{\Gamma}{}_{j_1}^2 \cap \cdots \cap \overset{\circ}{\Gamma}{}_{j_r}^2 \cap \overline{\Omega} \times S_{n-1}$, then the forms

(13.21)
$$d\overset{\circ}{\varphi}{}_{i_1}^1(X), \ldots, d\overset{\circ}{\varphi}{}_{i_s}^1(X),$$
$$d\overset{\circ}{\varphi}{}_{j_1}^2(X), \ldots, d\overset{\circ}{\varphi}{}_{j_r}^2(X)$$

are linearly independent (the set of indices $\{i_1, \ldots, i_r, j_1, \ldots, j_s\}$ can consist of one element).

The functions $\overset{\circ}{\varphi}{}_i^2$ can be constructed in the following way. First we construct functions satisfying condition (13.20) with $\Delta/2$ instead of Δ, as well as the condition $d\overset{\circ}{\varphi}{}_i^2(X) \neq 0 \; \forall X \in \Omega_\varepsilon \times S_{n-1}$; if condition (13.21) does not hold, then (13.13) implies that it can be satisfied with an arbitrarily small change of the functions $\overset{\circ}{\varphi}{}_i^2$. This will also result in (13.20) being satisfied.

We fix $\theta \in C_0^\infty(R^1)$, $0 \leqslant \theta \leqslant 1$, $\theta\big|_{|t|<\frac{1}{2}} = 1$, $\theta\big|_{|t|>1} = 0$ and let

$$\theta_c = 1 - \theta(c^{-1}|\cdot|), \quad \varphi_1 = \prod_i \overset{\circ}{\varphi}{}_i^1, \quad \varphi_2(x,\xi) = \theta_{1/2}(\xi) \prod_i \overset{\circ}{\varphi}{}_i^2(x,\xi),$$

$$\Gamma_j = \{X \mid \varphi_j(X) = 0\}.$$

Then $\varphi_1, \varphi_2 \in S(g,1)$, and if $\Delta > 0$ is sufficiently small, then conditions (13.20) and (13.12) guarantee existence of the matrix functions $e_i \in S(g, I, U_i)$, where $U_i = \{(x,\xi) \mid (x,|\xi|^{-1}\xi) \in \overset{\circ}{U}_i, \ |\xi| \geqslant 1\}$, such that

$$e_i^* e_i = e_i e_i^* = I, \qquad e_i^* a_0 e_i = [\lambda_j^0 \delta_{jk}]_{j,k=1}^m;$$

the order $\lambda_{j_1}^0 \leqslant \cdots \leqslant \lambda_{j_m}^0$ can vary for different i. By construction, $a_0 \in S(g, h^\ell, U_i)$; therefore, $\lambda'_{ij} = \lambda_j^0\big|_{U_i} \in S(g, h^\ell, U_i)$ [homogeneous of degree $-\ell$ with respect to ξ]. We can construct $\overset{\circ}{\varphi}{}_i^2$ such that $\overset{\circ}{U}_i$ are Lipschitz sets; hence, the functions λ'_{ij} can be analytically continued to functions of the class $S(g, h^\ell)$, homogeneous with respect to ξ of degree $-\ell$ on $\mathcal{M}^\varepsilon$. Then, denoting

$$\overset{\circ}{\lambda}_{t,ij}(x,\xi) = \psi_\varepsilon(x)\theta_1(\xi)(1 + t|\xi|^{-\ell})^{-1}(I - t\lambda'_{ij}(x,\xi)),$$

where $\psi_\varepsilon \in C_0^\infty(\Omega_\varepsilon)$ and $\psi_\varepsilon(x) = 1 \quad \forall x \in \Omega$, we see that all the conditions of Theorem 7.3 are satisfied.

Estimate (13.17) [with $\delta \in (0, \frac{2}{3})$] holds here, too; and in order to estimate $W_c^\delta(a, d)$, it suffices to estimate mes $W'_{c,j}$, where
(13.22)

$$W'_{c,j} = \{(x,\xi) \mid x \in \Omega_\varepsilon, \ 1 \leqslant |\xi| \leqslant ct^{1/\ell},$$

$$|\varphi_1(x)\varphi_2(x,\xi)(1 + t|\xi|^{-\ell})^{-1}(1 - t|\xi|^{-\ell}\lambda_j^0(x, |\xi|^{-1}\xi))| \leqslant C|\xi|^{-\delta}\}.$$

Now (13.9) implies that there exist $c_1, \varepsilon > 0$ such that

(13.23) $$|\lambda_j^0(x,\xi)| \geqslant c_1 \quad \forall(x,\xi) \in \Omega_\varepsilon \times S_{n-1};$$

therefore, if $(x,\xi) \in B_t = \{(x,\xi) \mid |\xi|^\ell \leqslant c_1 t/2\}$, then

$$W'_{c,j} \cap B_t \subset \{(x,\xi) \mid x \in \Omega_\varepsilon, \ 1 \leqslant |\xi|^\ell \leqslant c_1 t/2, \ |\varphi_1(x)\varphi_2(x, |\xi|^{-1}\xi)| \leqslant C|\xi|^{-\delta}\}.$$

In order to estimate the measure of the last set, we first change to the spherical coordinate system (r, θ) in R_ξ^n, and then (locally) to the system $(y_1, \ldots y_{2n-1}, r)$ on $\Omega_\varepsilon \times S_{n-1} \times [1, \infty)$, such that (also locally)

$$\Omega_\varepsilon \times S_{n-1} \backslash (\overset{\circ}{\Gamma}_1 \cup \overset{\circ}{\Gamma}_2) \subset \{y \mid y_i \geqslant 0, \ i = 1, \ldots, k\},$$

and $|\varphi_1 \varphi_2| \asymp \prod\limits_{1 \leqslant i \leqslant k} |y_i|$ [this is possible due to (13.21)]. Then

$$\mathrm{mes}(W'_{c,j} \cap B_t) \leqslant C \int\limits_1^{C\tau} dr \cdot r^{n-1} \int\limits_{\substack{|y_1 \cdots y_k| \leqslant Cr^{-\delta}, \\ |y_i| \leqslant 1}} dy_1 \cdots dy_k$$

$$\leqslant C \int\limits_1^{C\tau} dr \cdot r^{n-1} \left(C_1 r^{-\delta} + \int\limits_{\substack{|y_1 \cdots y_k| \leqslant Cr^{-\delta}, \\ r^{-\delta} \leqslant |y_i| \leqslant 1}} dy_1 \cdots dy_k \right)$$

(13.24)

$$\leqslant C \int\limits_1^{C\tau} dr \cdot r^{n-1-\delta} \prod\limits_{2 \leqslant i \leqslant k} \int\limits_{r^{-\delta} \leqslant |y_i| \leqslant 1} |y_i|^{-1} dy_i$$

$$\leqslant C_\varepsilon \int\limits_1^{C\tau} dr \cdot r^{n-1-\delta+\varepsilon} \leqslant C'_\varepsilon \tau^{n-\delta+\varepsilon},$$

where $\varepsilon > 0$ is arbitrary and $\tau = t^{1/\ell}$. Since $\delta \in (0, \frac{2}{3})$ is arbitrary, we can assume that

(13.25) $$\mathrm{mes}(W'_{c,j} \cap B_t) \leqslant Ct^{(n-\delta)/\ell}.$$

If $(x, \xi) \in W'_{c,j} \backslash B_t$, then $r \asymp \tau$. In $W'_{c,j} \backslash B_t$ we make the substitution $1 - tr^{-\ell} \lambda_j^0(x, \varphi) = \rho$. (13.23) yields $|1 - \rho| \geqslant c_1 > 0$; therefore, as in (13.24) and (13.25),

(13.26) $$\mathrm{mes}(W'_{c,j} \backslash B_t) \leqslant C\tau^n \int\limits_{\substack{|y_i \cdots y_k \rho| \leqslant C\tau^{-\delta}, \\ y_i, \rho \in [-1, 1]}} dy_1 \ldots dy_k d\rho \leqslant Ct^{(n-\delta)/\ell}.$$

Now (13.17), (13.25), and (13.26) yield estimate (13.10).

13.5. Here b^0 will be nonelliptic, i.e., condition (13.9) does not hold. If one of the functions λ_j^0 vanishes on a set of nonzero measure, then we assume that $\chi = 0$; otherwise, there exists $\chi \in [0, 1)$ such that for all j,

(13.27) $$\iint\limits_{\Omega \times S_{n-1}} |\lambda_j^0(x, \xi)|^{-\chi} dx \, dS(\xi) < \infty.$$

Theorem 13.2.
   a) *If conditions (13.1)–(13.3) and (13.27) hold, then for all $\delta \in (0, \frac{1}{3})$,*

(13.28) $$N_\pm(t) = c_\pm \, t^{n/\ell} + O(t^{\rho(\delta)}),$$

where $\rho(\delta) = \max\{(n - \delta)/\ell, \; n(1 - \chi)/(\ell + \delta - \ell\chi)\}$.

   b) *If conditions (13.2), (13.3), and (13.27) hold, if Ω is a Lipschitz domain, and if the symbol a_0 is microlocally smoothly diagonalizable, in the sense of part b) of Theorem 13.1, then estimate (13.28) is valid for all $\delta \in (0, \frac{1}{2})$.*

   c) *If the conditions of parts b) and c) of Theorem 13.1 and one of the following conditions hold:*

     1. *If $X \in \Gamma''_{j_1} \cap \cdots \cap \Gamma''_{j_s} \cap \Gamma_{(i)} \cap \overline{\Omega} \times S_{n-1}$, where $\Gamma''_j = \Gamma'_j \times S_{n-1}$, $\Gamma_{(i)} = \{X \mid \lambda_i^0(X) = 0\}$, then the vectors*

$$(13.29) \qquad (\nabla_x \varphi'_{j_1}(x), 0), \ldots, (\nabla_x \varphi'_{j_s}(x), 0), \qquad \nabla_X \lambda_i^0(X)$$

are linearly independent (the set $\{j_1, \ldots, j_s, i\}$ can consist of one element);

     2. *The matrix a_0 is globally smoothly diagonalizable on $\overline{\Omega} \times (R^n \backslash O)$, and condition (13.29) holds with $\partial\Omega \times S_{n-1}$ instead of $\overline{\Omega} \times S_{n-1}$;*

     3. *$n - \ell - \frac{2}{3} \geqslant 0$;*

then estimate (13.28) holds for all $\delta \in (0, \frac{2}{3})$.

Remark. If either condition 1 or condition 3 holds, then (13.28) is reduced to estimate (13.10) for all χ.

Proof. The proof of parts a) and b) differs from the proof of Theorem 13.1 only by the estimate of the right-hand side of inequality (13.19). We use the following lemma:

Lemma 13.2. *Let $\Sigma \subset R^m$, $\operatorname{mes} \Sigma < \infty$, $x \in [0, 1)$, $p, q, \delta > 0$, $q > \delta$,*

$$(13.30) \qquad \int_\Sigma |\lambda(\sigma)|^{-\chi} d\sigma < \infty, \qquad |\lambda(\sigma)| < C$$

($\chi = 0$ if $\lambda(\sigma) = 0$ on a set of nonzero measure), and $\mathcal{M}(\sigma) = \mathcal{M}(\sigma, t) = \{r > 1 \mid |1 - tr^{-p}\lambda(\sigma)| \leqslant Ctr^{-p-\delta}\}$.

   Then for $t \geqslant 1$,

$$(13.31) \qquad \int_\Sigma d\sigma \int_{\mathcal{M}(\sigma)} r^{q-1} dr \leqslant Ct^\beta,$$

where

$$(13.32) \qquad \beta = \max\{(q - \delta)/p, \; q(1 - \chi)/(p + \delta - p\chi)\}.$$

Proof. Let $\Sigma' = \{\sigma \mid \lambda(\sigma) = 0\}$. Then, by substituting Σ' for Σ in (13.31), we obtain (13.31) and (13.32) with $\chi = 0$. Hence, we can assume that λ does not vanish on Σ (and is even positive, since $|1 - tr^{-p}|\lambda(\sigma)|| \leqslant |1 - tr^{-p}\lambda(\sigma)|$).

After the substitution $r \to r^q$, we obtain

$$\int_{\mathcal{M}(\sigma)} dr \cdot r^{q-1} \leqslant \operatorname{mes} \mathcal{M}'(\sigma)/q,$$

where

$$\mathcal{M}'(\sigma) = \{r > 1 \mid |1 - tr^{-p'}\lambda(\sigma)| \leqslant Ctr^{-p'-\delta'}\}, \quad p' = p/q, \quad \delta' = \delta/q.$$

Let

$$\mathcal{M}_1(\sigma) = \{r \mid tr^{-p'}\lambda(\sigma) \leqslant \tfrac{1}{2}, \quad 1 \leqslant 2ctr^{-p'-\delta'}\},$$
$$\mathcal{M}_2(\sigma) = \{r \mid tr^{-p'}\lambda(\sigma) \geqslant \tfrac{3}{2}, \quad \lambda(\sigma) \leqslant 2cr^{-\delta'}\},$$
$$\mathcal{M}_3(\sigma) = \mathcal{M}'(\sigma)\backslash(\mathcal{M}_1(\sigma) \cup \mathcal{M}_2(\sigma)).$$

If $r \in \mathcal{M}_1(\sigma)$, then $t^s \lambda(\sigma)^s r^{-p's} \leqslant 1$ and $(t\lambda(\sigma))^s r^{p'+\delta'-p's} \leqslant C_1 t$ for any $s \in [0, 1)$. This yields $r \leqslant C_s t^{s_1} \lambda(\sigma)^{-s_2}$, where $s_1 = (1 - s)/(p' + \delta' - p's)$ and $s_2 = s/(p' + \delta' - p's)$. Let $s = \chi(p' + \delta')/(1 + \chi p')$. Then $s_2 = \chi$ and $s_1 = (q - \delta\chi)/(p + \delta)$; therefore, for $i = 1$,

$$(13.33) \qquad \operatorname{mes} \mathcal{M}_i(\sigma) \leqslant Ct^{(q-\delta\chi)/(p+\delta)}\lambda(\sigma)^{-\chi}.$$

Similarly, we prove (13.33) for $i = 2$. Finally, if $r \in \mathcal{M}_3(\sigma)$, then $r \asymp (t\lambda(\sigma))^{1/p'}$; and for any s,

$$(13.34) \qquad |r - (t\lambda(\sigma))^{1/p'}| \leqslant C_s t(t\lambda(\sigma))^{(1-p'-\delta'+s)/p'} r^{-s}.$$

If $(1 - p' - \delta')/p' \geqslant -\chi$, then, assuming $s = 0$ in (13.34), we have

$$(13.35) \qquad \operatorname{mes} \mathcal{M}_3(\sigma) \leqslant Ct^{(q-\delta)/p}\lambda(\sigma)^{-\chi}.$$

On the other hand, if $(1 - p' - \delta')/p' \leqslant -\chi$, then we denote

$$\mathcal{M}_{3,\gamma}^1(\sigma) = \{r \in \mathcal{M}_3(\sigma) \mid r \leqslant t^\gamma\} \qquad \mathcal{M}_{3,\gamma}^2(\sigma) = \mathcal{M}_3(\sigma)\backslash\mathcal{M}_{3,\gamma}^1(\sigma),$$

where $\gamma > 0$ will be chosen below. Obviously,

$$(13.36) \qquad \operatorname{mes} \mathcal{M}_{3,\gamma}^1(\sigma) \leqslant Ct^\gamma;$$

and if $s = -\chi p' - 1 + p' + \delta'$ in (13.34), then

$$(13.37) \qquad \operatorname{mes} \mathcal{M}_{3,\gamma}^2(\sigma) \leqslant C_1 \lambda(\sigma)^{-\chi} t^{1-\chi+\gamma(\chi p'+1-p'-\delta')}.$$

By solving the equation $\gamma = 1 - \chi + \gamma(\chi p' + 1 - p' - \delta')$, we determine that $\gamma = (1 - \chi)/(p' + \delta' - p'\chi)$; and, therefore, (13.30) and (13.33)–(13.37) yield (13.31) with

$$\beta = \max\{(q - \delta)/p, \ (q - \delta\chi)/(p + \delta), \ q(1 - \chi)/(p + \delta - p\chi)\}.$$

One can easily verify that if $q - p(1 - \chi) - \delta \geqslant 0$, then

$$(q - \delta)/p \geqslant (q - \delta\chi)/(p + \delta) \geqslant q(1 - \chi)/(p + \delta - p\chi);$$

and if $q - p(1 - \chi) - \delta \leqslant 0$, then the inequalities of opposite direction hold. This implies that β is defined by (13.32). ∎

Lemma 13.2, (13.17)–(13.19), and (13.27) yield estimate (13.28), provided that the conditions in parts a) and b) of the theorem hold.

13.6. We turn to the proof of part c) of Theorem 13.2. If (13.29) holds, then Γ_i' are smooth hypersurfaces and the functions $\overset{\circ}{\varphi}_i^2$ [see the proof of part c) of Theorem 13.1] can be constructed in such a way that for any i, the totality of the functions $\overset{\circ}{\varphi}_1^1, \ldots, \overset{\circ}{\varphi}_{N_0}^1, \overset{\circ}{\varphi}_1^2, \ldots, \overset{\circ}{\varphi}_{N_1}^2, \lambda_i^0$ satisfies condition (13.21). Now, as in Theorem 13.1, we change to spherical coordinates and break $\Omega_\varepsilon \times S_{n-1}$ into the sets $\omega_1, \ldots, \omega_{N_2}$ of sufficiently small diameter. If on ω_i we have $|\lambda_j^0(x, \xi)| \geqslant c$ or $\omega_1 \cap (\overset{\circ}{\Gamma}_1 \cap \overset{\circ}{\Gamma}_2) = \emptyset$ [the sets $\overset{\circ}{\Gamma}_j$ are the same as in the proof of Theorem 13.1], then the same arguments as in the proof of part c) of Theorem 13.1 and Lemma 13.2 yield the estimate

$$R_{ij} = \mathrm{mes}\{X \in \mathcal{M}^\varepsilon \cap \omega_i \times [1, \infty) \mid h(X)^{-\ell} \leqslant Ct, \ |(\varphi_1 \varphi_2 \lambda_{t,j}^+)(X)| \leqslant Ch(X)^\delta\}$$
$$\leqslant C_\varepsilon t^{\rho(\delta) + \varepsilon}, \quad \forall \varepsilon > 0,$$

where $\delta \in (0, \frac{2}{3})$. On the other hand, if $\omega_i \cap (\overset{\circ}{\Gamma}_1 \cup \overset{\circ}{\Gamma}_2) \neq \emptyset$, if $\mathrm{diam}\,\omega_i$ is sufficiently small, and if $\lambda_j^0(X) = 0$ for some $X = (x, \varphi, r) \in \omega_i \times (1, \infty)$, then by applying (13.29) and selecting a suitable local coordinate system, we obtain

$$R_{ij} \leqslant C \iint\limits_{\mathcal{M}} dy\,dr \cdot r^{n-1},$$

where

$$\mathcal{M} = \{(y, r) \in (0, 1)^{2n-1} \times [1, ct^{1/\ell}] \mid$$
$$|y_1 \cdots y_{2n-1}(1 + tr^{-\ell})^{-1}(1 - tr^{-\ell}y_1)| \leqslant Cr^{-\delta}\}.$$

As in (13.24) and (13.25),

$$R_{ij} \leqslant Ct^{(n-\delta)/\ell} + C \iint\limits_{\mathcal{M}'} dy\,dr \cdot r^{n-1},$$

where

$$\mathcal{M}' = \{(y,r) \in (0,1)^{2n-1} \times (1, Ct^{1/\ell}) \,|$$
$$|y_2 \cdots y_{2n-1}| \geqslant Cr^{-\delta},$$
$$|y_2 \cdots y_{2n-1}(1 + tr^{-\ell})^{-1}(1 - tr^{-\ell}y_1)| \leqslant Cr^{-\delta}\}.$$

If $(y,r) \in \mathcal{M}'$, then

$$t^{-1}r^\ell - Cr^{-\delta}(y_2 \cdots y_{2n-1})^{-1} < y_1 < t^{-1}r^\ell + Cr^{-\delta}(y_2 \cdots y_{2n-1})^{-1};$$

and, therefore,

$$R_{ij} \leqslant Ct^{(n-\delta)/\ell} + \int_0^{Ct^{1/\ell}} dr \cdot r^{n-1+\delta} \prod_{2 \leqslant i \leqslant 2n-1} \int_{cr^{-\delta} \leqslant y_i \leqslant 1} y_i^{-1} dy_i$$
$$\leqslant C_\varepsilon t^{(n-\delta)/\ell+\varepsilon} \quad \forall \varepsilon > 0.$$

Hence, estimate (13.38) holds in this case, too.

Provided that condition 1 holds, (13.38) yields (13.28). If condition 2 holds, then the proof becomes simpler, since there is no need to construct the functions $\overset{0}{\varphi}{}^2_i$.

13.7. Now let condition 3 hold. The same arguments as in the previous parts show that it suffices to estimate the integrals

$$(13.39) \qquad I_j = \iiint_{W_j} r^{n-1} dx \, d\theta \, dr, \quad 1 \leqslant j \leqslant m,$$

where

$$W_j = \{(r, x, \theta) \,\big|\, 1 \leqslant r \leqslant Ct^{1/\ell}, \ (x, \theta) \in \Omega \times S_{n-1},$$
$$(1 + tr^{-\ell})^{-1} \,\big|1 - tr^{-\ell}|\lambda^0_j(x, \theta)|\big| \,|f(x, \theta)| \leqslant Cr^{-\delta}\},$$

and the function f is locally a product of the coordinate functions in a suitable coordinate system on $\Omega \times S_{n-1}$.

We select subsets $W^i_j \subset W_j$, for $i = 1, 2$, by imposing the conditions

$$tr^{-\ell}|\lambda^0_j(x, \theta)| < \tfrac{1}{2}, \qquad tr^{-\ell}|\lambda^0_j(x, \theta)| > \tfrac{3}{2},$$

respectively; we denote $W^3_j = W_j \backslash (W^1_j \cup W^2_j)$; and then we define I^i_j through equality (13.39), replacing W_j with W^i_j. Then for $i = 1, 2$ we have

$$(13.40) \qquad I^i_j \leqslant c_1 \int_{1 \leqslant r \leqslant Ct^{1/\ell}} r^{n-1} dr \iint_{|f(x,\theta)| \leqslant C'tr^{-\ell-\delta}} dx \, d\theta + c_2 \int_{1 \leqslant r \leqslant c'_2 t^{1/(\ell+\delta)}} r^{n-1} dr.$$

Arguments similar to those of (13.24) allow us to estimate the inner integral in (13.40) by

$$Cr^{-\delta} + C_\varepsilon tr^{-\ell-\delta+\varepsilon} \quad \forall \varepsilon > 0.$$

Then, by integrating with respect to r and taking into account the fact that $n - \ell - \frac{2}{3} \geqslant 0$ and that $\delta \in (0, \frac{2}{3})$ is arbitrary such that $n/(\ell + \delta) < (n - \delta)/\ell$, we obtain

(13.41)
$$I_j^i \leqslant Ct^{(n-\delta)/\ell}$$

for $i = 1, 2$. In order to estimate I_j^3, we denote $\mu = |\lambda_j^0(x, \theta)|$, $r = \rho^{1/n} t^{1/\ell} \mu^{1/\ell}$. Then

(13.42)
$$I_j^3 \leqslant Ct^{n/\ell} \iiint\limits_{W^3} \mu^{n/\ell} dx\, d\theta\, d\rho,$$

where

$$W^3 = \{(x, \theta, \rho) \,\big|\, |f(x, \theta)(1 - \rho^{-\ell/n})| \leqslant c_1 t^{-\delta/\ell} \mu^{-(\ell+\delta)/\ell}, \quad c_1 < \rho < c_2\}.$$

Let

$$W' = \{(x, \theta, \rho) \,\big|\, c_1 < \rho < c_2, \; t^{-\delta/\ell} \mu^{-(\ell+\delta)/\ell} > c\}, \quad W'' = W^3 \backslash W',$$

and let us define I' and I'' through equality (13.42), replacing W^3 with W' and W'', respectively. On a strength of the condition $n - \ell - \frac{2}{3} \geqslant 0$ we have on W'

$$\mu^{n/\ell} \leqslant C\mu^{(\ell+\delta)/\ell} \leqslant C't^{-\delta/\ell};$$

thus, $I' \leqslant Ct^{(n-\delta)/\ell}$. In order to estimate I'', we notice that in a suitable local coordinate system in (x, θ, ρ)-space, we have

$$|f(x, \theta)(1 - \rho^{-\ell/n})| \asymp |y_1 \cdots y_k|;$$

therefore, as in (13.24), for any $\varepsilon > 0$,

$$I'' \leqslant c_\varepsilon t^{n/\ell} \iiint d\theta\, dx\, \mu^{n/\ell} (t^{-\delta/\ell} \mu^{-(\ell+\delta)/\ell})^{1-\varepsilon}.$$

The condition $n - \ell - \delta > 0$ implies that we can integrate with respect to (x, θ) to obtain

$$I'' \leqslant C_{\varepsilon_1} t^{(n-\delta)/\ell+\varepsilon_1} \quad \forall \varepsilon_1 > 0.$$

Since $\delta \in (0, \frac{2}{3})$ is arbitrary, we can assume that I''; and, therefore, I_j^3 admit estimate (13.41).

This proves the last statement of Theorem 13.2. ∎

13.8. We shall now consider a symmetric Douglis–Nirenberg elliptic PDO A of order $(\ell_1, \dots, \ell_m; \ell_1, \cdots, \ell_m)$ with $\ell_i > 0$, whose principle symbol a^0 satisfies condition (13.8). Let

$$\Lambda = [\delta_{ij} h^{-\ell_j}]_{i,j=1}^m.$$

Arguments similar to those of Section 13.2, show that we can assume that $A \in \mathcal{L}I^+(g, \Lambda, \Lambda)$. Lemma 8.1 and Theorem 4.2 yield

$$\mathcal{A}[u] = \langle Au, u \rangle \geqslant -c\|u\|^2 \quad \forall u \in C_0^\infty(\Omega)^m;$$

let $\widehat{A}$ be an operator associated with the variational triple $\mathcal{A}$, $C_0^\infty(\Omega)^m$, $L_2(\Omega)^m$. In order to formulate the theorem on asymptotics of the spectrum of the operator $\widehat{A}$, we need the following lemma:

Lemma 13.3. *The eigenvalues of the symbol a^0 admit the representation*

$$\text{(13.43)} \qquad\qquad \lambda_j^0 = \lambda_j^1 + \lambda_j^2,$$

where λ_j^1 are positive homogeneous with respect to ξ of degree $2\ell_j$,

$$\text{(13.44)} \qquad\qquad |\lambda_j^2(x,\xi)| \leqslant C|\xi|^{2\ell_j - 1}, \quad |\xi| \geqslant 1,$$

and

$$\text{(13.45)} \qquad\qquad \lambda_i^1(x,\xi) > 0 \quad \forall(x,\xi) \in \overline{\Omega} \times S_{n-1}.$$

Proof. By renumbering the rows and columns if necessary, we can assume that

$$\ell_1 = \cdots = \ell_{m_1} > \ell_{m_1+1} = \cdots = \ell_{m_1+m_2} > \cdots > \ell_{m-m_s+1} = \cdots = \ell_m.$$

We write a^0 in block form ,

$$a^0 = \begin{bmatrix} a_{01} & a_{12} \\ a_{12}^* & a_{22} \end{bmatrix},$$

where a_{01} is an $m_1 \times m_1$ matrix. Since a^0 satisfies condition (13.8), so does a_{01}. Let

$$e_1 = \begin{bmatrix} I_{m_1} & -a_{01}^{-1} a_{12} \\ 0 & I_{m-m_1} \end{bmatrix} = I_m + k.$$

Then

$$\text{(13.46)} \qquad\qquad \|k(x,\xi)\| \leqslant C|\xi|^{-1} \quad \text{for } |\xi| \geqslant 1;$$

and

$$e_1^* a^0 e_1 = \begin{bmatrix} a_{01} & 0 \\ 0 & a_0' \end{bmatrix}, \quad \text{where } a_0' = a_{22} - a_{12}^* a_{01}^{-1} a_{12}.$$

The matrix a_0' also satisfies condition (13.8); therefore, by iterating the previous construction, we can obtain a function e such that $k = I - e$ satisfies condition (13.46) and

$$(13.47) \qquad e^* a^0 e = [\delta_{jk} a_{0j}]_{j,k=1}^s, \quad \operatorname{ord} a_{0j} = 2\ell_j',$$

where $\ell_j' = \ell_{m_j'}$, $m_j' = m_1 + \cdots + m_j$. Thus, λ_j^0 are eigenvalues of the problem

$$(e^* a^0 e)(x,\xi) f = \lambda(I + k(x,\xi)) f, \quad f \in \mathbb{C}^m.$$

where $\|k(x,\xi)\| \leqslant C|\xi|^{-1}$ for $|\xi| \geqslant 1$. By applying the minimax principle, we find that

$$\lambda_j^-(x,\xi) \leqslant \lambda_j^0(x,\xi) \leqslant \lambda_j^+(x,\xi),$$

where $\lambda_j^\pm(x,\xi)$ are eigenvalues of the problem

$$(e^* a^0 e)(x,\xi) f = \lambda(1 \mp C|\xi|^{-1}) f, \quad f \in \mathbb{C}^m.$$

But (13.47) implies that a_{0j} satisfy condition (13.8), therefore $\lambda_j^\pm$ satisfy conditions (13.43)–(13.45). Hence, so do λ_j^0. ∎

For $j = 1, \ldots, s$, let

$$c_j = n^{-1} (2\pi)^{-n} \int_\Omega dx \int_{S_{n-1}} dS(\xi) \sum_{m_{j-1}' < i \leqslant m_j'} \lambda_i^0(x,\xi)^{-n/(2\ell_j')}.$$

We notice that by using formula (13.11), c_j can be obtained from the symbol a_{0j}^{-1}; and in turn, it defines the constant δ_j as in Theorem 13.1.

Theorem 13.3. *Under the conditions formulated above,*

$$(13.48) \qquad N(t, \widehat{A}) = \sum_{1 \leqslant j \leqslant s} \left(c_j t^{n/(2\ell_j')} + O(t^{(n-\delta_j)/(2\ell_j')}) \right).$$

Proof. From above, we can assume that $\hat{a}^0 = \sigma^w(A) \in SI^+(g, \Lambda, \Lambda)$. We write $\hat{a}^0$ in block form,

$$\hat{a}^0 = \begin{bmatrix} \hat{a}_{01} & \hat{a}_{12} \\ \hat{a}_{12}^* & \hat{a}_{22} \end{bmatrix},$$

notice that $\hat{a}_{01} \in SI^+(g, h^{-2\ell_1}) \subset SI(g, h^{-2\ell_1})$, and construct a paramatrix $r_w \in \mathcal{L}I(g, h^{2\ell_1})$ of the operator $\hat{a}_{01,w}$. Let

$$e_{1,w} = \begin{bmatrix} I_{m_1} & -r_w \hat{a}_{12,w} \\ 0 & I_{m-m_1} \end{bmatrix}.$$

Then

$$e_{1,w}^* \hat{a}_w^0 e_{1,w} - \begin{bmatrix} \hat{a}_{01,w} & 0 \\ 0 & \hat{a}_{0,w}' \end{bmatrix} \in \mathcal{L}^{-\infty}(g, I_m),$$

where

$$\hat{a}_{0,w}' = \hat{a}_{22,w} - \hat{a}_{12,w}^* r_w \hat{a}_{12,w} \in \mathcal{L}I^+(g, \Lambda', \Lambda',),$$
$$\Lambda' = [\delta_{jk} h^{-\ell_j}]_{j,k=m_1+1}^m.$$

As in Lemma 13.3, we can similarly transform $\hat{a}_{0,w}'$ further and finally construct a symbol e such that

$$e_w^* \hat{a}_{0,w} e_w - [\delta_{jk} \hat{a}_{0j,w}]_{j,l=1}^s \in \mathcal{L}^{-\infty}(g, I_m),$$
(13.49) $$k = I - e = [k_{ij}]_{i,j=1}^m, \qquad k_{ij} \in S(g, h^{|\ell_i - \ell_j|}),$$
$$k \in S(g, h).$$

By definition of principal symbol,

$$(\hat{a}^0 - a^0)\big|_{\mathcal{M}^\epsilon} \in S(g, h\Lambda, \Lambda);$$

and also

$$(\hat{a}_{0j} - a_{0j})\big|_{\mathcal{M}^\epsilon} \in S(g, h^{-2\ell_j'+1}), \quad j = 1, \ldots, s.$$

Hence, we can apply Theorem 7.4 with

$$\overset{\circ}{q}_t = t^{1/2} + \Lambda, \qquad q_{t,w} = e_w(\overset{\circ}{q}_t^{-1})_w, \qquad d_t = h,$$
$$a_t\big|_{\mathcal{M}^\epsilon} = [\delta_{jk}(a_{0j} - tI_{m_j})(h^{-\ell_j'} + t^{1/2})^{-1}\big|_{\mathcal{M}^\epsilon}]_{j,k=1}^s;$$

and the same arguments as in Theorem 13.1 here yield estimate (13.48) [the condition $q_t \in SI(g, \overset{\circ}{q}_t^{-1}, I)$ is not quite obvious from the conditions of Theorem 7.4; it follows from the inclusion $e \in SI(g, \overset{\circ}{q}_t^{-1}, \overset{\circ}{q}_t)$, which is implied by (13.49)].

Theorem 13.3 is proved.   ∎

§14 General Boundary Value Problems for Elliptic Operators

14.1. Let $\widehat{A}$ and $\widehat{B}$ be self-adjoint and symmetric operators, respectively, in a Hilbert space H with domain $D(\widehat{A})$.

Definition 14.1. The spectrum of the problem

(14.1) $$\widehat{A}u = t\widehat{B}u, \quad u \in D(\widehat{A}),$$

is the set of numbers $t \in \mathbb{C}$ such that the operator $\widehat{A} - t\widehat{B}$ does not have an inverse that would be defined on all of H.

Lemma 14.1. *Let one of the operators $\widehat{A}$, $\widehat{B}$ be positive-definite and let there exist $\mu \in \mathbb{C}$ such that $\widehat{A} - \mu I$ is invertible and*

(14.2) $\widehat{B}(\widehat{A} - \mu I)^{-1}$ *and* $(\widehat{A} - \mu I)^{-1}$ *are compact operators.*

Then spectrum of problem (14.1) *is real and consists of isolated eigenvalues of finite multiplicity.*

Proof. Condition (14.2) implies that the $\widehat{A}$-bound of operator $\widehat{B}$ is equal to zero; therefore (see Kato [1]), for any $\lambda \in \mathbb{C}$, the operator $\widehat{A} - \lambda \widehat{B}$ is closed and $(\widehat{A} - \lambda \widehat{B})^* = \widehat{A} - \overline{\lambda}\widehat{B}$. The equality

$$\widehat{A} - \lambda \widehat{B} = (I + (\mu - \lambda \widehat{B})(\widehat{A} - \mu)^{-1})(\widehat{A} - \mu)$$

implies that for any $\lambda \in \mathbb{C}$, $\text{Im}(\widehat{A} - \lambda \widehat{B})$ is closed and has finite codimension, and that for all λ from a complement to discrete set, we have $\text{Im}(\widehat{A} - \lambda \widehat{B}) = H$. But $\text{Ker}(\widehat{A} - \lambda \widehat{B}) \oplus \text{Im}(\widehat{A} - \overline{\lambda}\widehat{B}) = H$, and since nonreal λ cannot be eigenvalues, the lemma is proved. ∎

If the conditions of Lemma 14.1 do not hold, then we can use the following definition of the spectrum of problem (14.1).

Let $\widehat{A} \geqslant cI$, $D(\widehat{A}^{1/2}) = \{u \mid \|u\|_{D(\widehat{A}^{1/2})} = \langle \widehat{A}u, u \rangle_H^{1/2} < \infty\}$,

(14.3) $D(\widehat{A}^{1/2})$ is compactly imbedded in $H_1 \subset H$;

(14.4) $|\langle \widehat{B}u, u \rangle_H| \leqslant C\|u\|_{H_1}^2$ $\forall u \in D(\widehat{A})$.

Then the closure in $D(\widehat{A}^{1/2})$ of the form $\langle B\cdot, \cdot \rangle_H$ defines a bounded symmetric operator B_1.

Definition 14.2. Provided that conditions (14.3) and (14.4) hold, we define the generalized spectrum of problem (14.1) as the spectrum of the problem

(14.5) $u = t B_1 u$, $u \in D(\widehat{A}^{1/2})$.

Lemma 14.2. *Under conditions* (14.3) *and* (14.4), *the spectrum of problem* (14.5) *is discrete, and*

(14.6) $N_\pm(t) = \mathcal{N}(\widehat{A} \mp t\widehat{B}, d(\widehat{A}), H)$.

Proof. $D(\widehat{A})$ is dense in $D(\widehat{A}^{1/2})$; therefore, Glazman's lemma yields

$$N_\pm(t) = \mathcal{N}(I \mp t B_1, D(\widehat{A}), D(\widehat{A}^{1/2})) = \mathcal{N}(\widehat{A} \mp t\widehat{B}, D(\widehat{A}), H).$$

Conditions (14.3) and (14.4) yield that for any $\tau > 0$,

$$|\langle \widehat{B}u, u \rangle_H| \leqslant \tau \langle \widehat{A}u, u \rangle_H + C_\tau \|u\|_H^2;$$

and since $\langle \widehat{A}u, u \rangle \geqslant c\|u\|_H^2$, then

(14.7) $$\mathcal{N}(\widehat{A} \mp t\widehat{B}, D(\widehat{A})) \leqslant \mathcal{N}(c_1\widehat{A} - C(t)I, D(\widehat{A})).$$

But condition (14.3) implies that the spectrum of the operator $\widehat{A}$ is discrete, and due to Glazman's lemma, the right-hand side of (14.7) is finite for any t. Hence, the right-hand side of (14.6) is also finite; and due to Glazman's lemma, the spectrum of problem (14.5) is discrete. ∎

Lemma 14.3. *Under conditions (14.2)–(14.4) the generalized spectrum of problem (14.1) coincides with the usual one.*

Proof. Let $\{u_n\} \subset D(\widehat{A})$ converge to $u \in D(\widehat{A}^{1/2})$ in the $D(\widehat{A}^{1/2})$ norm. Then $\forall v \in D(\widehat{A})$,

$$\langle (I - tB_1)u_n, v \rangle_{D(\widehat{A}^{1/2})} \to \langle (I - tB_1)u, v \rangle_{D(\widehat{A}^{1/2})},$$

but $\widehat{A} - t\widehat{B}$ is self-adjoint due to (14.2); hence,

$$\langle (I - tB_1)u_n, v \rangle_{D(\widehat{A}^{1/2})} = \langle (\widehat{A} - t\widehat{B})u_n, v \rangle_H$$
$$= \langle u_n, (\widehat{A} - t\widehat{B})v \rangle_H \to \langle u, (\widehat{A} - t\widehat{B})v \rangle_H.$$

Therefore,

$$u = tB_1 u, \quad u \in D(\widehat{A}^{1/2}) \iff \langle (I - tB_1)u, v \rangle_{D(\widehat{A}^{1/2})} = 0 \quad \forall v \in D(\widehat{A})$$
$$\iff \langle u, (\widehat{A} - t\widehat{B})v \rangle = 0 \quad \forall v \in D(\widehat{A})$$
$$\iff u \in D((\widehat{A} - t\widehat{B})^*) = D(\widehat{A}) \quad \text{and}$$
$$\langle (\widehat{A} - t\widehat{B})u, v \rangle = 0 \quad \forall v \in D(\widehat{A}) \iff \widehat{A}u = t\widehat{B}u, \quad u \in D(\widehat{A}). \quad ∎$$

14.2. Let $\Omega \subset R^n$ be a bounded Lipschitz domain; let A and B be DO of order $2\ell_1$ and $2\ell_2$, respectively, with matrix C^∞-coefficients, $\ell = 2\ell_1 - 2\ell_2 > 0$; and let $\widehat{A}$ and $\widehat{B}$ be, respectively, self-adjoint and symmetric operators in $L_2(\Omega)^m$ defined by differential expressions A and B on $V = D(\widehat{A})$ such that

(14.8) $$C_0^\infty(\Omega)^m \subset V \subset H^{\ell_1}(\Omega)^m$$

and $\forall u \in V$,

(14.9) $$c\|u\|_{\ell_1}^2 \leqslant \langle \widehat{A}u, u \rangle_0, \qquad |\langle \widehat{B}u, u \rangle_0| \leqslant C\|u\|_{\ell_2}^2,$$

where $\langle \cdot, \cdot \rangle_S$ and $\| \cdot \|_S$ are, respectively, a scalar product and a norm in $H^S(\Omega)^m$. We shall denote a scalar product in $L_2(R^n)^m$ by $\langle \cdot, \cdot \rangle$.

Since $\Omega \subset R^n$ is a bounded Lipschitz domain, then $H^{\ell_1}(\Omega) \subset H^{\ell_2}(\Omega)$ is compact; therefore, condition (14.9) implies that we can define the generalized spectrum of the problem

$$\widehat{A}u = t\widehat{B}u, \quad u \in V;$$

and if $\ell_1 > 2\ell_2$, then the conditions of Lemma 14.3 are satisfied, so that the generalized spectrum coincides with the usual one.

In the particular case of $B = I$, we can weaken condition (14.9):

(14.9′) $$\langle \widehat{A}u, u \rangle_0 \geqslant c\|u\|_{\ell_1}^2 - C\|u\|_0^2;$$

the algebraic conditions that would guarantee (14.9′) are given by Grubb in [4], where she considered Douglis–Nirenberg elliptic operators.

Theorem 14.1. *Under the conditions formulated above,*

(14.10) $$N_\pm(t) = c_\pm t^{n/\ell} + O(t^{\rho(\delta)}),$$

where $c_\pm$ and $\rho(\delta)$ are defined as in Theorems 13.1 and 13.2.

14.3.

Proof. (14.6) and (14.8) yield

$$N_\pm(t) \geqslant \mathcal{N}_0(A - tB, \Omega);$$

therefore, the estimate from below for $N_\pm(t)$ follows from Theorems 13.1 and 13.2.

To verify the estimate from above for $N_+(t)$, we construct the metric g and the functions $a_0, \overset{\circ}{q}_t, a_t, \lambda_{t,j}$ as in Section 13.2, and we fix $C_0 > 0$ and denote by $\psi_{t,c}$ a cutoff function associated with $h^{-1/3}g$, c, and

(14.11) $$\mathcal{M}_{t,C_0} = \{X \mid h(X)^{-\ell} < C_0 t\}.$$

Starting with the functions a_t, $\psi_{t,c}$, $d_t = h$, we construct the functions $\lambda_{t,0}$, a_t^+, $\eta_{t,\pm c}$ as was done in §§9–12 [the construction is determined by the conditions satisfied by the function a_0 and, therefore, by the function a_t—see §13] and denote, for $c > 0$,

$$\overset{\circ}{\mu}_{t,c} = (\eta_{t,-c}^2 + (1 - \eta_{t,-c})^2)^{-1/2},$$

$$\mu_{t,c} = (1 - \eta_{t,-c})\overset{\circ}{\mu}_{t,c}, \quad \mu_{t,-c} = \eta_{t,-c}\overset{\circ}{\mu}_{t,c}.$$

Then

$$(14.12) \qquad \mu_{t,-c}^2 + \mu_{t,c}^2 = 1,$$

$$(14.13) \qquad \mu_{t,c}\mu_{t,2c} = \mu_{t,c}, \qquad \mu_{t,-c}\mu_{t,-2c} = \mu_{t,-2c},$$

$$(14.14) \quad \operatorname{supp}\mu_{t,-c}, \ \operatorname{supp}(1 - \mu_{t,c}) \text{ are compact subsets of } \Omega \times R^n;$$

and if a_t satisfies the conditions of Theorems 7.1 and 7.2 [i.e., the symbol a_0, defined in §13, satisfies the conditions in parts a) and b) of Theorems 13.1 and 13.2], then

$$(14.15) \qquad \eta_{t,\pm c}, \mu_{t,\pm c} \in S(h^{-2\delta}g, 1),$$

where $\delta \in (0, \frac{1}{3})$ under the conditions in part a) and $\delta \in (0, \frac{1}{2})$ under the conditions in part b). If the conditions in part c) of either Theorems 13.1 or 13.2 hold, then a_t satisfies the conditions of Theorem 7.3; therefore, instead of (14.15), we have

$$(14.15') \qquad \eta_{t,\pm c}, \mu_{t,\pm c} \in S(G^{(1)}, 1),$$

where $G^{(1)}$ is the metric constructed in §11 for the proof of the estimate from above [if $m = 1$, then $G^{(1)}$ coincides with the metric in §10].

Similar to the transformation of the functions in §11 [see (11.19)–(11.25) and the following], we can transform the functions $\mu_{t,\pm c}$ such that conditions (14.13)–(14.15) would still hold, and, additionally,

$$(14.16) \qquad \mu_{t,-c,r}\mu_{t,-c,\ell} + \mu_{t,c,r}\mu_{t,c,\ell} = I + k_{t,c,w},$$

where

$$(14.16') \qquad \operatorname{supp}_\infty k_{t,c} \subset \operatorname{supp}\mu_{t,-c}, \quad k_t \in S(h^{-2\delta}g, h)(S(G^{(1)}, h)).$$

If (14.15) holds, then $\lambda_0 = h^{\delta-\omega}$ and

$$(14.17) \qquad [A - tB, \mu_{t,\pm c,r}] \in \mathcal{L}(h^{-2\delta}g, \lambda_0 h^\omega \overset{\circ}{q}_t, \overset{\circ}{q}_t);$$

and on the other hand, if (14.15') holds, then by construction, $\lambda_0 \asymp h^{\delta/2-\omega}$ on $\operatorname{supp}_d \mu_{t,\pm c}$, and since for any $s > 0$,

$$n_s(G^{(1)}, \lambda_0 \overset{\circ}{q}_t, \overset{\circ}{q}_t, \sigma^w(A - tB)) \leqslant C_s n_s(g, \overset{\circ}{q}_t, \overset{\circ}{q}_t, \sigma^w(A - tB)),$$

then

$$(14.17') \qquad [A - tB, \mu_{t,\pm c,r}] \in \mathcal{L}(G^{(1)}, \lambda_0 h^\omega \overset{\circ}{q}_t, \overset{\circ}{q}_t).$$

All of the following arguments are the same for condition (14.15′) as for condition (14.15); therefore, we shall write $G^{(1)}$ throughout, implying the metric $h^{-2\delta}g$, provided that (14.15) holds.

We shall obtain the estimate from below for the form $\langle(\widehat{A}-t\widehat{B})\cdot,\cdot\rangle_0$ through the sum of two forms, using (14.16) [compare this with the proof of the estimate from above in §11]; therefore, we must prove that

$$(14.18) \qquad r_+\mu_{t,\pm c,\ell}e_+u \in V \quad \forall u \in V.$$

Condition (14.14) yields that $\forall u \in L_2(R^n)$,

$$(14.19) \qquad \text{supp}\,\mu_{t,-c,\ell}u, \ \text{supp}(1-\mu_{t,c,\ell})u \ \subset \Omega.$$

Since A is elliptic, we have $V \subset H_{loc}^{2\ell_1}(\Omega)^m$; and since a self-adjoint operator is closed and $\overset{\circ}{C}_0^{\infty}(\Omega)^m \subset V$, then $\overset{\circ}{H}{}^{2\ell_1}(\Omega)^m \subset V$. Hence, (14.19) implies that it suffices to verify the inclusions

$$(14.20) \qquad \mu_{t,-c,\ell}e_+u, \ (1-\mu_{t,-c,\ell})e_+u \ \in H^{2\ell_1}(R^n)^m \quad \forall u \in V.$$

Let $\tilde{\ell}: H^{2\ell_1}(\Omega) \to H^{2\ell_1}(R^n)$ be a bounded linear extension operator from $H^s(\Omega)$ to $H^s(R^n)$, $s = 0,1,\ldots,2\ell_1$ ($\tilde{\ell}$ exists, since Ω is a bounded Lipschitz domain–see Stein[1]), and let $\psi \in C_0^{\infty}(R^n)$, $\psi(x) = 1 \quad \forall x \in \Omega$. Then $\ell = \psi\tilde{\ell}$ has the same properties as $\tilde{\ell}$. Condition (14.14) yields

$$(14.21) \qquad \mu_{t,-c,r}e_+u = \mu_{t,-c,r}\ell u \quad \forall u \in L_2(\Omega);$$

and condition $(14.13)_2$, combined with Theorems 3.1–3.3, implies

$$(14.22) \qquad \mu_{t,-c,\ell} = \mu_{t,-c,\ell}\mu_{t,-c/2,r} + T_t,$$

where $T_t \in \mathcal{L}^{-\infty}(g,1)$. Lemma 5.6 implies that for an integer $s \geqslant 0$, the norm in $\mathcal{H}(h^{-s},\mathbf{C}^1)$ can be defined by the equality

$$(14.23) \qquad \|u\|_{h^{-s}}^2 = \sum_{|\alpha|\leqslant s} \|\langle x\rangle^{s-|\alpha|}D^\alpha u\|_{L_2}^2;$$

and since supp ψ is bounded, $\ell : H^s(\Omega) \to \mathcal{H}(h^{-s},\mathbf{C}^1)$ is also bounded $[s = 0,1,\ldots,2\ell_1]$. Theorem 5.2 implies that $\mu_{t,\pm c,\ell}$ and $\mu_{t,\pm c,r}$ are operators bounded in $\mathcal{H}(h^{-s},\mathbf{C}^1)$ and that T_t is an operator bounded from $L_2(R^n)$ to $\mathcal{H}(h^{-s},\mathbf{C}^1)$, all uniformly with respect to t. Hence, by substituting (14.22) into (14.20) and taking (14.21) and (14.23) into account, we see that $\mu_{t,-c,\ell}e_+u \in H^{2\ell_1}(R^n)^m$.

Since (14.14) also holds for $1 - \mu_{t,c}$ instead of $\mu_{t,-c}$, the proof above also gives the second inclusion in (14.20). This proves (14.18).

Let $u_{-2} = \mu_{t,-2,\ell}e_+u$ and $u_2 = r_+\mu_{t,2,\ell}e_+u$. As we showed above,

$$\forall u \in V \quad u_{-2} \in \mathcal{H}(g, h^{-2\ell_1}) \quad \text{and} \quad u_2 \in V;$$

therefore, (14.16) and (14.19) yield

(14.24) $\langle\langle(\widehat{A} - t\widehat{B})u, u\rangle_0 = \langle\langle(\widehat{A} - t\widehat{B})u_2, u_2\rangle_0 + \langle(A - tB)u_{-2}, u_{-2}\rangle$

$$+ \sum_{j=\pm 2} \langle r_+[A - tB, \mu_{t,j,r}]\mu_{t,j,\ell}e_+u, u\rangle_0 + \langle r_+(A - tB)k_{t,2,w}e_+u, u\rangle_0.$$

By construction, $(1 - \mu_{t,2})\mu_{t,-1} = (1 - \mu_{t,2})$; therefore, Theorems 3.1–3.3 yield

(14.22′) $$(1 - \mu_{t,2,r})\mu_{t,-1,\ell} + T_t = 1 - \mu_{t,2,r},$$

and (14.16′) and (14.13) imply

$$k_{t,1,w} = \mu_{t,-1,\ell}k_{t,w}\mu_{t,-1,r} + T'_t,$$

where $T_t \in \mathcal{L}^{-\infty}(g,1)$ and $T'_t \in \mathcal{L}^{-\infty}(g,t)$; hence, using (14.21), we can write the sum of the last three terms in (14.24) as

(14.25) $$\langle\mu_{t,-1,\ell}K_t\mu_{t,-1,r}\ell u, \ell u\rangle + \langle r_+T_t e_+u, u\rangle,$$

where $K_t \in \mathcal{L}(G^{(1)}, \lambda_0 h^w \overset{\circ}{q}_t, \overset{\circ}{q}_t)$ and $T_t \in \mathcal{L}^{-\infty}(g,t)$.

Let $\mathcal{L}^{-\infty}(\Omega, g, t)$ be a linear span of operators of the form $D_{1,t}T_tD_{2,t}$, where

$$D_{1,t} : L_2(R^n)^m \to L_2(\Omega)^m, \qquad D_{2,t} : L_2(\Omega)^m \to L_2(R^n)^m$$

are operators bounded uniformly with respect to $t \geqslant 1$ and where $T_t \in \mathcal{L}^{-\infty}(g, t^s)$ for some $s = s(T_t)$ [also uniformly with respect to $t \geqslant 1$]. Then the last term in (14.25) can be estimated with the help of the following lemma:

Lemma 14.4. *If $K_t \in \mathcal{L}^{-\infty}(\Omega, g, t)$, then for any $\varepsilon, N > 0$, there exist $C > 0$ and a subspace $L_t \subset L_2(\Omega)^m$ of codimension not larger than Ct^ε such that $\forall u \in L_t$,*

(14.26) $$\|K_t u\|_0 \leqslant Ct^{-N}\|u\|_0.$$

Proof. Let $K_t = D_{1,t}T_tD_{2,t}$, where $D_{j,t}$ and T_t satisfy the conditions formulated above. Since the metric g satisfies the condition $h(X) \leqslant \langle X\rangle^{-1}$, then by applying Theorem 7.1 to the operator $t^{-2N} - T_t^*T_t \in \mathcal{L}(g, t^{-2N} + t^{2s}h^M)$ $\forall M, N$, we obtain for any $N, \varepsilon > 0$,

$$\mathcal{N}(t^{-2N} - T_t^*T_t, L_2(R^n)^m) \leqslant C_{N,\varepsilon}t^\varepsilon.$$

Therefore, there exists a subspace $L_t' \subset L_2(R^n)^m$ of codimension not larger than $C_{N,\varepsilon} t^\varepsilon$ such that

$$\|T_t u\| \leqslant t^{-N} \|u\| \quad \forall u \in L_t',$$

and we can assume that $L_t = D_{2,t}^{-1}(L_t')$. ∎

14.4. Let

$$A_t' = (\overset{\circ 2}{q_t} \lambda_0 h^{\omega/2})_w, \qquad A_t'' = \mu_{t,-1,\ell} A_t' \mu_{t,-1,r}.$$

Using (14.18) and (14.21) and repeating all the above arguments, we obtain
(14.27)
$$\langle (\widehat{A} - t\widehat{B})u, u \rangle_0$$
$$= \langle (\widehat{A} - t\widehat{B} - r_+ A_t'' e_+ - t^{-1})u, u \rangle_0 + \langle A_t'' \ell u, \ell u \rangle + t^{-1} \|u\|_0^2$$
$$= \langle (\widehat{A} - t\widehat{B} - r_+ A_t'' e_+ - t^{-1})u_2, u_2 \rangle_0 + \langle (A - tB - A_t'' - t^{-1})u_{-2}, u_{-2} \rangle$$
$$+ t^{-1} \|u\|_0^2 + \langle T_t' e_+ u, e_+ u \rangle + \langle \mu_{t,-1,\ell}(A_t' + K_t')\mu_{t,-1,r}\ell u, \ell u \rangle,$$

where $K_t' \in \mathcal{L}(G^{(1)}, \lambda_0 h^\omega, \overset{\circ 2}{q_t})$ and $T_t \in \mathcal{L}^{-\infty}(g, t)$. By construction, $A_t' + K_t' \in \mathcal{L}I^+(G^{(1)}, \lambda_0 h^{\omega/2}\overset{\circ 2}{q_t})$; therefore, Lemma 8.1 yields $A_t' + K_t' \geqslant T_t''$, where $T_t'' \in \mathcal{L}^{-\infty}(g, t)$, and the sum of the last two terms in (14.27) is estimated by $\langle T_t u, u \rangle_0$, where $T_t \in \mathcal{L}^{-\infty}(\Omega, g, t)$.

We shall estimate the first term in (14.27). Let

$$P_j = \sum_{|\alpha| \leqslant \ell_j} D^\alpha \langle x \rangle^{2\ell_j - 2|\alpha|} D^\alpha, \qquad \widehat{P}_t = P_1 + tP_2.$$

Then

$$\varepsilon \widehat{P}_t - A_t'' \in \mathcal{L}I^+(G^{(1)}, \overset{\circ 2}{q_t}) \quad \forall \varepsilon > 0;$$

therefore, by using Lemma 8.1 and the equality $\langle r_+ A_t'' e_+ u, u \rangle_0 = \langle A_t'' \ell u, \ell u \rangle$, we obtain

$$-\langle r_+ A_t'' e_+ u_2, u_2 \rangle \geqslant -\varepsilon \langle \widehat{P}_t \ell u_2, \ell u_2 \rangle - \langle T_{t,\varepsilon} u, u \rangle_0,$$

where $T_{t,\varepsilon} \in \mathcal{L}^{-\infty}(\Omega, g, t)$. But

(14.28)
$$\langle \widehat{P}_t \ell u_2, \ell u_2 \rangle \asymp (\|\ell u_2\|_{h-\ell_1}^2 + t\|\ell u_2\|_{h-\ell_2}^2)$$
$$\asymp (\|\ell u_2\|_{\ell_1}^2 + t\|\ell u_2\|_{\ell_2}^2);$$

therefore, by choosing sufficiently small $\varepsilon > 0$ and using (14.9), we obtain (14.29)

$$(14.29) \quad \langle(\widehat{A} - t\widehat{B} - r_+ A_t'' e_+ - t^{-1})u_2, u_2\rangle \geqslant c\|u_2\|_{\ell_1}^2 - Ct\|u_2\|_{\ell_2}^2 - \langle T_t u, u\rangle_0,$$

where $T_t \in \mathcal{L}^{-\infty}(\Omega, g, t)$.

Let $\eta_{t,c}$ be the functions in §§9–12 constructed from the function a_t^+. Since $\eta_{t,c}(x) = 1 \quad \forall x \in \overline{\Omega}$, then

$$\ell_t = \eta_{t,2,t}\ell : H^s(\Omega) \to H(g, h^{-s}), \quad s = 0, \dots, 2\ell_1,$$

is a bounded linear operator extension. By denoting $u_1 = \ell_t u_2$ and using (14.28), we obtain

$$(14.30) \qquad c\|u_2\|_{\ell_1}^2 - C\|u_2\|_{\ell_1}^2 \geqslant \langle(c_1 P_1 - C_1 t P_2)u_1, u_1\rangle.$$

Furthermore, let $\mu_{t,w}' = \eta_{t,8,w}\mu_{t,8,w}$ and recall that

$$\eta_{t,2c}\eta_{t,c} = \eta_{t,c}, \qquad \mu_{t,2c}\mu_{t,c} = \mu_{t,c}, \qquad (1 - \mu_{t,2c})\mu_{t,-c} = 1 - \mu_{t,2c}.$$

Thus,

$$(\eta_{t,8,w} - I)\mu_{t,8,w}\eta_{t,2,\ell} \in \mathcal{L}^{-\infty}(g, 1);$$

and

$$
(14.31) \quad
\begin{aligned}
\mu_{t,w}' u_1 &= T_{1,t}\ell u_2 + u_1 + (\mu_{t,8,w} - I)\eta_{t,2,\ell}r_+\mu_{t,2,\ell}e_+ u \\
&= T_{2,t}\ell u_2 + u_1 + (\mu_{t,8,w} - I)\eta_{t,2,\ell}\mu_{t,-4,r}\ell r_+\mu_{t,2,\ell}e_+ u \\
&= T_{2,t}\ell u_2 + u_1 + (u_{t,8,w} - I)\eta_{t,2,\ell}\mu_{t,-4,r}\mu_{t,2,\ell}e_+ u \\
&= u_1 + T_{2,t}\ell u_2 + T_{3,t}e_+ u,
\end{aligned}
$$

where $T_{j,t} \in \mathcal{L}^{-\infty}(g, 1)$. (27) and (29)–(31) yield the estimate

$$\langle(\widehat{A} - t\widehat{B})u, u\rangle_0 \geqslant \sum_{j=\pm 1}\langle Q_{t,j}u_j, u_j\rangle + \langle T_t u, u\rangle_0 + t^{-1}\|u\|_0^2,$$

where $T_t \in \mathcal{L}^{-\infty}(\Omega, g, t)$, $u_{-1} = u_{-2}$ and

$$Q_{t,1} = cP_1 - Ct\mu_{t,w}'P_2\mu_{t,w}', \qquad Q_{t,-1} = A - tB - A_t'' - t^{-1}.$$

Lemma 14.4 enables us to find the space $L_t \subset V$ of codimension t^χ, where $\chi = (n-1)/\ell$, such that for $t \geqslant t_0$ and all $u \in L_t$,

$$(14.32) \qquad \langle(\widehat{A} - t\widehat{B})u, u\rangle_0 \geqslant \sum_{j=\pm 1}\langle Q_{t,j}u_j, u_j\rangle.$$

Now (14.16) yields

$$\sum_{j=\pm 1} \|u_j\|^2 = \|u\|_0^2 + \langle k_{t,w} e_+ u, e_+ u \rangle + \langle T_t u, u \rangle_0,$$

where $T_t \in \mathcal{L}^{-\infty}(\Omega, g, t)$ and $k_t \in S(G^{(1)}, h)$. By applying Lemma 8.2 to the second term and Lemma 14.4 to the third, we see that for $t \geqslant t_0$, the dimension of the kernel of the operator

$$V \ni u \to (u_1, u_{-1}) \in \mathcal{H}(h^{-\ell_1}, C^m) \oplus \overset{\circ}{H}{}^{2\ell_1}(\Omega)^m$$

does not exceed t^χ; therefore, (14.32) and Lemmas of E.6 imply the estimate

$$(14.33) \qquad N_+(t) \leqslant t^\chi + \mathcal{N}(Q_{t,1}; \mathcal{H}(h^{-\ell_1}, C^m)) + \mathcal{N}(Q_{t,-1}, \overset{\circ}{H}{}^{2\ell_1}(\Omega)^m).$$

14.5. In order to estimate the second term, we notice that $A - tB : \overset{\circ}{H}{}^{2\ell_1}(\Omega)^m \to L_2(\Omega)^m$ is continuous; the same also holds for A_t'', since Theorem 5.2 implies that the latter is a continuous operator from $\mathcal{H}(h^{-2\ell_1}, C^m)$ to $L_2(R^n)^m$, and supp $A_t'' u \subset \Omega \quad \forall u \in \overset{\circ}{H}{}^{2\ell_1}(\Omega)^m$. Thus,

$$\mathcal{N}(Q_{t,-1}, \overset{\circ}{H}{}^{2\ell_1}(\Omega)^m) = \mathcal{N}_0(Q_{t,-1}, \Omega);$$

and since $A_t'' + t^{-1} \in \mathcal{L}(G^{(1)}, \lambda_0 h^{\omega/2} \overset{\circ}{q}{}_t^2)$, the proofs of Chapter 2 give the same estimates for both $\mathcal{N}_0(Q_{t,-1}, \Omega)$ and $\mathcal{N}_0(A - tB, \Omega)$. Therefore, the arguments from §13 yield

$$(14.34) \qquad \mathcal{N}(Q_{t,-1}, \overset{\circ}{H}{}^{2\ell_1}(\Omega)^m) \leqslant c_+ t^{n/\ell} + O(t^{\rho(\delta)}).$$

In order to estimate the first term in (14.33), we use Theorem 5.3 to construct the isomorphism

$$(\mathcal{L}I(g, h^{\ell_1}) \ni) Q : L_2(R^n)^m \to \mathcal{H}(h^{-\ell_1}; C^m),$$

and note that

$$(14.35) \qquad \mathcal{N}(Q_{t,1}, \mathcal{H}(h^{-\ell_1}, C^m)) = \mathcal{N}(Q'_{t,1}, L_2(R^n)^m),$$

where $Q'_{t,1} = Q^* Q_{t,1} Q$. To estimate the last function, we denote by $\widetilde{\psi}_t$ a cutoff function associated with $h^{-1/3}g$, 1, $M_{t,c_0/2}$, where C_0 is the same constant as in (14.11), and let

$$\chi_t = \eta_{t,16}\mu_{t,16}\widetilde{\psi}_t, \qquad \mathcal{E}_t = \chi_{t,w}, \qquad \mathcal{E}_{1,t} = \mathcal{E}_t^2(3 - 2\mathcal{E}_t).$$

Then

$$\text{(14.36)} \qquad \text{supp}_\infty \, \sigma^w(\mathcal{E}_{1t}) \subset \text{supp} \, \eta_{t,16} \cap \text{supp} \, \mu_{t,16} \cap \mathcal{M}_{t,c_0},$$

and the arguments in §9 show that $\mathcal{E}_{1t}$ is an operator of the trace class whose nonzero eigenvalues, counted with their multiplicity and numbered arbitrarily, satisfy the estimate

$$\text{(14.37)} \qquad \text{card}\{i \mid \nu_{1ti} \geqslant \tfrac{1}{2}\} \leqslant C \, \text{mes} \, \mathcal{M},$$

where $\mathcal{M}$ is the intersection on the right-hand side of (14.36). The arguments in §13 give the estimate mes $\mathcal{M} \leqslant Ct^{\rho(\delta)}$; therefore, we can rewrite (14.37) as

$$\text{(14.38)} \qquad \text{card}\{i \mid \nu_{1ti} \geqslant \tfrac{1}{2}\} \leqslant Ct^{\rho(\delta)}.$$

Let $L_t = \text{span}\{u \mid \mathcal{E}_{1t}u = \nu_{1ti}u, \ \nu_{1ti} \geqslant \tfrac{1}{2}\}$. If $v \in {}^\perp L_t$, then $u = (I - \mathcal{E}_{1t})v \in {}^\perp L_t$, $\|u\| \geqslant \|v\|/2$, and $\forall \varepsilon > 0$,

$$\text{(14.39)} \qquad \langle Q'_{t,1}u, u \rangle \geqslant \langle Q''_{t,\varepsilon}v, v \rangle,$$

where

$$Q''_{t,\varepsilon} = (I - \mathcal{E}_{1t})(Q'_{t,1} - 4\varepsilon I)(I - \mathcal{E}_{1,t}) + \varepsilon I.$$

Since $\eta_{t,16}\eta_{t,8} = \eta_{t,8}$ and $\mu_{t,16}\mu_{t,8} = \mu_{t,8}$, we have

$$\text{(14.40)}$$
$$(I - \mathcal{E}_{1t})(-Q^* C t \mu'_{t,w} P_2 \mu'_{t,w} Q)(I - \mathcal{E}_{1t})$$
$$= (I - \tilde{\psi}^2_{t,w}(3 - 2\tilde{\psi}_{t,w}))(-Q^* C t \mu'_{t,w} P_2 \mu'_{t,w} Q)(I - \tilde{\psi}^2_{t,w}(3 - 2\tilde{\psi}_{t,w})) + T_t,$$

where $T_t \in \mathcal{L}^{-\infty}(g,t)$; and since $th(X)^{-2\ell_2} \leqslant C_0^{-1}h(X)^{-2\ell_1}$ for every $X \in \text{supp}_\infty(I - \tilde{\psi}^2_{t,w}(3 - 2\psi_{t,w}))$, then the operator in (14.40) belongs to $\mathcal{L}(G^{(1)}, C_0^{-1}) + \mathcal{L}^{-\infty}(g,t)$ uniformly with respect to $C_0 \geqslant 1$ and $t \geqslant 1$.

By applying Theorem 4.2, we obtain

$$Q''_{t,\varepsilon} \geqslant c(I - \mathcal{E}_{1,t})Q^* P_1 Q (I - \mathcal{E}_{1,t}) + (\varepsilon - C_1 C_0^{-1})I + T_t,$$

where $T_t \in \mathcal{L}^{-\infty}(g,t)$ and where C_1 does not depend on C_0, t. But $Q^* P_1 Q \in \mathcal{L}I^+(g,1)$, and Lemma 8.1 yields

$$Q''_{t,\varepsilon} \geqslant (\varepsilon - C_1 C_0^{-1})I + T'_t, \quad \text{where } T'_t \in \mathcal{L}^{-\infty}(g,t).$$

Then by applying Lemma 14.4, we see that if C_0 is sufficiently large and $t \geqslant t_0$, and if $v \in L'_t \subset {}^\perp L_t$, where codim $L'_t \leqslant t^\chi$, then $\langle Q''_{t,\varepsilon}v, v \rangle \geqslant 0$.

Now (14.39) and (14.38) yield the estimate

$$\mathcal{N}(Q'_{t,1}, L_2(R^n)^m) \leqslant Ct^{\rho(\delta)},$$

and (14.33)–(14.35) give estimate (14.34) for $N_+(t)$.
This proves Theorem 14.1. ∎

14.6. Now let $\widehat{A}$ be a self-adjoint, elliptic DO of order $\ell > 0$ without fixed sign; let

$$(14.41) \qquad C_0^\infty(\Omega) \subset V = D(\widehat{A}) \subset H^\ell(\Omega)^m;$$

and let there exist c_1, c_2 such that

$$(14.42) \qquad \|u\|_\ell \leqslant c_1 \|\widehat{A}u\|_0 + c_2 \|u\|_0 \quad \forall u \in V.$$

These conditions imply invertibility of the principal symbol a of the operator A on $\overline{\Omega} \times S_{n-1}$; the proof is the same as that of Lemma 13.1.

The spectrum of the operator $\widehat{A}$ is discrete, since condition (14.42) ensures that $(\widehat{A}+iI)^{-1}$ is compact as an operator in $L_2(\Omega)^m$. In order to formulate the theorem on asymptotics of spectrum distribution functions on the operator $\widehat{A}$, we denote $a_0 = a^{-1}$ and define $c_\pm$ and δ through a_0 and $\partial\Omega$, as was done in Theorem 13.1.

Theorem 14.2. *As $t \to \infty$,*

$$(14.43) \qquad N_\pm(t) = c_\pm t^{n/\ell} + O(t^{(n-\delta)/\ell}).$$

Proof. Assume that $t_0 \in R$ does not belong to the spectrum. The substitution $\widehat{A} \to \widehat{A} - t_0 I$ does not change the form of formula (14.43); therefore, we can assume that $\widehat{A}$ is invertible, and condition (14.42) is satisfied with $c_2 = 0$. But then the spectrum of the operator $\widehat{A}$ coincides with the spectrum of the problem

$$u = t\widehat{A}^{-1}u, \quad u \in L_2(\Omega)^m.$$

By applying Glazman's lemma and (14.41), we obtain

$$N_\pm(t) = \mathcal{N}(I \mp t\widehat{A}^{-1}, L_2(\Omega)^m) = \mathcal{N}(\|\widehat{A}\cdot\|_0^2 \mp t\langle\widehat{A}\cdot,\cdot\rangle_0, D(\widehat{A})) \geqslant \mathcal{N}_0(A^2 \mp tA, \Omega);$$

therefore, Theorem 13.1 yields

$$(14.44) \qquad N_\pm(t) \geqslant c_\pm t^{n/\ell} - Ct^{(n-\delta)/\ell}.$$

In order to obtain the estimate from above, we notice that if $N(t)$ is a distribution function of the spectrum of the problem

$$(14.45) \qquad \widehat{A}^2 u = t^2 u, \quad u \in D(\widehat{A}^2),$$

then

$$(14.46) \qquad N(t) = N_+(t) + N_-(t),$$

and we also notice that condition (14.42) with $c_2 = 0$ means that problem (14.45) satisfies the conditions of Theorem 14.1, so we have

$$(14.47) \qquad N(t) = c_+(a_0^2)t^{n/\ell} + O(t^{(n-\delta)/\ell}).$$

But $c_+(a_0^2) = c_+(a_0) + c_-(a_0)$; therefore, (14.44), (14.46), and (14.47) yield (14.43).

Remark. Algebraic conditions that ensure (14.42) (Shapiro–Lopatinsky conditions) are well-known—see, for example, Lions and Magenes [1] (scalar operators), Eskin [1] (matrix, including pseudodifferential operators).

14.7. The technique in this section enables us to analyze linear pencils for PDO with the transmission property and are perturbed by singular operators, with a leading operator without fixed sign and with $B \geqslant cI$, and with Douglas–Nirenberg elliptic operators (see the author's works [7, 8, 13, and 15]).

§15 Problems with Resolvable Constraints

15.1. Let $\ell_1, \ell_2, m_1, m_2 \geqslant 1$ be integers, let $\Omega \subset R^n$ be a bounded Lipschitz domain, and let $V_j \subset H^{\ell_j}(\Omega)^{m_j}$, $j = 1, 2$, be subspaces such that

$$(15.1) \qquad C_0^\infty(\Omega)^{m_j} \subset V_j, \quad j = 1, 2.$$

On V_j, $j = 1, 2$, let us define the forms

$$(15.2) \qquad \mathcal{A}[u_j, v_j] = \sum_{|\alpha|, |\beta| \leqslant \ell_j} \langle a_{j\alpha\beta}(x)D^\alpha u_j, D^\beta v_j \rangle_0,$$

which satisfy the conditions

$$(15.3) \qquad a_{j\alpha\beta} = a_{j\beta\alpha}^* \in C^\infty(\overline{\Omega}; \operatorname{End} C^{m_j}),$$

$$(15.4) \qquad \mathcal{A}[u_j, u_j] \geqslant c\|u_j\|_{\ell_j}^2, \quad \forall u_j \in V_j,$$

where $c > 0$ does not depend on u_j; and on $V_1 \oplus V_2$, let us define the forms

$$\mathcal{A}_3[u_1, u_2] = \sum_{\substack{|\alpha| \leqslant \ell_1, \\ |\beta| \leqslant \ell_2}} \langle a_{3\alpha\beta}(x)D^\alpha u_1, D^\beta u_2 \rangle_0,$$

where $a_{3\alpha\beta} \in C^\infty(\overline{\Omega}; \operatorname{Hom}(C^{m_1}, C^{m_2}))$.

Let us consider the problem

(15.5)
$$\begin{cases} \mathcal{A}_1[u_1, w_1] + \mathcal{A}_3[w_1, u_2] = t\langle u_1, w_1\rangle_0, & \forall w_1 \in V_1, \\ \mathcal{A}_3[u_1, w_2] - \mathcal{A}_2[u_2, w_2] = 0, & \forall w_2 \in V_2. \end{cases}$$

Such, for example, is the eigenfrequency problem of electroelastic bodies (see Belokon and Vorovich [1]). Here, u_1 is a displacement vector and u_2 is an electrostatic potential; $\ell_1, \ell_2 = 2$, $m_1 = 3$, $m_2 = 1$; and condition (15.3) is satisfied.

We define the spectrum of problem (15.5) as a set of numbers $t \in \mathbb{C}$ such that system (15.5) has a nontrivial solution. In order to analyze the spectrum, we rewrite (15.5) as

(15.6)
$$\begin{cases} \widehat{A}_1 u_1 + \widehat{A}_3 u_2 = t u_1, \\ \widehat{A}_3^* u_1 - \widehat{A}_2 u_2 = 0, \end{cases}$$

where $\widehat{A}_j : V_j \to V_j^*$ $[j = 1, 2]$ and $\widehat{A}_3 : V_2 \to V_1^*$ are bounded operators. (15.4) implies that the operator $\widehat{A}_2$ is invertible, and problem (15.6) is equivalent to the problem

(15.7)
$$\widehat{A}_0 u_1 \overset{\text{def.}}{=} (\widehat{A}_1 + \widehat{A}_3 \widehat{A}_2^{-1} \widehat{A}_3^*) u_1 = t u_1;$$

moreover,

(15.8)
$$c\|u_1\|_{\ell_1}^2 \leqslant \langle \widehat{A}_0 u_1, u_1 \rangle \leqslant C\|u_1\|_{\ell_1}^2 \quad \forall u \in V_1,$$

where $\langle \cdot, \cdot \rangle$ is a pairing between V_1^* and V_1. (15.8) implies that we can introduce a new scalar product $\langle \cdot, \cdot \rangle_H = \langle \widehat{A}_0 \cdot, \cdot \rangle$ in V_1. Then (15.7) is equivalent to the equation

(15.7')
$$u_1 = t K u_1, \qquad u_1 \in H,$$

where $K = \widehat{A}_0^{-1}\big|_H$ is a symmetrical operator in H. Hence, the spectrum of problems (15.7'), (15.7), and (15.5) is real [and positive due to (15.8)].

Glazman's lemma, applied to (15.7'), yields the following formula for the distribution function of the spectrum:

(15.9)
$$N(t) = \mathcal{N}(\langle \widehat{A}_0 \cdot, \cdot \rangle - t\| \cdot \|_0^2, V_1).$$

Furthermore, we shall show that the right-hand side of (15.9) is finite for any t; therefore, the spectrum is discrete.

The operator $\widehat{A}_0$ is not local [and is not even pseudodifferential]; hence, analysis of the asymptotics of the function $N(t)$ is somewhat more difficult than in §14; however, we succeed in obtaining a similar result.

Let

$$a_j(x, \xi) = \sum_{|\alpha|=|\beta|=\ell_j} a_{j\alpha\beta}(x)\xi^{\alpha+\beta} \qquad [j = 1, 2],$$

$$a_3(x, \xi) = \sum_{|\alpha|=\ell_1, |\beta|=\ell_2} a_{3\alpha\beta}(x)\xi^{\alpha+\beta}, \quad a_0 = (a_1 + a_3 a_2^{-1} a_3^*)^{-1}.$$

Conditions (15.4) and (15.1) and Lemma 13.1 imply that a_1, a_2 are positive-definite on $\overline{\Omega} \times S_{n-1}$; therefore, the same holds for a_0.

Theorem 15.1. a) *As* $t \to \infty$, $\forall \delta \in (0, \frac{1}{3})$, *we have*

$$(15.10) \qquad N(t) = c_+ t^{n/\ell} + O(t^{(n-\delta)/\ell}),$$

where $\ell = 2\ell_1$ *and where* c_+ *is defined by equality* (13.11).

b) *If* a_0 *is microlocally smoothly diagonalizable, in the sense of part* b) *of Theorem 13.1, then estimate* (15.10) *holds for any* $\delta \in (0, \frac{1}{2})$.

c) *In addition, if* $\partial\Omega$ *is piecewise smooth, in the sense of part* c) *of Theorem 13.1, then estimate* (15.10) *holds for any* $\delta \in (0, \frac{2}{3})$.

15.2.

Proof. Given the function a_0, we shall construct the function a_t as in §10; and then, given the functions a_t, $d_t = h$ and the metric g [the same as in §13], we shall construct the functions $\eta_{t,\pm c}$ and $\lambda_{t,0}$ and the ASP $\mathcal{E}_{1,t}$ as in the Chapter 2 proof of the estimate from below. The arguments in Chapter 2 and in §13 yield the following estimate for the eigenvalues ν_{1ti} and the operator $\mathcal{E}_{1,t}$:

$$(15.11) \qquad \mathrm{card}\{i \mid \nu_{1ti} \geqslant \tfrac{1}{2}\} = c_+ t^{3/2} + O(t^{(3-\delta)/2}).$$

Now (15.4) allows us to change the coefficients of the DO

$$A_j = \sum_{|\alpha|,|\beta| \leqslant \ell_j} D^\beta a_{j\alpha\beta}(x) D^\alpha \quad [j = 1, 2], \qquad A_3 = \sum_{|\alpha| \leqslant \ell_1, |\beta| \leqslant \ell_2} D^\beta a_{3\alpha\beta}(x) D^\alpha$$

outside Ω such that

$$A_j \in \mathcal{L}I^+(g, h^{-2\ell_j} I_{m_j}) \quad [j = 1, 2], \qquad A_3 \in \mathcal{L}(g, h^{-\ell_2} I_{m_2}, h^{-\ell_1} I_{m_1}).$$

Then Lemma 5.5 implies that there exists a paramatrix $R \in \mathcal{L}I(g, g^{2\ell_2})$ of the operator A_2. By replacing R with $(R + R^*)/2$, if necessary, we can assume that $R \in \mathcal{L}I^+(g, h^{2\ell_2})$ and that $A_0 = A_1 + A_3 R A_3^* \in \mathcal{L}I^+(g, h^{-2\ell_1})$. As shown in the process of proving Lemma 8.1, there exists $b \in \mathcal{L}I(g, h^{\ell_1})$ such that

$$b_w^* A_0 b_w = I + T, \quad T \in \mathcal{L}^{-\infty}(g, I_{m_1}).$$

Let $\pi_t = (\lambda_{t,0} + d_t)^{-1/2}$ and $q_{t,w} = b_w((1 + th^{2\ell_1})^{-1/2})_w$, and let

$$L_t' = \mathrm{span}\{u_i \mid \mathcal{E}_{1t} u_i = \nu_{1ti} u_i, \; \nu_{1ti} \geqslant \tfrac{1}{2}\},$$

$$L_t' = \eta_{t,-1,\ell} q_{t,w} \pi_{t,w}(L_t').$$

Then the arguments in Chapter 2 yield $L_t \subset C_0^\infty(\Omega)^{m_1}$; and due to (15.11),

$$(15.12) \qquad \dim L_t \geqslant \dim L_t' - C = c_+ t^{n/\ell} - C t^{(n-\delta)/\ell}.$$

Since $V_1 \supset C_0^\infty(\Omega)^{m_1}$, then in order to verify the estimate from below, it suffices to find a subspace $L_t^- \subset L_t$ of codimension t^χ, where $\chi = (n-1)/\ell$, such that for $t \geqslant t_0$,

$$(15.13) \qquad \langle \widehat{A}_0 u, u \rangle - t\|u\|_0^2 < 0 \quad \forall (0 \neq) u \in L_t^-.$$

We notice that if we replace $\widehat{A}_0$ with A_0 in (15.13), then we obtain the statement proved in Chapter 2. We shall transform the proof of Chapter 2 into the proof of estimate (15.13) with the help of Lemmas 14.4 and 11.1 and the following lemma:

Lemma 15.1. *Let G_t be a σ-temperate splitting metric and let*

$$g \leqslant G_t \leqslant Ch_g^{-C}g, \quad h_{G_t} \leqslant h_g^\varepsilon, \quad \eta_{t,-c} \in S(G_t, 1), \quad \varepsilon > 0$$

uniformly with respect to t. Then

$$(15.14) \qquad \widehat{A}_2^{-1}r_+\eta_{t,-c,\ell} = r_+R\eta_{t,-c,\ell} + \widehat{A}_2^{-1}r_+T_{2,t} + r_+T_{3,t},$$

where $T_{j,t} \in \mathcal{L}^{-\infty}(g, 1)$.

Proof. Since supp $\eta_{t,-c} \subset \Omega \times R^n$, then

$$(15.15) \qquad \widehat{A}_j r_+\eta_{t,-c,\ell} = r_+A_j\eta_{t,-c,\ell},$$

but $\eta_{t,-c/2}\eta_{t,-c} = \eta_{t,-c}$; therefore, (15.15) yields the equality

$$\begin{aligned}
\widehat{A}_2^{-1}r_+\eta_{t,-c,\ell} &= \widehat{A}_2^{-1}r_+A_2R\eta_{t,-c,\ell} + \widehat{A}_2^{-1}r_+T_{1,t} \\
&= \widehat{A}_2^{-1}r_+A_2\eta_{t,-c/2,\ell}R\eta_{t,-c,\ell} + \widehat{A}_2^{-1}r_+T_{2,t} \\
&= r_+\eta_{t,-c/2,\ell}R\eta_{t,-c,\ell} + \widehat{A}_2^{-1}r_+T_{2,t} \\
&= r_+R\eta_{t,-c,\ell} + r_+T_{3,t} + \widehat{A}_2^{-1}r_+T_{2,t}.
\end{aligned}$$

∎

Let $v \in L_t'$ and $v_1 = \mathcal{E}_{1,t}v \in L_t'$ and let $u = \eta_{t,-1,\ell}q_{t,w}\pi_{t,w}v_1$. Then (15.15), (15.14), and the equality $\eta_{t,-c/2}\eta_{t,-c} = \eta_{t,-c}$ yield

$$\begin{aligned}
\langle \widehat{A}_0 u, u \rangle &= \langle A_1 u, u \rangle_0 + \langle \widehat{A}_2^{-1}r_+A_3^*u, r_+A_3^*u \rangle \\
&= \langle A_0 u, u \rangle + \langle T_{1,t}v, v \rangle + \langle \widehat{A}_2^{-1}r_+T_{2,t}v, r_+K_t v \rangle,
\end{aligned}$$

where $T_{j,t} \in \mathcal{L}^{-\infty}(g, 1)$ and $K_t \in \mathcal{L}(G_t, h^{-2\ell_1})$.

By construction, $\text{supp}_\infty \sigma^w(\mathcal{E}_{1t}) \subset \{X \mid h(X)^{-2\ell_1} < Ct\}$; therefore, the same holds for K_t. Hence, $K_t \in \mathcal{L}(G_t, t)$; and due to Theorem 4.2, $\|K_t v\| \leqslant C_1 t \|v\|$. Therefore, Lemma 14.4 shows that, on a subspace $\widehat{L}_t \subset L_t$ of codimension t^χ, we have

$$(15.16) \qquad \langle \widehat{A}_0 u, u \rangle - t\|u\|_0^2 \leqslant \langle (A_0 - t)u, u \rangle + Ct^{-1}\|v\|_0^2.$$

Arguments similar to those of Chapter 2 show that on a subspace $L_t^- \subset \widehat{L}_t$ of finite codimension, the right-hand side of (15.16) is negative if $v \neq 0$; therefore, both (15.13) and the estimate from below are proved.

15.3. Proof of the estimate from above almost coincides with the proof of the estimate from above in §14. Estimate (15.8) replaces estimates (14.9); and if

$\eta_{t,-c}$ and $\mu_{t,\pm c}$ are the functions constructed from the function a_0 as in §14, then

$$\eta_{t,-c/2}\mu_{t,-c} = \mu_{t,-c}, \qquad \eta_{t,-c/2}(1 - \mu_{t,c}) = 1 - \mu_{t,c}.$$

Therefore, the same arguments as in Lemma 15.1 yield

$$[\widehat{A}_0, r_+\mu_{t,\pm c,r}e_+] = r_+[A_0, \mu_{t,\pm c,r}]e_+ + \widehat{A}_3\widehat{A}_2^{-1}r_+T_{1,t} + r_+T_{2,t},$$
$$\widehat{A}_0 r_+\mu_{t,-c,r}e_+ = r_+A_0\mu_{t,-c,r}e_+ + \widehat{A}_3\widehat{A}_2^{-1}r_+T_{3,t}e_+ + r_+T_{4,t}e_+;$$

and if k_t is a symbol from (14.16), then (14.16) yields

$$\widehat{A}_0 r_+k_{t,w} = r_+A_0k_{t,w} + \widehat{A}_3\widehat{A}_2^{-1}r_+T_{5,t} + r_+T_{6,t},$$

where $T_{j,t} \in \mathcal{L}^{-\infty}(g, 1)$.

These equalities and Lemma 14.4 allow us, as in §14, to obtain estimate (14.33) with

$$Q_{t,1} = c \sum_{|\alpha| \leqslant \ell_1} D^\alpha \langle x\rangle^{2\ell_1 - 2|\alpha|} D^\alpha - Ct(\mu'_{t,w})^2,$$
$$Q_{t,-1} = A_0 - t - \mu_{t,-1,\ell}((h^{-2\ell_1} + t)\lambda_{t,0}h^{\omega/2})_w\mu_{t,-1,r} - t^{-1},$$

and, thus, the estimate

$$N(t) \leqslant c_+ t^{n/\ell} + Ct^{(n-\delta)/\ell}.$$

§16 Electromagnetic Resonator

In their works, Vekk [1], Birman [2], Birman and Solomyak [8–10] formulated boundary value problems for Maxwell operators in domains with nonsmooth boundaries (in the last work, tensors ε, μ of electric and magnetic permeability are not constant and are anisotropic) and showed that domains of corresponding operators are not always subspaces of $H'(\Omega)^6$ (for example, this is not the case if a domain has incoming edges). In [9, 10] Birman and Solomyak calculated the principal term of the asymptotics of a spectrum for very weak conditions of regularity of the boundary and of ε, μ. In [5] Safarov obtained a sharp estimate for the remainder for a Maxwell operator with $\varepsilon = \mu = I$ in a domain with $\partial\Omega \in C^\infty$; and by making additional global assumptions, he calculated the second term of the asymptotics.

In this section we obtain an estimate for the remainder in an asymptotic formula for the distribution function of the squares ω_n^2 of eigenfrequencies of an electromagnetic resonator of $\Omega \subset R^3$, where Ω is a convex bounded Lipschitz domain with piecewise smooth boundary, and $\varepsilon, \mu \in C^\infty(\overline{\Omega}; \mathrm{End}\,\mathbf{C}^3)$.

Let

(16.1) $$V_1 = \{u \in H^1(\Omega)^3 \mid u^\tau|_\Gamma = 0\},$$

where Γ is a union of smooth parts of the boundary and u^τ is a tangent component of the vector u. Results of the quoted works imply that ω_n^2 are eigenvalues of the problem

(16.2) $$\langle \mu^{-1}(x)\,\text{rot}\,u, \text{rot}\,w\rangle_0 = t\langle \varepsilon(x)u, w\rangle_0 \quad \forall w \in V_1,$$

where $u \in V_1$ satisfies the constraint condition

(16.3) $$\nabla^* \varepsilon u = 0.$$

We fix a positive smooth scalar function ρ, let

$$\mathcal{A}_\rho[u, w] = \langle \mu^{-1}\,\text{rot}\,u, \text{rot}\,w\rangle_0 + \langle \rho\nabla^*\varepsilon u, \nabla^*\varepsilon w\rangle_0,$$

and denote by $V \subset V_1$ a subspace singled out by condition (16.3). We shall show that in (16.2), V_1 can be replaced with V. For this purpose, we denote by $F : \overset{\circ}{H}{}^1(\Omega) \to \overset{\circ}{H}{}^1(\Omega)^*$ the operator of the problem

(16.4) $$\langle \varepsilon\nabla v, \nabla v_1\rangle_0 = \langle f, v_1\rangle \quad \forall v_1 \in \overset{\circ}{H}{}^1(\Omega)$$

and notice that F has a bounded inverse F^{-1}. Since the operator $\nabla^*\varepsilon : L_2(\Omega)^3 \to \overset{\circ}{H}{}^1(\Omega)^3$ is bounded, so is the operator $P = I_3 - \nabla F^{-1}\nabla^*\varepsilon : L_2(\Omega)^3 \to L_2(\Omega)^3$. Since Ω is a bounded convex Lipschitz domain, then $F^{-1}(L_2(\Omega)) \subset H^2(\Omega)$ and $P(V_1) \subset V_1$ (see Grisvard [1]). Obviously, P projects V_1 on V.

Furthermore, $\text{rot}\,\nabla = 0$; hence, $\text{rot}\,Pu = \text{rot}\,u \quad \forall u \in V_1$, .and we can replace $w \in V_1$ with $Pw \in V$ in (16.2).

Therefore, problem (16.2) with (16.3) is equivalent to the problem

(16.5) $$\langle \mu^{-1}(x)\,\text{rot}\,u, \text{rot}\,w\rangle_0 = t\langle \varepsilon(x)u, w\rangle_0 \quad \forall w \in V;$$

and since $\nabla^*\varepsilon u = 0 \quad \forall u \in V$, we can replace the form on the left-hand side with the form $\mathcal{A}_\rho[u, w]$.

We need the estimate

(16.6) $$\mathcal{A}_\rho[u, u] \geqslant c\|u\|_1^2 - C\|u\|_0^2 \quad \forall u \in V_1.$$

Standard arguments, which apply a partition of unity, allow us to reduce the proof of this estimate to the case where $\text{supp}\,u \subset \overline{\Omega}'$, where $\Omega' \subset \Omega$ is a subdomain of a small diameter. Therefore, we can assume that the coefficients are constant. We integrate by parts, with the help of the condition in (16.1), and select an appropriate cartesian coordinate system to obtain

$$A_\rho[u, u] = \sum_{|\alpha|=|\beta|=1} \langle b_{\alpha\beta} D^\alpha u, D^\beta u \rangle_{L_2} + R(u),$$

where

$$|R(u)| \leqslant \omega \|u\|_1^2 + C_\omega \|u\|_0^2 \quad \forall \omega > 0,$$

$$\sum_{|\alpha|=|\beta|=1} \langle b_{\alpha\beta} \xi_\alpha, \xi_\beta \rangle_{\mathbf{C}^3} \geqslant c \sum_{|\alpha|=1} |\xi_\alpha|_{\mathbf{C}^3}^2 \quad \forall \xi_\alpha \in \mathbf{C}^3.$$

This yields (16.6).

Let $L_2(\varepsilon, \Omega)$ be a space $L_2(\Omega)^3$ with the norm $\| \cdot \|_{\varepsilon,2} = \langle \varepsilon \cdot, \cdot \rangle_0^{1/2}$, let $L_2(\varepsilon, V, O)$ be a closure of V in $L_2(\varepsilon, \Omega)$, and let $\widehat{A}_\rho$ be an operator associated with the variational triple A_ρ, V, $L_2(\varepsilon, V, \Omega)$. Estimate (16.6) and standard arguments used to analyze the Neumann problem for Laplace operators show that the spectrum of problem (16.5) coincides with the spectrum of the operator $\widehat{A}_\rho$. Applying Glazman's lemma to $\widehat{A}_\rho$, we obtain the following equality for the distribution function of the spectrum:

$$(16.7) \qquad N(t) = \mathcal{N}(A_\rho[\cdot] - t \langle \varepsilon P \cdot, \cdot \rangle_0, V).$$

Let

$$F_t(V) = \mathcal{N}(A_\rho[\cdot] - t \langle \varepsilon P \cdot, \cdot \rangle_0, V),$$

and let us prove that $F_t(V) = F_t(V_1)$. Since $V \subset V_1$, then $F_t(V) \leqslant F_t(V_1)$; and in order to prove the inequality $F_t(V_1) \leqslant F_t(V)$, we denote $u_1 = Pu \in V$ and $u_2 = (I - P)u$ for $u \in V_1$. Then $\operatorname{rot} u_2 = 0$,

$$\langle \varepsilon Pu, u \rangle_0 = \langle \varepsilon Pu_1, u_1 \rangle_0, \qquad \nabla^* \varepsilon u = \nabla^* \varepsilon u_2, \qquad \nabla^* \varepsilon u_1 = 0;$$

and, hence,

$$A_\rho[u] - t \langle \varepsilon Pu, u \rangle_0 = A_\rho[u_1] - t \langle \varepsilon Pu_1, u_1 \rangle_0 + \langle \rho \nabla^* \varepsilon u_2, \nabla^* \varepsilon u_2 \rangle_0.$$

Noticing that $\langle \varepsilon u, u \rangle_0 = \langle \varepsilon u_1, u_1 \rangle_0 + \langle \varepsilon u_2, u_2 \rangle_0$, we see that the last equality and Lemma E.3 yield the estimate

$$F_t(V_1) \leqslant F_t(V) + \mathcal{N}(\langle \rho \nabla^* \varepsilon \cdot, \nabla^* \varepsilon \cdot \rangle), V_1).$$

Since the last term is equal to zero, we have shown that

$$N(t) = \mathcal{N}(A_\rho - t \langle \varepsilon P \cdot, \cdot \rangle_0, V_1).$$

Therefore, the asymptotics of the function $N(t)$ can be calculated in exactly the same way as in §15. Indeed, the problems in this section and in §15 differ only by the fact that in §15, the higher operator is nonlocal; and here, it is the subordinate one.

The asymptotic coefficient is defined as follows [see §13]: Let $r(\xi)$ be a symbol of the operation rot, let $\xi \in \mathbf{C}^3$ be a column vector, and let

$$\tilde{a}_\rho(x,\xi) = r(\xi)\mu^{-1}(x)r(\xi) + \rho(x)\varepsilon(x)[\xi_i\xi_j]_{i,j=1}^3\varepsilon(x),$$
$$b(x,\xi) = \varepsilon(x) - (\xi^T\varepsilon(x)\xi)^{-1}\varepsilon(x)[\xi_i\xi_j]_{i,j=1}^3\varepsilon(x),$$
$$a_\rho = \tilde{a}_\rho^{-1/2}b\tilde{a}_\rho^{-1/2},$$
$$c = \frac{1}{24\pi^3}\int_\Omega dx \int_{S_2} [\lambda_1(x,\xi)^{3/2} + \lambda_2(x,\xi)^{3/2}]dS(\xi),$$

where λ_1, λ_2 are positive eigenvalues of the matrix a_ρ.

We remark that c does not depend on the choice of ρ, since $\lambda_1(x,\xi)^{-1}$ and $\lambda_2(x,\xi)^{-1}$ are eigenvalues of the problem

$$(16.8) \qquad \begin{cases} r(\xi)\mu(x)^{-1}r(\xi)u = \lambda\varepsilon(x)u, \\ \xi^T\varepsilon(x)u = 0, \quad u \in \mathbf{C}^3 \end{cases}$$

[and are positive eigenvalues of the first equation in (16.8)].

Thus, we proved the following theorem:

Theorem 16.1. a) *As* $t \to \infty$ *and* $\forall\delta \in (0,\frac{1}{3})$,

$$(16.9) \qquad N(t) = ct^{3/2} + O(t^{(3-\delta)/2}).$$

b) *If the function* ρ *is chosen such that* a_ρ *is microlocally smoothly diagonalizable, then estimate* (16.9) *holds for any* $\delta \in (0,\frac{1}{2})$.

c) *In addition, if* $\partial\Omega$ *is piecewise smooth, in the sense of part* c) *of Theorem 13.1, then estimate* (16.9) *holds for any* $\delta \in (0,\frac{2}{3})$.

Remark. If the resonator is isotropic [i.e., ε and μ are scalar functions], then the condition of microlocal smooth diagonalizability can be satisfied by assuming that $\rho = \mu^{-1}\varepsilon^{-2}$. Indeed, in this case, $\tilde{a}_\rho(x,\xi) = \mu^{-1}(x)|\xi|^2$ is a scalar function; and the function

$$b(x,\xi) = \varepsilon(x)(I_3 - |\xi|^{-2}[\xi_i\xi_j]_{i,j=1}^3)$$

is microlocally smoothly diagonalizable, since one of its eigenvalues $\lambda_1 = 0$ is simple and two others are mutually equal:

$$\lambda_2(x,\xi) = \lambda_3(x,\xi) = \varepsilon(x).$$

§17 Asymptotics of the Discrete Spectrum of Douglis–Nirenberg Operators with a Totally Disconnected Essential Spectrum

17.1. The spectral problem for equations of the multigroup diffusion model has the form (see Zweiffel [1])

$$(17.1) \qquad \begin{bmatrix} A_{11} & A_{12} \\ A_{21} & A_{22} \end{bmatrix} \begin{bmatrix} u_1 \\ u_2 \end{bmatrix} = t \begin{bmatrix} u_1 \\ u_2 \end{bmatrix},$$

where $A_{21} = A_{12}^*$ and A_{12} are operators of multiplication by matrix functions and where

$$A_{11} = [-\delta_{jk}\nabla^* d_j(x)\nabla]_{j,k=1}^{m_1}, \qquad A_{22} = [\delta_{jk}\mu_j]_{j,k=1}^{m_2}, \quad \mu_j \in R.$$

We shall consider a natural extension of problem (17.1), show that a discrete spectrum can accumulate at $\mu_j - 0$, $j = 1, \ldots, m_2$ [but not at $\mu_j + 0$] and does accumulate at $+\infty$, and then calculate the asymptotics of a corresponding series of eigenvalues.

17.2. Let m_1, m_2, $\ell \geqslant 1$ be integers and $\Omega \in R^n$ be a bounded Lipschitz domain; let $a_{\alpha\beta} = a_{\beta\alpha}^* \in C^\infty(\overline{\Omega}; \operatorname{End} C^{m_1})$ $[|\alpha|, |\beta| \leqslant \ell]$; let

$$A_{21} = A_{12}^* \in C^\infty(\overline{\Omega}; \operatorname{Hom}(C^{m_1}; C^{m_2})), \qquad V_1 \subset H^\ell(\Omega)^{m_1}$$

be a closed subspace such that

$$(17.2) \qquad\qquad C_0^\infty(\Omega)^{m_1} \subset V_1;$$

and let the form

$$A_1[u_1] = \sum_{|\alpha|,|\beta| \leqslant \ell} \langle a_{\alpha\beta}(x)D^\alpha u_1, D^\beta u_1 \rangle$$

be coercive on V_1:

$$(17.3) \qquad\qquad A_1[u_1] \geqslant c\|u_1\|_\ell^2 - C\|u_1\|_0^2 \quad \forall u_1 \in V_1.$$

Let $u = (u_1, u_2)$. Condition (17.3) implies that the form

$$A[u] = A_1[u_1] + \langle u_1, A_{12}u_2 \rangle_0 + \langle u_2, A_{12}^*u_1 \rangle_0 + \langle A_{22}u_2, u_2 \rangle_0$$

is semibounded from below on $V = V_1 \oplus L_2(\Omega)^{m_2}$. Let $\widehat{A}$ and $\widehat{A}_{11}$ be operators associated with the variational triples A, V, $L_2(\Omega)^m$ $[m = m_1 + m_2]$ and A_1, V_1, $L_2(\Omega)^{m_1}$, respectively. Then $\widehat{A} = [\widehat{A}_{ij}]_{i,j=1,2}$, where $\widehat{A}_{21} = \widehat{A}_{12}^*$ and $\widehat{A}_{ij}$, with $i + j > 2$, are operators of multiplication by matrix functions.

Lemma 17.1.   *The essential spectrum* $\sigma_{\text{ess}}\widehat{A} = \omega \overset{\text{def.}}{=} \{\mu_1, \ldots, \mu_{m_2}\}$.

Proof.   If $\lambda \notin \omega$, then the equation $\widehat{A}u = \lambda u$ is equivalent to the equation $F_\lambda u = 0$, where $F_\lambda = \widehat{A}_{11} - F_\lambda'$ and $F_\lambda' = \lambda + \widehat{A}_{12}(\widehat{A}_{22} - \lambda)^{-1}\widehat{A}_{12}^*$. But F_λ' is a bounded operator function analytic in $\mathbb{C}\backslash\omega$, and $\widehat{A}_{11}$ is an operator with a discrete spectrum, due to (17.3); therefore, in $\mathbb{C}\backslash\omega$ we can only have eigenvalues of the pencil F_λ [of finite multiplicity] that accumulate at $+\infty$ and possibly at ω. Hence, $\omega \supset \sigma_{\text{ess}}\widehat{A}$. On the other hand, $\sigma_{\text{ess}}\widehat{A} \supset \omega$, since $\dim \text{Ker}(\widehat{A} - \mu_j I) = \infty$ for all j.

Lemma 17.2.   *For any* $j = 1, \ldots, m_2$, *the eigenvalues of the operator* $\widehat{A}$ *do not accumulate at* $\mu_j + 0$.

Proof.   Assume that $\lambda > \mu_j$ does not belong to the spectrum of the operator $\widehat{A}$, and let $\text{dist}(\lambda, \omega) = \lambda - \mu_j$. Then F_λ is invertible and

$$N((\mu_j, \lambda), \widehat{A}) = N((-\infty, 0), (\lambda - \mu_j)^{-1} + (\widehat{A} - \lambda)^{-1}),$$

where $N((a, b), A)$ denotes a number of eigenvalues [counting multiplicity] of the operator A on the interval (a, b) [here $N((a, b), A) = +\infty$ if there are points of the essential spectrum in (a, b)].

   Let $\pi : \mathbb{C}^m \to \mathbb{C}^{m_2}$ be an operator such that $\pi u = u_2$, and let us consider it as an operator from $L_2(\Omega)^m$ to $L_2(\Omega)^{m_2}$. Then

$$(17.4) \qquad (\widehat{A} - \lambda)^{-1} = \pi^*(\widehat{A}_{22} - \lambda)^{-1}\pi + \widehat{A}_\lambda',$$

where

$$\widehat{A}_\lambda' = \begin{bmatrix} P_\lambda & -P_\lambda \widehat{A}_{12}(\widehat{A}_{22} - \lambda)^{-1} \\ -(\widehat{A}_{22} - \lambda)^{-1}\widehat{A}_{12}^* P_\lambda & (\widehat{A}_{22} - \lambda)^1 \widehat{A}_{12}^* P_\lambda \widehat{A}_{12}(\widehat{A}_{22} - \lambda)^{-1} \end{bmatrix}$$

with $P_\lambda = F_\lambda^{-1}$. But the choice of λ implies that

$$(\lambda - \mu_j)^{-1}I_m + \pi^*(\widehat{A}_{22} - \lambda)^{-1}\pi \geqslant 0;$$

therefore,

$$(17.5) \qquad (\lambda - \mu_j)^{-1}I_m + (\widehat{A} - \lambda)^{-1} \geqslant \widehat{A}_\lambda' = C_\lambda B_\lambda C_\lambda^*,$$

where

$$C_\lambda = \begin{bmatrix} I_{m_1} & 0 \\ 0 & -(\widehat{A}_{22} - \lambda)^{-1}\widehat{A}_{12}^* \end{bmatrix}, \qquad B_\lambda = \begin{bmatrix} P_\lambda & P_\lambda \\ P_\lambda & P_\lambda \end{bmatrix}.$$

Now (17.3) implies that the operator F_λ is semibounded from below and has a discrete spectrum; therefore, the number of negative eigenvalues of the operator P_λ is finite. But if $\mu \neq 0$ is an eigenvalue of the operator B_λ of multiplicity k, then $\mu/2$ is an eigenvalue of the operator P_λ of multiplicity $k/2$; and if μ is a point of the essential spectrum of the operator B_λ, then $\mu/2$ is a point of the essential spectrum of the operator P_λ. Hence, the negative spectrum of the operator B_λ, and, therefore, of the operator $\widehat{A}'_\lambda$, consists of a finite number of eigenvalues of finite multiplicity; and (17.5) yields the statement of the lemma. ∎

17.3. Let $m'_j = \dim \text{Ker}(A_{22} - \mu_j)$, let $\pi_j : C^{m_2} \to C^{m'_j}$ be an orthoprojection on $\text{Ker}(A_{22} - \mu_j)$, and $\forall (x, \xi) \in \overline{\Omega} \times (R^n \backslash O)$, let

$$a_0(x,\xi) = \sum_{|\alpha|=|\beta|=\ell} a_{\alpha\beta}(x)\xi^{\alpha+\beta},$$

$$a_j(x,\xi) = \pi_j A^*_{12}(x) a_0^{-1}(x,\xi) A_{12}(x)\pi_j^*$$

[the function a_0 is invertible due to (17.3) and Lemma 13.1].

Theorem 17.1. *Let $\lambda < \mu_j$ and let* $\text{dist}(\lambda, \omega) = \mu_j - \lambda$. *Then*
  a) *As $\tau \to +0$, $\forall \delta \in (0, \frac{1}{3})$,*

$$(17.6) \qquad N((\lambda, \mu_j - \tau), \widehat{A}) = c_j \tau^{-n/(2\ell)} + O(\tau^{-\rho(\delta)}),$$

where

$$(17.7) \qquad c_j = (2\pi)^{-n} n^{-1} \int_\Omega dx \int_{|\xi|=1} \text{Tr}\, a_j(x,\xi)_+^{n/(2\ell)} dS(\xi),$$

$$\rho(\delta) = \max\{(n-\delta)/(2\ell),\ n/(2\ell+\delta)\};$$

  b) *if $m'_j = 1$, then formula (17.6) holds for any $\delta \in (0, \frac{1}{2})$; and*
  c) *if $m'_j = 1$ and $n > 2\ell$ and if $\partial\Omega$ is piecewise smooth, in the sense of part c) of Theorem 13.1, then formula (17.6) holds with $\rho(\delta) = (n-\delta)/2\ell$, where $\delta \in (0, \frac{2}{3})$ is arbitrary.*

Corollary 17.1. *If $\partial\Omega$ is piecewise smooth, in the sense of part c) of Theorem 13.1 and if, as is usually assumed in a multigroup diffusion model, all eigenvalues of the matrix A_{22} are single, then for all j,*

$$N((\lambda, \mu_j - \tau), \widehat{A}) = c_j \tau^{-3/2} + O(\tau^{-7/6-\epsilon}) \quad \forall \epsilon > 0,$$

for the operator of the indicated model.

17.4.

Proof of Theorem 17.1. Without loss of generality, we can assume that λ does not belong to the spectrum of the operator $\widehat{A}$. Then

$$(17.8) \qquad N((\lambda, \mu_j - \tau), \widehat{A}) = N((-\infty, 0), \widehat{A}_{\lambda, t}),$$

where

$$\widehat{A}_{\lambda, t} = (\mu_j - \lambda)^{-1} - (\widehat{A} - \lambda)^{-1} + t^{-1}(\mu_j - \lambda)^{-2},$$
$$t^{-1} = (\mu_j - \lambda)^2((\mu_j - \lambda - \tau)^{-1} - (\mu_j - \lambda)^{-1}) = \tau + O(\tau^2).$$

We consider $\tilde{\pi}_2 = \pi_j \pi$ as an orthoprojection in $L_2(\Omega)^m$ and denote $\tilde{\pi}_1 = I - \tilde{\pi}_2$ and $\widehat{A}_{\lambda, t}^{ij} = \tilde{\pi}_i \widehat{A}_{\lambda, t} \tilde{\pi}_j^*$. Then

$$\widehat{A}_{\lambda, t} = [\widehat{A}_{\lambda, t}^{ij}]_{i,j=1,2};$$

and due to (17.4),

$$(17.9) \qquad \widehat{A}_{\lambda, t}^{22} = (\mu_j - \lambda)^{-2}(t^{-1} - \pi_j \widehat{A}_{12}^* P_\lambda \widehat{A}_{12} \pi_j^*);$$

$$(17.10) \qquad \widehat{A}_{\lambda, t}^{11} = M + t^{-1}(\mu_j - \lambda)^{-2} + \widehat{A}_\lambda^{111};$$

$(17.11) \qquad M$ is a diagonal matrix with positive constants on diagonal;

$(17.12) \qquad \widehat{A}_{\lambda, t}^{12} = \widehat{A}_\lambda^{12} = (\widehat{A}_\lambda^{21})^* = (A_{\lambda, t}^{21})^*$ and $\widehat{A}_\lambda^{111}$ are products of the operators of multiplication by matrix functions and the operator P_λ.

The exact form of the matrices M, $\widehat{A}_\lambda^{12}$, $\widehat{A}_\lambda^{21}$, $\widehat{A}_\lambda^{111}$ is not essential for further discussion.

Lemma 17.3. *There exists an operator $R \in \mathcal{L}(g, h^{2\ell})$ such that*

$$(17.13) \qquad \|u\|_{V_1^*}^2 \leqslant C \leqslant \langle Re_+ u, e_+ u \rangle \quad \forall u \in L_2(\Omega)^{m_1}$$

[the metric g and the function h are the same as in §§13–15].

Proof. Let $\widetilde{\Omega} \supset \overline{\Omega}$ be a bounded open set. Then there exists a bounded linear operator extension $\ell : H^\ell(\Omega)^{m_1} \to H^\ell(R^n)^{m_1}$ such that supp $\ell\varphi \subset \widetilde{\Omega}$; therefore,

$$\|u\|_{V_1^*}^2 = \sup_{0 \neq \varphi \in V_1} \langle u, \varphi \rangle_0^2 / \|\varphi\|_\ell^2 \leqslant C_1 \sup_{\ell\varphi} \langle e_+ u, \ell\varphi \rangle^2 / \|\ell\varphi\|_\ell^2$$

$$\leqslant C_2 \sup_{\ell\varphi \in \mathcal{H}(g, h^{-\ell})^{m_1}} \langle e_+ u, \ell\varphi \rangle^2 / \|\ell\varphi\|_{\mathcal{H}(g, h^{-\ell})^{m_1}}^2$$

$$\leqslant C_2 \|e_+ u\|_{(\mathcal{H}(g, h^{-\ell})^*)^{m_1}}^2 \leqslant C_1 \|e_+ u\|_{\mathcal{H}(g, h^\ell)^{m_1}}^2,$$

since Lemma 5.6 and Theorem 5.2 yield

$$\|\ell\varphi\|_{\mathcal{H}(g,h^{-\ell})^{m_1}} \leqslant C\|\ell\varphi\|_\ell \quad \text{and} \quad \mathcal{H}(g,h^{-\ell})^* \cong \mathcal{H}(g,h^\ell).$$

Therefore, in (17.13) we can assume that $R = Q^*Q$, where $(\mathcal{L}(g,h^\ell) \ni)Q :$ $\mathcal{H}(g,h^\ell) \to L_2(R^n)$ is an isomorphism. ∎

Since

$$|\langle P_\lambda u_1, v_1\rangle| \leqslant C\|u_1\|_{V_1^*}\|v_1\|_{V_1^*},$$

then (17.9)–(17.13) yield the estimate

(17.14) $$\langle \widehat{A}_{\lambda,t}u, u\rangle_0 \geqslant ct^{-1}\|u\|_0^2 - C\langle Re_+u, e_+u\rangle,$$

which allows us, as in §14 and §15, to analyze the asymptotics of the function $N((-\infty,0), \widehat{A}_{\lambda,t})$ almost the same way as was done for the distribution functions of the spectrum of pseudodifferential pencils in §13.

17.5. By applying (17.3), it is easy to prove [see Lemma 13.1] that

$$a_0(x,\xi) > 0 \quad \forall (x,\xi) \in \overline{\Omega} \times (R^n\backslash 0);$$

therefore, we can continue the coefficients $a_{\alpha\beta}$ in R^n such that the operator

$$A_{11} = \sum_{|\alpha|,|\beta|\leqslant\ell} D^\beta a_{\alpha\beta}(x)D^\alpha$$

will belong to $\mathcal{L}I^+(g,h^{-2\ell})$. Also, the operator

$$\widetilde{F}_\lambda = A_{11} - \lambda - A_{12}(A_{22} - \lambda)^{-1}A_{12}^* \in \mathcal{L}I^+(g,h^{-2\ell}) \subset \mathcal{L}I(g,h^{-2\ell})$$

[here we continue A_{12} to a function of the class $C_0^\infty(R^n, \text{Hom}(C^{m_2}, C^{m_1}))$]; and, hence, F_λ has a parametrix $R_\lambda \in \mathcal{L}I^+(g,h^{2\ell})$. By replacing R_λ with $(R_\lambda + R_\lambda^*)/2$, if necessary, we can assume that $R_\lambda = R_\lambda^*$.

Considering the function a_0 to be defined on $\mathcal{M}^\varepsilon = \Omega_\varepsilon \times \{\xi \mid |\xi| > 1\}$ for some $\varepsilon > 0$, we have

(17.15) $$(\sigma^w(R_\lambda) - a_0)\big|_{m^\varepsilon} \in S(g, h^{2\ell+1}, \mathcal{M}^\varepsilon).$$

We define an operator $A_{\lambda,t}$ by taking the definition of the operator $\widehat{A}_{\lambda,t}$ and substituting A_{ij} for $\widehat{A}_{ij}$ [with $i + j > 2$] and R_λ for P_λ. Then, similarly to (17.9)–(17.12),

(17.16) $$A_{\lambda,t}^{22} = (\mu_j - \lambda)^{-1}t^{-1}(I - ta_{\lambda,w}^{22}), \quad a_\lambda^{22} \in S(g, h^{2\ell}),$$

(17.17) $$A_{\lambda,t}^{11} = M + t^{-1}(\mu_j - \lambda)^{-2} + A_\lambda^{111},$$

where M is the same as in (17.10) and (17.11);

$$(17.18) \qquad A^{12}_{\lambda,t} = A^{12}_{\lambda} = (A^{21}_{\lambda})^* = (A^{21}_{\lambda,t})^*, \qquad A^{111}_{\lambda} \in \mathcal{L}(g, h^{2\ell}),$$

and due to (17.15),

$$(17.19) \qquad (a^{22}_{\lambda} - a_j)\big|_{\mathcal{M}^\epsilon} \in S(g, h^{2\ell+1}).$$

Now (17.19) implies that we can construct a function $\tilde{a}^{22}_{\lambda} \in S(g, h^{2\ell+1})$ such that $\tilde{a}^{22}_{\lambda} = a_j$ and $a^{22}_{\lambda} - \tilde{a}^{22}_{\lambda} \in S(g, h^{2\ell+1})$ on $\mathcal{M}^\epsilon$. Let

$$a_t(x, \xi) = (1 + th(x, \xi)^{2\ell})^{-1}(I_{m'_j} - t\tilde{a}^{22}_{\lambda}(x, \xi)), \qquad d_t = h,$$

and, starting with the functions a_t and d_t, construct functions $\lambda_{t,0}$ and $\eta_{t,-c}$ as was done in Chapter 2 [the way of construction is defined by the properties of the symbol a_j and of the boundary $\partial\Omega$—see §13]. We shall further construct the ASP $\mathcal{E}_{1t}$ as in Chapter 2 for the estimate from below. The arguments in Chapter 2 and in §13 show that its nonzero eigenvalues satisfy the condition

$$(17.20) \qquad \operatorname{card}\{i \mid \nu_{1ti} \geqslant \tfrac{1}{2}\} = c_j t^{n/(2\ell)} + O(t^{\rho(\delta)}).$$

Let $L_t = \operatorname{span}\{u \mid \mathcal{E}_{1,t}u = \nu_{1ti}u, \ \nu_{1ti} \geqslant \tfrac{1}{2}\}$, and let

$$q_t = (t^{-1} + h^{2\ell})^{-1/2}, \qquad \pi_t = (\lambda_{t,0} + d)^{-1/2}, \qquad Q_t = q_{t,w}\pi_{t,w}.$$

In a natural way we identify $L_t \subset \mathcal{S}(R^n; \operatorname{Ker}(A_{22} - \mu_j))$ with a subspace of $\mathcal{S}(R^n)^m$ and denote

$$L^-_t = \eta_{t,-1,\ell}Q_t(L_t) \subset C^\infty_0(\Omega)^m.$$

The arguments in Chapter 2 and (17.20) show that

$$(17.21) \qquad \dim L^-_t \geqslant \dim L_t - C = c_j t^{n/(2\ell)} + O(t^{\rho(\delta)}).$$

If $v \in L_t$ and $u = \eta_{t,-1,\ell}Q_t\mathcal{E}_{1,t}v \in L^-_t$, then by applying Lemmas 15.1 and 14.4 as in §15, we obtain

$$(17.22) \qquad \langle \widehat{A}_{\lambda,t}u, u \rangle \leqslant \langle A^{22}_{\lambda,t}\tilde{\pi}_2 u, \tilde{\pi}_2 u \rangle + t^{-1}\|\tilde{\pi}_2 v\|^2_0,$$

for $v \in L'_t \subset L_t$, where the codimension of L'_t does not exceed Ct^χ for $\chi = (n-1)/(2\ell)$. The arguments in Chapter 2 show that on a subspace $L''_t \subset L_t$ of finite codimension, the first term on the right-hand side of inequality (17.22) is estimated from above by $-\|\tilde{\pi}_2 \cdot v\|^2_0/2$; therefore, for $t \geqslant t_0$, we have $\langle \widehat{A}_{\lambda,t}u, u \rangle < 0$ for all nonzero u belonging to a subspace $\widetilde{L}_t \subset L^-_t$ of codimension $C_1 t^\chi$, and (17.21) yields the estimate

$$N((-\infty, 0), \widehat{A}_{\lambda,t}) \geqslant c_j t^{c/(2\ell)} + O(t^{\rho(\delta)}).$$

17.6. In order to obtain an estimate from above, we use the constructions in §14 and §15 with $\ell = e_+$ and estimate (17.14) instead of estimates (14.9) and (15.8). We obtain

$$(17.23) \quad \begin{aligned} N((-\infty,0), \widehat{A}_{\lambda,t}) &\leqslant \mathcal{N}_0(t^{-1} - \mu'_{t,w} K_1 \mu'_{t,w}, R^n) + \\ &\mathcal{N}_0(A_{\lambda,t} - \tilde{\pi}_2 K_{2,t} \tilde{\pi}_2 - (1-\tilde{\pi}_2) K_{3,t}(1-\tilde{\pi}_2), \Omega), \end{aligned}$$

where μ'_t is the same function as in Section 14.5; where $K_1 \in \mathcal{L}(g, h^{2\ell})$, $K_{2,t} \in \mathcal{L}(G_t^{(1)}, \lambda_{t,0} h^{\omega/2}(t^{-1} + h^{2\ell}) + t^{-2})$, $K_{3,t} \in \mathcal{L}(G_t^{(1)}, h^{\omega} + t^{-1})$; and where the function $\lambda_{t,0}$ and the metric $G_t^{(1)}$ are constructed from a_t as in Chapter 2 when we verified the estimates from above for the function $\mathcal{N}_0(\overset{\circ}{a}_w, \Omega)$.

Exactly the same arguments as in Section 14.5 yield

$$\mathcal{N}_0(t^{-1} - \mu'_{t,w} K_1 \mu'_{t,w}, R^n) \leqslant C t^{\rho(\delta)}.$$

In order to estimate the second term in (17.23), we recall that $A_{\lambda,t} = [A^{ij}_{\lambda,t}]_{i,j=1,2}$; and for any $\varepsilon > 0$, due to (17.18),

$$\begin{aligned} |\langle A^{12}_{\lambda,t} u_2, u_1 \rangle| &\leqslant C \|A^{12}_\lambda u_2\| \, \|u_1\| \\ &\leqslant \varepsilon \|u_1\|^2 + C_\varepsilon \|A^{12}_\lambda u_2\|^2 = \varepsilon \|u_1\|^2 + \langle D_{\lambda,\varepsilon} u_2, u_2 \rangle, \end{aligned}$$

where $D_{\lambda,\varepsilon} \in \mathcal{L}(g, h^{4\ell})$. Therefore, (17.17) and (17.18) yield

$$\begin{aligned} \langle A_{\lambda,t} u, u \rangle &\geqslant \langle (M + t^{-1}(\mu_j - \lambda)^{-2} - \varepsilon + A^{111}_\lambda) u_1, u_1 \rangle \\ &+ (\mu_j - \lambda)^{-2} \langle (t^{-1} - a^{22}_{\lambda,w} - D'_{\lambda,\varepsilon}) u_2, u_2 \rangle. \end{aligned}$$

Condition (17.11) allows us to choose $0 < \varepsilon I < M$; therefore, the second term in (17.23) is estimated from above by

$$\mathcal{N}(Q_t; C_0^\infty(R^n)^{m_1}) + \mathcal{N}(t^{-1} - a^{22}_{\lambda,w} - K_{t,w}; C_0^\infty(\Omega)^{m_2}),$$

where

$$Q_t \in \mathcal{L}I^+(G_t^{(1)}, 1), \qquad K_t \in \mathcal{L}(G_t^{(1)}, \lambda_{t,0} h^{\omega/2}(t^{-1} + h^{2\ell}) + t^{-2}).$$

Lemma 8.2 yields $\mathcal{N}(Q_t, C_0^\infty(R^n)^{m_1}) \leqslant C$; and the arguments in Chapter 2 give the same estimate for both $\mathcal{N}(t^{-1} - a^{22}_{\lambda,w} - K_{t,w}; C_0^\infty(\Omega)^{m_2})$ and $\mathcal{N}_0(t^{-1} - a^{22}_{\lambda,w}, \Omega)$. Recalling (17.19) and the arguments in §13, we see that the last term in (17.23) and, hence, $N((-\infty,0), \widehat{A}_{\lambda,t})$ are both estimated from above by $c_j t^{n/(2\ell)} + O(t^{\rho(\delta)})$, where c_j is defined by equality (17.7). But $t^{-1} = \tau + O(\tau^2)$; therefore, (17.8) and the estimates just obtained yield (17.6), and the theorem is proved. ∎

17.7. We shall calculate the asymptotics of a series of eigenvalues that accumulates at $+\infty$.

Theorem 17.2. *Let* $t > \lambda = \max \mu_j$. *Then there exists* C *such that*

$$(17.24) \qquad -C + N(t - C, \widehat{A}_{11}) \leqslant N((\lambda, t), \widehat{A}) \leqslant N(t + C, \widehat{A}_{11}) + C.$$

Proof. In Lemmas 17.1 and 17.2 we have shown that the spectrum of the operator $\widehat{A}$ in (λ, ∞) is discrete and accumulates only at $+\infty$; therefore, the form of formula (17.24) would not change if we took $\lambda > \max \mu_j$ not belonging to the spectrum. It also would not change if we replaced $\widehat{A}_{11}$, $\widehat{A}$, t with $\widehat{A}_{11} - \lambda$, $\widehat{A} - \lambda$, $t - \lambda$, respectively; therefore, we can assume in (17.24) that $\lambda = 0 > \max \mu_j$ and that $\widehat{A}$ is invertible. But then

$$(17.25) \qquad N((\lambda, t), \widehat{A}) = N(0, t^{-1} - \widehat{A}^{-1}).$$

Recalling (17.4) and noticing that for any $\varepsilon > 0$,

$$|\langle P_0 \widehat{A}_{12} \widehat{A}_{22}^{-1} u_2, u_1 \rangle| \leqslant \varepsilon \|u_2\|^2 + C_\varepsilon \langle P_0^2 u_1, u_1 \rangle,$$

we obtain

$$(17.26) \qquad \begin{aligned} &\langle (t^{-1} - P_0 - C_\varepsilon P_0^2) u_1, u_1 \rangle + \langle (t^{-1} - \widehat{A}_{22}^{-1} - \varepsilon - K_0) u_2, u_2 \rangle \\ &\qquad \leqslant \langle (t^{-1} - \widehat{A}^{-1}) u, u \rangle \\ &\qquad \leqslant \langle (t^{-1} - P_0 + C_\varepsilon P_0^2) u_1, u_1 \rangle + \langle (t^{-1} - \widehat{A}_{22}^{-1} + \varepsilon - K_0) u_2, u_2 \rangle, \end{aligned}$$

where $K_0 = \widehat{A}_{22}^{-1} \widehat{A}_{12}^* P_0 \widehat{A}_{12} \widehat{A}_{22}^{-1}$. But P_0 and, therefore, K_0 are compact operators in $L_2(\Omega)^{m_1}$ and $L_2(\Omega)^{m_2}$, respectively, and $-\widehat{A}_{22}^{-1} > 0$ by assumption; hence, if $\varepsilon > 0$ is sufficiently small, then the second forms on the left- and right-hand sides of inequality (17.26) are positive on a subspace $L \subset L_2(\Omega)^{m_2}$ of finite codimension.

Therefore,

$$(17.27) \qquad \begin{aligned} -C + N(0, t^{-1} - P_0 + C_\varepsilon P_0^2) &\leqslant N(0, t^{-1} - \widehat{A}^{-1}) \\ &\leqslant C + N(0, t^{-1} - P_0 - C_\varepsilon P_0^2). \end{aligned}$$

P_0 is an inverse of the operator F_0 semibounded from below with a discrete spectrum; therefore, for fixed $\varepsilon > 0$, there exists a subspace $L \subset L_2(\Omega)^{m_1}$ of finite codimension that is spanned over a part of the eigenvectors of the operator P_0 such that $C_\varepsilon \langle P_0 u, u \rangle \leqslant \langle P_0 u, u \rangle / 2 \quad \forall u \in L$. Let $\tilde{P} : L \to L$ be a restriction of P_0 to L, and let $\tilde{F}$ be an inverse of $\tilde{P}$. Then

$$(17.28)$$
$$N(0, t^{-1} - P_0 - C_\varepsilon P_0^2)$$
$$\leqslant \operatorname{codim} L + N(0, T^{-1} - \tilde{P} - C_\varepsilon \tilde{P}^2) = C_1 + N(t, (\tilde{P} + C_\varepsilon \tilde{P}^2)^{-1})$$
$$\leqslant C_1 + N(t, \tilde{F} - C_2) \leqslant C_1 + N(t, F_0 - C_2) \leqslant C_1 + N(t + C_3, \widehat{A}_{11}),$$

since $\widehat{A}_{11} - F_0 = \widehat{A}_{12}\widehat{A}_{22}^{-1}\widehat{A}_{12}^*$ is a bounded operator. Similarly,

$$(17.29) \qquad N(0, t^{-1} - P_0 + C_\epsilon P_0^2) \geqslant -C_1 + N(t - C_3, \widehat{A}_{11});$$

and (17.25)–(17.29) imply (17.24). ∎

Remark. If $\partial\Omega \in C^\infty$, then both the formula

$$N(t, \widehat{A}_{11}) = c_0 t^{n/(2\ell)} + O(t^{(n-1)/(2\ell)}),$$

and, with additional global assumptions, the second term of the asymptotics

$$N(t, \widehat{A}_{11}) = c_0 t^{n/(2\ell)} + c_1 t^{(n-1)/(2\ell)} + o(t^{(n-1)/(2\ell)})$$

are well-known [see the bibliography review]. Theorem 17.2 implies the same formulas for $N((\lambda, t), \widehat{A})$.

§18 Linearized Stationary Navier–Stokes System

18.1. In §15 we studied general problems with solvable constraints; in this section, for the sake of brevity, we consider the simplest model as an example of a problem with an unsolvable constraint. The scheme of this section enables us to analyze general problems with an unsolvable constraint as well; but in the latter case, the conditions and the proofs of some auxiliary propositions become more cumbersome to formulate.

18.2. Let $\Omega \subset R^3$ be a bounded Lipschitz domain, let $V_1 \subset H^1(\Omega)^3$ be a closed subspace containing $\overset{\circ}{H}{}^1(\Omega)^3$, let

$$V = \{u \in V_1 \mid \nabla^* u = 0\}, \qquad A[u] = \sum_{i=1,2,3} \|\nabla u_i\|^2_{L_2(\Omega)^3},$$

where $u = (u_1, u_2, u_3)$ is a form in V_1, and let L_0 be a closure V in $L_2(\Omega)^3$ and $\widehat{A}$ be an operator associated with the variational triple $\mathcal{A}$, V, L_0. If $V_1 = \overset{\circ}{H}{}^1(\Omega)^3$ and if $\partial\Omega$ is piecewise smooth, then spectrum of the operator $\widehat{A}$ is a spectrum of the problem

$$\begin{cases} -\Delta u = tu + \nabla p, \\ \nabla^* u = 0, \\ u\big|_{\partial\Omega} = 0. \end{cases}$$

Since the embedding $H^1(\Omega) \subset L_2(\Omega)$ is compact and

$$(18.1) \qquad A[u] \geqslant c\|u\|_1^2 - C\|u\|_0^2,$$

the spectrum of the operator $\widehat{A}$ is discrete [and obviously non-negative].

Theorem 18.1. a) *As $t \to \infty$, $\forall \delta \in (0, \frac{1}{2})$,*

$$(18.2) \qquad N(t, \widehat{A}) = \frac{1}{3\pi^2} \operatorname{mes} \Omega t^{3/2} + O(t^{(3-\delta)/2}).$$

b) *If $\partial\Omega$ is piecewise smooth, in the sense of part c) of Theorem 13.1, then estimate (18.2) holds for any $\delta \in (0, \frac{2}{3})$.*

Proof. If V_1 has the form (16.1), then we have a particular case of the problem with a solvable constraint that we considered in §16; but in the general case, the constraint $\nabla^* u = 0$ is unsolvable, and the asymptotics of the spectrum must be calculated differently.

We choose a function $f \in C^\infty(R^3)$ such that $f(x) = 0 \quad \forall x \in \Omega_\varepsilon$ for some $\varepsilon > 0$, and $f(x) = |x|^2$ outside $\Omega_{2\varepsilon}$; construct the paramatrix of the operator $-\Delta + f(x) : R = R^* \in \mathcal{L}(g, h^2)$; and denote $P = I_3 - \nabla R \nabla^* \in S(g, I_3)$,

$$\overset{o}{q}_t = (h^{-2} + t)^{1/2}, \qquad \overset{o}{a}_{t,w} = (-\Delta + f)I_3 - tP, \qquad q_t = \overset{o}{q}_t^{-1},$$
$$a^0_{t,w} = q_{t,w} \overset{o}{a}_{t,w} q_{t,w}, \qquad d_t = h,$$

where h and g are the same throughout this chapter.

As in §15, we shall reduce the analysis of $N(t, \widehat{A})$ to that of $\mathcal{N}_0(\overset{o}{a}_{t,w}, \Omega)$ in Chapter 2; therefore, it is useful to know the properties of the symbol $\overset{o}{a}_t$ in $\mathcal{M}^\varepsilon = \Omega_\varepsilon \times \{\xi \mid |\xi| \geqslant 1\}$. The principal symbol of the operator p^t in $\mathcal{M}^\varepsilon$ has the form

$$(18.3) \qquad p(x, \xi) = I_3 - |\xi|^{-2} [\xi_j \xi_j]^3_{i,j=1};$$

hence, we can construct a function $a_t = a_t^* \in S(g, I_3)$ such that $a_t - a_t^0 \in S(g, h)$, and for $(x, \xi) \in \mathcal{M}^\varepsilon$,

$$(18.4) \qquad a_t(x, \xi) = \tilde{f}_t(x, \xi)(1 + t|\xi|^{-2})^{-1} (I_3 - t|\xi|^{-2} p(x, \xi)),$$

where $\tilde{f}_t \asymp 1$. As shown at the end of §16, the function p is microlocally smoothly diagonalizable in $\mathcal{M}^\varepsilon$, therefore a_t satisfies condition B in §7 [see §13], and if $\partial\Omega$ is piecewise smooth, in the sense of part c) of Theorem 13.1, then a_t also satisfies condition C in §7.

Starting with the functions a_t and $d_t = h$, we shall construct the functions $\eta_{t,\pm c}$ and $\mu_{t,\pm c}$ as in Section 14.3; and, also as in §14, we shall use the notation $G_t^{(1)}$ not only for the metric constructed in §11, but also for the metric $h^{-2\delta}g$ [if the boundary is not piecewise smooth].

We notice that, similar to (14.17) and (14.17'),

$$[P, \mu_{t,-c,\ell}], [P, \mu_{t,-c,r}] \in \mathcal{L}(G_t^{(1)}, \lambda_{t,0} h^{\omega/2});$$

and, therefore, (14.22) yields that for $u \in V$,

(18.5)
$$
\begin{aligned}
P\mu_{t,-2,\ell}e_+u &= P\mu_{t,-2,\ell}\mu_{t,-1,r}\ell u + T_{1,t}e_+u \\
&= \mu_{t,-2,\ell}\mu_{t,-1,r}P\ell u \\
&\quad + K_t\mu_{t,-1,r}e_+u + T_{2,t}e_+u \\
&= \mu_{t,-2,\ell}e_+u - \mu_{t,-2,\ell}\mu_{t,-1,r}\nabla R\mu_{t,-1,r}\nabla^*\ell u \\
&\quad + K_t\mu_{t,-1,r}e_+u + T_{3,t}e_+u + T_{4,t}\ell u \\
&= \mu_{t,-2,\ell}e_+u \\
&\quad + K_t\mu_{t,-1,r}e_+u + T_{3,t}e_+u + T_{4,t}\ell u,
\end{aligned}
$$

where $T_{j,t} \in \mathcal{L}^{-\infty}(g,1)$ and $K_t \in \mathcal{L}(G_t^{(1)}, \lambda_{t,0}h^{\omega/2})$.

Condition (18.1) allows us to repeat the arguments from §14; by also applying (18.5), we obtain instead of (14.33) [see also the arguments following this formula]

$$
N(t,\widehat{A}) \leqslant t^{\chi} + \mathcal{N}(Q_{t,1}, \mathcal{H}(h^{-1}, C^3)) + \mathcal{N}_0(Q_{t,-1}, \Omega),
$$

where $\chi = (3-1)/2 = 1$, where $Q_{t,1}$ is defined as in §14 for $\ell_1 = 1$ and $\ell_2 = 0$, and where $Q_{t,-1} = -\Delta - tP + K_t - t^{-1}$ for $K_t \in \mathcal{L}(G_t^{(1)}, \lambda_{t,0}h^{\omega/2}\overset{\circ}{q}_t^2)$.

As we showed in §14,

$$
\mathcal{N}(Q_{t,-1}, \mathcal{H}(h^{-1}; C^3)) \leqslant Ct^{(3-\delta)/2};
$$

and the arguments in Chapter 2 give the same estimate for $\mathcal{N}_0(Q_{t,-1}, \Omega)$ as for $\mathcal{N}_0(-\Delta - tP, \Omega)$, but the asymptotics of the latter function are calculated in Theorem 13.2 and have the form

$$
\mathcal{N}_0(-\Delta - tP, \Omega) = c_0 t^{3/2} + O(t^{(3-\delta)/2}),
$$

since $3 > 2 + \delta$. Since, by (18.4), the symbol a_t is defined in $\mathcal{M}^\varepsilon$ and since, by (18.3), $\rho(x,\xi)$ is a projector with a two-dimensional image, then formula (13.11) yields $c_0 = \operatorname{mes} \Omega/(3\pi^2)$. Thus,

$$
N(t,\widehat{A}) \leqslant \frac{1}{3\pi^2} \operatorname{mes} \Omega t^{3/2} + Ct^{(3-\delta)/2}.
$$

18.3. In order to obtain a similar estimate from below, we denote $d_t = h$, $\overset{\circ}{q}_t = (h^{-2} + t)^{1/2}$, $a_t(x,\xi) = (|\xi|^2 + f(x) - t)\overset{\circ}{q}_t(x,\xi)^{-2}$, $q_t = \overset{\circ}{q}_t^{-1}$; and starting with the functions a_t and d_t, we construct the functions $\lambda_{t,0}$, $\eta_{t,-c}$, χ_t, the operator $\mathcal{E}_t = \eta_{t,-2,w}\chi_{t,w}\eta_{t,-2,w}$, and the metric G_t as in §9 and §10 when we were proving the estimate from below [if $\partial\Omega$ is not piecewise smooth, i.e., if it does not satisfy the conditions of Theorem 7.3, then $G_t = h^{-2\delta}g$ and $\delta \in (0, \frac{1}{2})$]. Furthermore, we denote $\widetilde{\mathcal{E}}_t = P\mathcal{E}_t I_3 P$ and $\widetilde{\mathcal{E}}_{1,t} = \widetilde{\mathcal{E}}_t^2(3 - 2\widetilde{\mathcal{E}}_t)$. The same arguments as in §9 and §10 show that $\widetilde{\mathcal{E}}_t$ and $\widetilde{\mathcal{E}}_{1,t}$ are operators of the trace class and that "almost all" nonzero eigenvalues ν_{1ti}, of the operator $\widetilde{\mathcal{E}}_{1,t}$, counted with their multiplicity and numbered in any order, belong to $[0,1]$:

$$\text{card}\{i \mid \nu_{1ti} \notin [0,1]\} \leqslant C,$$

where C does not depend on t.

By calculating $\sigma^w(\widetilde{\mathcal{E}}_{1,t})$ as in §9 and §10, we obtain [cf. (9.7)]

$$\sigma^w(\widetilde{\mathcal{E}}_{1,t}) = p(\eta_{t,-2}^2\chi_t)^2(3 - 2\eta_{t,-2}^2\chi_t) + f_t + r_t,$$

where f_t and r_t satisfy the same conditions as in §9 and §10. Since $(\eta_{t,-2}^2\chi_t)^2 \times (3 - 2\eta_{t,-2}^2\chi_t)$ is a scalar function and since $p(x,\xi)$, for $x \in \Omega$ and $|\xi| \geqslant 1$, is a projection with a two-dimensional image, then the arguments in §9, §10, and §13 show that

$$(18.6) \qquad \text{card}\{i \mid \nu_{1ti} \geqslant \tfrac{1}{2}\} = 2c_1 t^{3/2} + O(t^{(3-\delta)/2}),$$

where c_1 is an asymptotical coefficient for a Laplace operator in Ω. Thus, $2c_1 = (3\pi^2)^{-1} \operatorname{mes} \Omega$.

18.4. As in §9 and §10, we shall now use the ASP $\widetilde{\mathcal{E}}_{1,t}$ to construct a subspace $L_t' \subset V_0 = V \cap W$, where $W = \overset{\circ}{H}{}^1(\Omega)^3$, satisfying the condition

$$(18.7) \qquad \mathcal{A}[u] - t\|u\|_0^2 < 0 \quad \forall (0 \neq)u \in L_t',$$

Then we shall estimate $\dim L_t'$ and obtain the estimate from below.

As in Chapter 2, we denote $\pi_t = \lambda_{t,0}^{-1/2}$, $Q_t = q_{t,w}\pi_{t,w}$, and

$$L_{2,t} = \text{span}\{u \mid \widetilde{\mathcal{E}}_{1,t}u = \nu_{1ti}u, \ \nu_{1ti} \geqslant \tfrac{1}{2}\},$$
$$L_{1,t} = \eta_{t,-1,t}PQ_t(L_{2,t}) \subset C_0^\infty(\Omega)^3.$$

In order to obtain a subspace $L_t' \subset V_0$, we need a projection $P_0 : W \to V_0$. The following lemma will help us to construct it.

Lemma 18.1. *There exists a bounded operator $R_0 : L_2(\Omega) \to W$ such that*

$$(18.8) \qquad \nabla^* R_0\big|_{\nabla^\bullet(W)} = I, \qquad \nabla^* R_0\big|_{L_2(\Omega)\ominus\nabla^\bullet(W)} = 0.$$

Proof. It suffices to prove that the codimension of $\nabla^*(W)$ is finite in $L_2(\Omega)$. Let us consider an operator $\nabla : L_2(\Omega) \to W^* = H^{-1}(\Omega)^3$, adjoint to $\nabla^* : W \to L_2(\Omega)$. The well-known estimate

$$\|f\|_{L_2(\Omega)} \leqslant \gamma(\|\nabla f\|_{H^{-1}(\Omega)^3} + \|f\|_{H^{-1}(\Omega)})$$

implies (see, for example, Lions and Magenes [1]) that the codimension of $\nabla^*(W)$ is finite in $L_2(\Omega)$. ∎

Then (18.8) implies that $P_0 = I - R_0\nabla^* : W \to W$ is bounded and $P_0 W = V_0$.

Lemma 18.2. *Let $L_{0,t} = P_0 L_{1,t}$. Then, for $t \geqslant t_0$,*

$$(18.9) \qquad\qquad |\dim L_{0,t} - \dim L_{2,t}| \leqslant Ct.$$

Proof. Let $v \in L_{2,t}$ and $u = P_0 \eta_{t,-1,\ell} P Q_t \widetilde{\mathcal{E}}_{1,t} v \in L_{0,t}$. Provided that $\partial\Omega$ is piecewise smooth [and that $\partial\Omega$ satisfies the conditions of Theorem 7.3], let b_3 be the function constructed in §10; otherwise, let $b_3 = 1$. Then in either case,

$$(18.10) \qquad \lambda_{t,0} = h^{\delta-w} b_3^{-1} \in S(b_3^{-2} g, \lambda_{t,0}), \quad ch^{\delta/2} \leqslant b_3 \leqslant 1.$$

Furthermore, by construction,

$$(18.11) \qquad \sigma^w(\nabla^* P)\big|_{\Omega \times R^n} \in S^{-\infty}(g, 1, I_3, \Omega \times R^n),$$

$$(18.12) \qquad \mathrm{supp}_\infty \sigma^w(\widetilde{\mathcal{E}}_{1,t}) \subset \mathrm{supp}\, \eta_{t,-2} \subset \{X \mid h(X)^{-1} < Ct^{1/2}\} \cap \Omega \times R^n,$$

and (18.12) implies that $\widetilde{\mathcal{E}}_{1,t}$ admits the representation

$$(18.13) \qquad\qquad \widetilde{\mathcal{E}}_{1,t} = \widetilde{\widetilde{\mathcal{E}}}_{1,t} + \widetilde{\widetilde{\mathcal{E}}}'_{1,t},$$

where $\widetilde{\widetilde{\mathcal{E}}}'_{1,t} \in \mathcal{L}^{-\infty}(g,1)$, and $\widetilde{\widetilde{\mathcal{E}}}_{1,t} \in \mathcal{L}(G_t, h^{2s} t^s) \quad \forall s \geqslant 0$.
Now (18.10) and (18.13) yield

$$\partial_i \eta_{t,-1,\ell} P Q_t \widetilde{\mathcal{E}}_{1,t} \in \mathcal{L}(G_t, h^{-\delta/2} h^{\delta/2} t^{\delta/4}) + \mathcal{L}^{-\infty}(g,1) = \mathcal{L}(G_t, t^{\delta/4});$$

therefore, due to Theorem 4.2,

$$(18.14) \qquad\qquad \|\eta_{t,-1,\ell} P Q_t \widetilde{\mathcal{E}}_{1,t} v\|_1 \leqslant Ct^{\delta/4} \|v\|_0.$$

The equality $\eta_{t,-c} \eta_{t,-2c} = \eta_{t,-2c}$ yields

$$(18.15) \qquad \eta_{t,-1,\ell} P Q_t \widetilde{\mathcal{E}}_{1,t} - P Q_t \widetilde{\mathcal{E}}_{1,t} \in \mathcal{L}^{-\infty}(g,1);$$

therefore, (18.11) and (18.12) imply that

$$\nabla^* \eta_{t,-1,\ell} P Q_t \widetilde{\mathcal{E}}_{1,t} \in \mathcal{L}^{-\infty}(g,1),$$

and

$$(18.16) \qquad\qquad u = \eta_{t,-1,\ell} P Q_t \widetilde{\mathcal{E}}_{1,t} v - R_0 r_+ T_t v,$$

where $T_t \in \mathcal{L}^{-\infty}(g,1)$. Since $R_0 : L_2(\Omega) \to H^1(\Omega)^3$ is bounded, Lemma 14.4 yields the estimate

$$(18.17) \qquad\qquad \|R_0 r_+ T_t v\|_1 \leqslant Ct^{-2} \|v\|_0,$$

for v belonging to some subspace $L_t \subset L_{2,t}$ of codimension not higher than Ct.

Now (18.14), (18.16), and (18.17) yield that for the same v, (18.18)
$$C\|u\|_1^2 \geqslant A[u] = \langle Q_t^* P \eta_{t,-1,r}(-\Delta)\eta_{t,-1,\ell} PQ_t \tilde{\mathcal{E}}_{1,t} v, \tilde{\mathcal{E}}_{1,t} v \rangle + O(t^{-2+\delta/4}\|v\|^2).$$

Then (18.10) implies that $Q_t \in \mathcal{L}(b_3^{-2}g, q_t\lambda_{t,0}^{-1/2})$, and since $P \in \mathcal{L}(g, I)$, we have
$$[P, Q_t] \in \mathcal{L}(b_3^{-2}g, q_t\pi_t b_3^{-1}h) \subset \mathcal{L}(b_3^{-2}g, q_t\pi_t h^{\delta+\omega}).$$

Furthermore,
$$\sigma^w(PP - P)\big|_{\Omega \times R^n} \in S^{-\infty}(g, I, \Omega \times R^n);$$

therefore,

$$(18.19) \qquad PQ_t\tilde{\mathcal{E}}_{1,t} = Q_t\tilde{\mathcal{E}}_{1,t} + K_t\tilde{\mathcal{E}}_{1,t},$$

where $K_t \in \mathcal{L}(b_3^{-2}g, q_t\pi_t h^{\delta+\omega})$.

Finally, we can replace $-\Delta$ with $-\Delta + f(x)$ in (18.18) by adding a term of class $\mathcal{L}^{-\infty}(g, 1)$; hence, (18.15) and (18.19) and Lemma 14.4 allow us to rewrite (18.18) as:

$$(18.20) \;\; C\|u\|_1^2 \geqslant A[u] = \langle(Q^*(-\Delta+f)Q+K_t)\tilde{\mathcal{E}}_{1,t} v, \tilde{\mathcal{E}}_{1,t} v\rangle + O(t^{-2+\delta/4}\|v\|_0^2),$$

where $K_t \in \mathcal{L}(b_3^{-2}g, \pi_t^2 h^{\delta+\omega})$ [and v belongs to a space $L_t \subset L_{2,t}$ of codimension not exceeding Ct].

But
$$Q^*(-\Delta + f)Q + K_t - 4I \in \mathcal{L}I^+(b_3^{-2}g, \pi_t^2);$$

therefore, Lemma 8.4 allows us to deduce from (18.20) the estimate

$$C\|u\|_1^2 \geqslant 4\|\tilde{\mathcal{E}}_{1,t} v\|_0^2 - Ct^{-2+\delta/4}\|v\|_0^2 \geqslant \|v\|_0^2(1 - Ct^{-2+\delta/4}),$$

which yields estimate (18.9) [for $t \geqslant t_0$].   ∎

The idea used in the proof of estimate (18.9) allows us to reduce the proof of estimate (18.7) to those in Chapter 2.

Now (18.16), (18.17), (18.15), and (18.19) yield for $t \geqslant t_0$

$$
\begin{aligned}
(18.21) \quad \|u\|_0^2 &\geqslant (\|(Q_t + K_t)\tilde{\mathcal{E}}_{1,t} v + T_{1,t} v\| - C\|T_{2,t} v\|)^2 \\
&= \langle(Q_t^* Q_t + K_{1,t})\tilde{\mathcal{E}}_{1,t} v, \tilde{\mathcal{E}}_{1,t} v\rangle \\
&\quad + \langle T_{3,t} v, v\rangle - 2C\|((Q_t + K_t)\tilde{\mathcal{E}}_{1,t} + T_{1,t})v\| \, \|T_{2,t} v\| \\
&\geqslant \langle(Q_t^* Q_t + K_{1,t})\tilde{\mathcal{E}}_{1,t} v, \tilde{\mathcal{E}}_{1,t} v\rangle \\
&\quad - t^{-4}\langle(\tilde{\mathcal{E}}_{1,t}(Q_t^* Q_t + K_{1,t})\tilde{\mathcal{E}}_{1,t} v, v\rangle - Ct^8\langle T_{4,t} v, v\rangle,
\end{aligned}
$$

where $T_{j,t} \in \mathcal{L}^{-\infty}(g,1)$ and $K_{1,t} \in \mathcal{L}(b_3^{-2}g, q_t^2 \pi_t^2 h^{\delta+\omega})$. (18.13) implies

$$\widetilde{\mathcal{E}}_{1,t}(Q_t^* Q_t + K_{1,t})\widetilde{\mathcal{E}}_{1,t} \in \mathcal{L}(G_t, h^{-2-\delta}h^{2+\delta}t^{1+\delta/2});$$

and due to Theorem 4.2, the norm of this operator in L_2 is estimated by $Ct^{1+\delta/2}$. This gives us an estimate for the second term in (18.21); by estimating the third one with the help of Lemma 14.4, we see that for v from the subspace $L_t \subset L_{2,t}$ of codimension not higher than Ct, the following estimate holds:

$$t\|u\|_0^2 \geqslant t(\langle (Q_t^* Q_t + K_{1,t})\widetilde{\mathcal{E}}_{1,t}v, \widetilde{\mathcal{E}}_{1,t}v\rangle - Ct^{-1}\|v\|_0^2.$$

For the same v, (18.20) implies the estimate

(18.22)
$$\mathcal{A}[u] - t\|u\|_0^2 \leqslant \langle (Q_t^*(-\Delta + f - t)Q_t\widetilde{\mathcal{E}}_{1,t}v, \widetilde{\mathcal{E}}_{1,t}v) + \langle K_t v, v\rangle + C_1 t^{-1}\|v\|_0^2$$
$$= \langle Q_t^* \overset{\circ}{a}_{t,w}^- Q_t \widetilde{\mathcal{E}}_{1,t}v, \widetilde{\mathcal{E}}_{1,t}v\rangle - 4\langle Q_t^*(q_t^{-2}\lambda_{t,0})_w Q_t\widetilde{\mathcal{E}}_{1,t}v, \widetilde{\mathcal{E}}_{1,t}v\rangle$$
$$+ \langle K_t v, v\rangle + C_1 t^{-1}\|v\|_0^2,$$

where $\overset{\circ}{a}_{t,w}^- = -\Delta + f - t + 4(q_t^{-2}\lambda_{t,0})_w$, and $K_t \in \mathcal{L}(G_t, h^\omega)$, where $\omega > 0$, since $\pi_t^2 h^{\delta+\omega} \leqslant h^\omega$.

As we have shown in Chapter 2,

$$Q_t^*(q_t^{-2}\lambda_{t,0})_w Q_t = I + K_{2,t}, \qquad Q_t^* \overset{\circ}{a}_{t,w}^- Q_t = \pi_{t,w} a_{t,w}^- \pi_t + K_{3,t},$$

where $K_{j,t} \in \mathcal{L}(G_t, h^\omega)$ and $a_t^- = a_t + 4\lambda_{t,0}$; therefore, (18.22) implies the estimate

(18.23)
$$\mathcal{A}[u] - t\|u\|_0^2 \leqslant \langle \widetilde{\mathcal{E}}_t(3 - 2\widetilde{\mathcal{E}}_t)P\eta_{t,-2,w}\chi_{t,w}\eta_{t,-2,w}P\pi_{t,w}a_{t,w}^- \pi_{t,w}$$
$$\times P\eta_{t,-2,w}\chi_{t,w}\eta_{t,-2,w}P\widetilde{\mathcal{E}}_t(3 - 2\widetilde{\mathcal{E}}_t)v, v\rangle$$
$$- \|v\|_0^2(1 - C_1 t^{-1}) + \langle K_t v, v\rangle,$$

where $K_t \in \mathcal{L}(G_t, h^\omega)$. As shown in §9 and §10, by calculating $[\pi_{t,w}P\eta_{t,-2,w}, \chi_{t,w}]$, we can assume that $a_{t,w}^- \in \mathcal{L}(G_t, \lambda_{t,0})$; therefore, in the first term of (18.23), we can obtain the factor $\chi_{t,w}a_{t,w}^-\chi_{t,w}$ [here, of course, $K_t \in \mathcal{L}(G_t, h^\omega)$ will also change]. But in §9 and §10, we have also shown that $\chi_{t,w}a_{t,w}^-\chi_{t,w} \leqslant K_{1,t} \in \mathcal{L}(G_t, h^\omega)$; therefore, (18.23) yields the estimate

$$\mathcal{A}[u] - t\|u\|_0^2 \leqslant \langle (-I + C_1 t^{-1} + K_t)v, v\rangle,$$

where $K_t \in \mathcal{L}(G_t, h^\omega)$.

By applying Lemma 8.2 to the operator $K_t \in \mathcal{L}(G_t, h^\omega)$, we see that for $t \geqslant t_0$, (18.7) holds for v belonging to a subspace $L_t \subset L_{2,t}$ of codimension not higher than Ct, and the estimate from below for $N(t, \widehat{A})$ follows from (18.9) and (18.6).

§19 Asymptotics for Eigenfrequencies of a Shell in a Vacuum

We shall consider a thin shell of a relative thickness h [to avoid confusion with
the notation from previous sections, the function h characterizing the metric g
will now be denoted by h_g]. On the middle surface S of this shell, we introduce
an orthogonal system of coordinates (x_1, x_2) such that the coordinate lines
coincide with the lines of curvature. Let $\Omega \times R^2$ [a range of parameters (x_1, x_2)]
be a bounded Lipschitz domain. We shall assume that the coefficients A, B
of the first quadratic form of the surface S and its principle curvatures k_1, k_2
belong to $C^\infty(\overline{\Omega})$.

Let $L(\Omega)$ be a Hilbert space with the scalar product $\langle \cdot, \cdot \rangle_1 = \langle AB \cdot, \cdot \rangle_{L_2}$,
and let $\mathcal{J}_h$ be a quadratic functional of the shell fixed along the edge. In [1],
Aslanyan and Lidskiĭ have shown that $\mathcal{J}_h$ defines a self-adjoint operator $\widetilde{\mathcal{J}}_h$
in $L(\Omega)^3$ that is semibounded from below and coincides with the Friedrichs
extension of the shell theory operator defined in $C_0^\infty(\Omega)^3$.

Let $n_h(\lambda)$ be a distribution function of the spectrum of the operator $\widetilde{\mathcal{J}}_h$.
Then

$$(19.1) \qquad n_h(\lambda) = \mathcal{N}_0(\widehat{L}_h - \lambda; \Omega),$$

where (see Aslanyan and Lydsky [1]) $\widehat{L}_h$ is a DO with the Weyl symbol

$$(19.2) \qquad L_h = \begin{bmatrix} L_{11} + h^2 L'_{11} + L''_{11} & L_{12} + h^2 L'_{12} + L''_{12} \\ L^*_{12} + h^2 L'^*_{12} + L''^*_{12} & L_{22,h} + h^2 L'_{22} \end{bmatrix},$$

where

$$L_{11}(x, \xi) = \begin{bmatrix} \eta_1^2 + \frac{1-\sigma}{2}\eta_2^2 & \frac{1+\sigma}{2}\eta_1\eta_2 \\ \frac{1+\sigma}{2}\eta_1\eta_2 & \frac{1-\sigma}{2}\eta_1^2 + \eta_2^2 \end{bmatrix}$$

$$L^*_{12}(x, \xi) = -i(\eta_1(k_1(x) + \sigma k_2(x)), \eta_2(\sigma k_1(x) + k_2(x))),$$
$$L_{22,h}(x, \xi) = \frac{h^2}{12}|\eta|^4 + K_1(x)^2 + 2\sigma k_1(x)k_2(x) + k_2(x)^2,$$
$$(19.3) \qquad \eta_1 = A(x)^{-1}\xi_1, \qquad \eta_2 = B(x)^{-1}\xi_2,$$

and where L'_{11}, L''_{11}, L'_{12}, L''_{12}, L'_{22} are polynomials with respect to ξ of degree
not higher than 2, 1, 3, 0, 2, respectively, and $\sigma < \frac{1}{2}$ is a Poisson coefficient.

We shall demonstrate how to analyze asymptotics of the function $n_h(\lambda)$ as
$h \to 0$, for fixed λ, by using Theorem 7.4. Since Ω is a Lipschitz domain, the
functions A, B, k_1, k_2 can be continued to real functions of the class $C^\infty(R^2)$,
which are constant outside some circle, such that $A, B > 0$ everywhere. Let

us construct the function $p \in C^\infty(R^2)$ such that $p(x) = |x|^2$ if $|x| \geqslant C$ and $p(x) = 0$ if $x \in \Omega$, and denote $t = h^{-2}$,

$$\overset{\circ}{q}_t(x,\xi)^2 = \begin{bmatrix} (1 + |x|^2 + |\xi|^2)I_2 & 0 \\ 0 & h^2|\xi|^4 + |x|^2 + 1 \end{bmatrix},$$

$$e(x,\xi) = \begin{bmatrix} I_2 & -(L_{11}(x,\xi) + |x|^2 + 1)^{-1}L_{12}(x,\xi) \\ 0 & 1 \end{bmatrix},$$

$q_t = e\overset{\circ}{q}_t^{-1}$. Then $e \in SI(g, \overset{\circ}{q}_t^{-1}, \overset{\circ}{q}_t)$ and, hence, $q_t \in SI(g, \overset{\circ}{q}_t^{-1}, I)$, where g is the same metric as throughout this chapter. Obviously, the conditions of Theorem 7.4 hold with $d_t = d_{t,j} = h_g + h$, $a_{t,1} \in SI^+(g, I_2)$, $a_{t,1} \geqslant cI_2 > 0$, and

(19.4) $\quad a_{t,2}|_{\mathcal{M}^\epsilon}(x,\xi) = (1 + h^2|\xi|^4)^{-1}(\frac{h^2}{12}|\eta|^4 + G(x,\varphi) - \lambda),$

where $\mathcal{M}^\epsilon = \Omega_\epsilon \times \{\xi \mid |\xi| \geqslant 1\}$, $\varphi = \arcsin \eta_2/|\eta|$, and

$$G(x,\varphi) = (1 - \sigma^2)[k_1(x)\sin^2 \varphi + k_2(x)\cos^2 \varphi]^2.$$

Since $a_{t,1} \geqslant cI_2 > 0$, we have

(19.5) $\quad V(\Omega \times R^n, a_{t,1}) + O((RV)_c^{\delta_1}(\Omega, a_{t,1}, d_t + d_{t,1})) = O(1).$

If $G(x,\varphi) > \lambda$ for all $(x,\varphi) \in \overline{\Omega} \times S_1$, then $a_{t,2}(x,\xi) \geqslant c > 0$ for all $(x,\xi) \in \mathcal{M}^\epsilon$ [for small $\epsilon > 0$], and estimate (19.5) also holds for $a_{t,2}$. Therefore, in this case, $n_h(\lambda) = O(1)$.

If, on the other hand, the last condition does not hold, then by rewriting (19.4):

$$a_{t,2}|_{\mathcal{M}^\epsilon}(x,\xi) = (t + |\xi|^4)^{-1}(1 - t|\eta|^{-4} \cdot 12(\lambda - G(x,\varphi))),$$

we see that the second term in formula (7.14) has already been estimated in Theorem 13.2. (19.5) yields the following theorem:

Theorem 19.1.   a) *Let $\mathcal{X} = 0$, provided that $\lambda = G(x,\varphi)$ on a set of nonzero measure. Otherwise, let $\mathcal{X} \in [0, 1)$ be a number such that*

(19.6) $\quad \displaystyle\int\limits_{\Omega \times S_1} |\lambda - G(x,\varphi)|^{-\mathcal{X}} dx\, d\varphi < \infty$

Then, as $h \to 0$, $\forall \delta \in (0, \frac{1}{2})$,

(19.7) $\quad n_h(\lambda) = \dfrac{\sqrt{3}}{4\pi^2 h} \displaystyle\iint\limits_{\Omega} \int\limits_0^{2\pi} \mathrm{Re}\,\sqrt{\lambda - G(x,\varphi)}\, d\varphi\, dS(x) + O(h^{-\rho(\delta)}),$

where $dS(x) = (AB)(x)dx$ and

(19.8) $\rho(\delta) = 2 \max\{(2 - \delta)/4,\ 2(1 - \mathcal{X})/4 + \delta - 4\mathcal{X})\}.$

b) *If $\partial\Omega$ is a union of smooth transversally intersecting arcs $\Gamma_i = \{x \mid f_i(x) = 0\}$, $f_i \in C^\infty(R^2)$, $df_i(x) \neq 0$ for $x \in \Gamma_i$, and if $G(x, \varphi) - \lambda = 0$ and $x \in \Gamma_{1_i} \cap \cdots \cap \Gamma_{i_s}$, then the vectors $\nabla_{x,\varphi} G(x, \varphi), (\nabla_x f_{i_1}(x), 0), \ldots, (\nabla_x f_{i_s}(x), 0)$ are linearly independent; and estimate (19.7) holds for any $\delta \in (0, \frac{2}{3})$ $[\rho(\delta)$ and $\mathcal{X}$ are defined by formulas (19.8) and (19.6)].*

c) *If $G(x, \varphi) > \lambda \quad \forall(x, \varphi) \in \overline{\Omega} \times S_1$, then $n_h(\lambda) = O(1)$.*

Remark. The statements of parts a) and b) are also valid for other shell theory functionals; differences between the proofs are the same as in §13 and §14.

CHAPTER 4

OPERATORS IN UNBOUNDED DOMAINS

In this chapter, for the sake of brevity, we shall only consider Schrödinger and Dirac model operators. Of course, this scheme also allows us to apply the theorems in §7 in order to analyze general operators in unbounded domains. (See the Review of the Bibliography.)

§20 Schrödinger Operators With Increasing Potential

20.1. Let $\Omega \times R^n$ be a Lipschitz domain conical at infinity and let $A_{\Omega,h}$ be a Friedrichs extension of the operator

$$A_h^0 = -h^2 \Delta I_m + q(x) : C_0^\infty(\Omega)^m \to L_2(\Omega)^m.$$

We assume that there exist $C, N, k > 0$ and $\gamma \in (-k, 1)$ such that

$$(20.1) \qquad \forall \varepsilon > 0, \quad c_\varepsilon \langle x \rangle^{2k-\varepsilon} \leqslant q(x) \leqslant C \langle x \rangle^N;$$

$$(20.2) \qquad \forall \alpha, \quad |\langle q^{(\alpha)}(x)u, u \rangle_{C^\ell}| \leqslant C_\alpha \langle x \rangle^{-\gamma|\alpha|} \langle q(x)u, u \rangle_{C^\ell}.$$

Let $\mathcal{X} = (k+1)/(2k)$, $\mathcal{X}_1 = (k+\gamma)/(2k)$,

$$\widehat{V}(t, q, \Omega) = (2\pi)^{-n} |V_n| \int_\Omega \mathrm{Tr}(t - q(x))_+^{n/2} dx,$$

where $|V_n|$ is a volume of a unit ball in R^n, and let

$$(20.3) \qquad \forall \varepsilon > 0, \quad c_\varepsilon t^{n\mathcal{X}-\varepsilon} \leqslant \widehat{V}(t, q, \Omega).$$

We notice that inequality (20.3) holds if, for example, $\gamma = 1$ and there exists a ray $\ell \subset \Omega$ such that

$$(20.4) \qquad \forall \varepsilon > 0, \quad \|q(x)\| \leqslant C_\varepsilon \langle x \rangle^{2k+\varepsilon}, \quad \forall x \in \ell.$$

139

Indeed, by using (20.2) and an argument of the type (1.11) with $g_x(y) = \langle x \rangle^{-2}|y|^2$, one can easily prove that (20.4) holds in some conic neighborhood of the ray ℓ.

For the sake of simplicity, we shall formulate and prove the theorem on asymptotics of the function $N(\lambda, A_{\Omega,h})$ under the conditions

$$(20.5) \qquad\qquad n \geqslant 2, \qquad n \geqslant 2(\gamma + k)/3.$$

Theorem 20.1. a) *For any* $\delta \in (0, \frac{1}{3})$,

$$(20.6) \qquad N(\lambda, A_{\Omega,h}) = h^{-n}\widehat{V}(\lambda, q, \Omega)(1 + O((h\lambda^{-\mathcal{X}_1})^\delta)$$

uniformly with respect to $\lambda \geqslant 1$ *and* $h \in (0, 1]$.

b) *If the potential* q *is locally smoothly diagonalizable and, in addition, if there exist a constant* C *and, for any* $x \neq 0$, *a conical neighborhood* U_x *of a point* x *and a function* $e_x = (e_x^*)^{-1} \in C^\infty(U_x; \mathrm{End}\, C^m)$ *that is positively homogeneous of degree 0 such that the matrix* $(e_x^* q e_x)(y)$ *is diagonalizable for* $y \in U_x$ *and* $|y| > C$, *then estimate (20.6) holds for any* $\delta \in (0, \frac{1}{2})$.

c) *If the conditions in part b) hold, if* $\partial\Omega$ *is locally piecewise smooth, in the sense of part c) of Theorem 13.1, and if, for sufficiently large* C, $\partial\widetilde{\Omega}_C = \partial\Omega \cap \{x \mid |x| = C\}$ *is piecewise as smooth as the subset* $\{x \mid |x| = C\}$, *then estimate (20.6) holds for any* $\delta \in (0, \frac{2}{3})$.

Proof. We notice that in Theorems 7.1–7.4 we did not use the fact that the parameter t is scalar, i.e., that t can be considered to be elements of some set. Let

$$t \in \{(\lambda, h) \mid \lambda \geqslant 1,\ h \in (0, 1]\}.$$

We fix a small $\varepsilon > 0$ and denote $k_\varepsilon = k(1 - \varepsilon)$,

$$g_{t,x,\xi}(y, \eta) = (|\xi|^2 + h^{-2}\langle x \rangle^{2k_\varepsilon})^{-1}|\eta|^2 + \langle x \rangle^{-2\gamma}|y|^2,$$

$$\overset{\circ}{q}_t(x, \xi) = (h^2|\xi|^2 + q(x) + \lambda)^{1/2}, \qquad \overset{\circ}{q}_t = \overset{\circ}{q}_t^{-1}, \qquad d_t = h_t = h_{g_t},$$

$$a_t(x, \xi) = q_t(x, \xi)^*(h^2|\xi|^2 + q(x) - \lambda)q_t(x, \xi).$$

Since

$$(20.7) \qquad\qquad h_t(X) \leqslant C \min\{\langle X \rangle^{-c}, h\langle x \rangle^{-k_\varepsilon - \gamma}\},$$

where $c = \min\{1, k_\varepsilon + \gamma\} > 0$, then (20.1), (20.2), and Lemmas 1.2, 1.3, 3.2, and 3.3 imply that, uniformly with respect to t, g_t is a σ-temperate metric, $\overset{\circ}{q}_t \in O(C^m; g_t)$, $A_h^0 - \lambda \in \mathcal{L}(g_t, \overset{\circ}{q}_t^*, \overset{\circ}{q}_t)$, and conditions (7.1), (7.2), (7.4), and (7.5) hold; obviously, condition (7.3) holds, too.

Therefore, Theorem 7.1 applies, and the principal term of the asymptotics of the function $N(\lambda, A_{\Omega,h})$ is equal to

$$(20.8) \qquad\qquad V(\Omega \times R^n, a_t) = h^{-n}\widehat{V}(\lambda, q, \Omega).$$

Now we estimate the remainder in formula (7.6). (20.1) and (20.7) yield

(20.9)
$$\mathcal{J}_c^\delta(a_t, d_t) \leqslant C_1 \iint\limits_{|h\xi|^2 + \langle x \rangle^{2k_\epsilon} < C\lambda} \langle x \rangle^{-\gamma\delta} h^\delta (h|\xi| + \langle x \rangle^{k_\epsilon})^{-\delta} dx\, d\xi$$

$$\leqslant C_2 h^{\delta - n} \int\limits_{|x|^{k_\epsilon} < C\lambda^{1/2}} \langle x \rangle^{nk_\epsilon - \gamma\delta - \delta k_\epsilon} dx \leqslant C_3 h^{\delta - n} \lambda^{\omega + \epsilon'},$$

where $C_j = C_j(\varepsilon)$, $\omega = n\mathcal{X} - \delta\mathcal{X}_1$, and $\varepsilon' \to 0$ as $\varepsilon \to 0$.

Since Ω is a Lipschitz domain conical at infinity, then for the unit ball $B_0 = V_n$ and for any spherical layer $B_j = \{x \mid 2^{j-1} < |x| < 2^j\}$, $j \geqslant 1$, we have

$$V((\partial\Omega \times R^n)(c, h_t^{-2\delta} g_t) \cap B_j, a_t - c(d_t + h_t^\delta))$$
$$\leqslant Ch^\delta 2^{-j\delta(k_\epsilon + \gamma)} V((\Omega \times R^n)(c, h_t^{-2\delta} g_t) \cap B_j, a_t - c(d_t + h_t^\delta));$$

hence, by summing with respect to j, we obtain

(20.10) $$V((\partial\Omega \times R^n)(c, h_t^{-2\delta} g_t), a_t - c(d_t + h_t^\delta)) \leqslant C(\varepsilon) h^{\delta - n} \lambda^{\omega + \epsilon'}.$$

Furthermore,

(20.11) $$W_c(\Omega \times R^n, a_t, d_t + h_t^\delta) \leqslant \sum_{1 \leqslant j \leqslant m} \operatorname{mes} \mathcal{M}_{j,\delta,t}^\epsilon,$$

where
(20.12)
$$\mathcal{M}_{j,\delta,t}^\epsilon = \{(x, \xi) \in \Omega \times R^n \mid |x|^{2k_\epsilon} + |h\xi|^2 < C(\varepsilon)\lambda, \big||h\xi|^2 + \lambda_j(x) - \lambda\big|$$
$$\leqslant C(\varepsilon) h^\delta \langle x \rangle^{-\delta(\gamma + k_\epsilon)}(|h\xi|^2 + \lambda_j(x) + \lambda)\},$$

and λ_j are eigenvalues of the matrix q. If C is large and $|h\xi|^2 + \lambda_j(x) > C\lambda$, then $(x, \xi) \notin \mathcal{M}_{j,\delta,t}^\epsilon$; therefore, λ can replace the last factor in (20.12).
Let

$$\mathcal{M}_{j,\delta,t}^{1,\epsilon} = \{(x, \xi) \in \mathcal{M}_{j,\delta,t}^\epsilon \mid |\lambda_j(x) - \lambda| \leqslant C_1 \lambda h^\delta \langle x \rangle^{-\delta(\gamma + k_\epsilon)}\},$$
$$\mathcal{M}_{j,\delta,t}^{2,\epsilon} = \mathcal{M}_{j,\delta,t}^\epsilon \backslash \mathcal{M}_{j,\delta,t}^{1,\epsilon}.$$

Then

(20.13)
$$\operatorname{mes} \mathcal{M}_{j,\delta,t}^{1,\epsilon} \leqslant C'(\varepsilon) h^{-n + n\delta/2} \lambda^{n/2} \int\limits_{|x|^{k_\epsilon} < C\lambda^{1/2}} dx\, \langle x \rangle^{-n\delta(\gamma + k_\epsilon)/2}$$

$$\leqslant C''(\varepsilon) h^{-n + \delta} \lambda^{\omega + \epsilon'};$$

and if C_1 is large and $(x, \xi) \in \mathcal{M}_{j,\delta,t}^{2,\varepsilon}$, then

$$(|\lambda - \lambda_j(x)|(1 \pm C(\varepsilon)\lambda h^\delta \langle x \rangle^{-\delta(\gamma+k_\varepsilon)}|\lambda - \lambda_j(x)|^{-1}))^{n/2}$$
$$= |\lambda - \lambda_j(x)|^{n/2} + O(|\lambda - \lambda_j(x)|^{-1+n/2}\lambda h^\delta \langle x \rangle^{-\delta(\gamma+k_\varepsilon)}),$$

$$\operatorname{mes} \mathcal{M}_{j,\delta,t}^{2,\varepsilon} \leqslant C_1(\varepsilon) h^{-n+\delta} \int\limits_{|x|^{k_\varepsilon} < C\lambda^{1/2}} |\lambda - \lambda_j(x)|^{-1+n/2}\lambda \langle x \rangle^{-\delta(\gamma+k_\varepsilon)} dx$$

$$\leqslant C_2(\varepsilon) h^{-n+\delta} \int\limits_{|x|^{k_\varepsilon} < C\lambda^{1/2}} \lambda^{n/2} \langle x \rangle^{-\delta(\gamma+k_\varepsilon)} dx \leqslant C_3(\varepsilon) h^{-n+\delta}\lambda^{\omega+\varepsilon'}.$$

Here we took into account the fact that $(20.5)_1$ yields $-1 + n/2 \geqslant 0$.

Now (20.8)–(20.14) and (20.3) imply (20.6) under the conditions in part a).

If the conditions in part b) hold, then we can apply Theorem 7.2, having constructed a set $\Gamma\Omega$ in the form $\Gamma' \times R_\xi^n$, where Γ' is a union of Lipschitz surfaces that are conic outside some compact set. Then estimate (20.10) holds with $\Gamma\Omega$ instead of $\partial\Omega \times R^n$, and we obtain (20.6) for any $\delta \in (0, \frac{1}{2})$.

Finally, under the conditions in part c), we can construct Γ' as a union of transversally intersecting smooth surfaces that are conic outside some compact set and can assume that $\Gamma\Omega = \Gamma' \times R_\xi^n$; and we can construct a function $f \in C^\infty(R^n)$ that vanishes on Γ', coincides with functions positively homogeneous of degree zero [outside some compact set], and is locally representable by a product of coordinate functions. Then in order to estimate the remainder in Theorem 7.3, we must estimate measures of the sets

$$\mathcal{N}_{j,\delta,t}^\varepsilon = \{(x,\xi) \, \big| |h\xi| + \langle x \rangle^{k_\varepsilon} < C\lambda^{1/2},$$

$$|f(x)|\big||h\xi|^2 + \lambda_j(x) - \lambda\big| < C(\varepsilon)h^\delta \lambda \langle x \rangle^{-\delta(\gamma+k_\varepsilon)}\}.$$

For $|x| > C$, let $f(x) = \tilde{f}(\theta)$, where $\theta = x/|x|$. Let

$$\tilde{\tilde{\mathcal{N}}}_{j,\delta,t}^\varepsilon = \mathcal{N}_{j,\delta,t}^\varepsilon \cap \{(x,\xi) \mid |x| < C\}, \qquad \tilde{\mathcal{N}}_{j,\delta,t}^\varepsilon = \mathcal{N}_{j,\delta,t}^\varepsilon \backslash \tilde{\tilde{\mathcal{W}}}_{j,\delta,t},$$

$$\tilde{\mathcal{N}}_{j,\delta,t}^{\varepsilon,3} = \{(x,\xi) \in \tilde{\mathcal{N}}_{j,\delta}^\varepsilon \mid |\tilde{f}(\theta)| < h^\delta |x|^{-\delta(\gamma+k_\varepsilon)}\},$$

$$\tilde{\mathcal{N}}_{j,\delta,t}^{\varepsilon,1} = \{(x,\xi) \in \tilde{\mathcal{N}}_{j,\delta,t}^\varepsilon \backslash \tilde{\mathcal{W}}_{j,\delta,t}^{\varepsilon,3} \mid |\lambda_j(x) - \lambda| < C(\varepsilon)h^\delta \lambda |x|^{-\delta(\gamma+k_\varepsilon)}\},$$

$$\tilde{\mathcal{N}}_{j,\delta,t}^{\varepsilon,2} = \tilde{\mathcal{N}}_{j,\delta,t}^\varepsilon \backslash (\tilde{\mathcal{N}}_{j,\delta,t}^{\varepsilon,1} \cup \tilde{\mathcal{N}}_{j,\delta,t}^{\varepsilon,3}).$$

By changing to the spherical coordinate system and by locally representing $\tilde{f}$ by a product of coordinate functions, we obtain, similar to (13.24), $\forall \varepsilon_1 > 0$,

$$\operatorname{mes} \tilde{\mathcal{N}}_{j,\delta,t}^{\varepsilon,3} \leqslant C(\varepsilon_1, \varepsilon) \int\limits_{|h\xi| + |x|^{k_\varepsilon} < C\lambda^{1/2}} h^{\delta-\varepsilon_1} \langle x \rangle^{-\delta(\gamma+k_\varepsilon)+\varepsilon_1} dx \, d\xi$$

$$(20.15)$$

$$\leqslant C'(\varepsilon, \varepsilon') h^{-n+\delta+\varepsilon'}\lambda^{\omega+\varepsilon'},$$

where $\varepsilon' \to 0$ as $\varepsilon, \varepsilon_1 \to 0$.

Furthermore, similar to (20.13),

$$\text{mes}\, \widetilde{\mathcal{N}}_{j,\delta,t}^{\varepsilon,1}$$

$$\leqslant C(\varepsilon) h^{-n+n\delta/2} \lambda^{n/2} \int\limits_{|x|^{k_\varepsilon}<C\lambda^{1/2}} dx \cdot \langle x \rangle^{-n\delta(\gamma+k_\varepsilon)/2} \int\limits_{|\tilde{f}(\theta)|>|x|^{-\delta'(\varepsilon)}h^\delta} |\tilde{f}(\theta)|^{-n/2} d\theta,$$

where $\delta'(\varepsilon) = \delta(\gamma + k_\varepsilon)$.

Let $r = h^\delta |x|^{-\delta'(\varepsilon)}$. Then for any $\varepsilon_2 > \varepsilon_1 > 0$, the inner integral is estimated by

$$C \int\limits_{\substack{r<y_1\cdots y_n<1 \\ 0<y_i<1}} \cdots \int (y_1 \cdots y_n)^{-n/2} dy_1 \cdots dy_n$$

$$\leqslant C(\varepsilon_1) \int\limits_{\substack{r<y_2\cdots y_n<1 \\ 0<y_i<1}} \cdots \int (y_2 \cdots y_n)^{-1+\varepsilon_1} dy_2 \cdots dy_n\, r^{1-n/2+\varepsilon_1}$$

$$\leqslant C(\varepsilon_1) r^{1-n/2+\varepsilon_1} \left(\int\limits_r^1 y_2^{1-\varepsilon_1} dy_2 \right)^{n-1}$$

$$\leqslant C_1(\varepsilon_2) r^{1-n/2-\varepsilon_2};$$

therefore, estimate (20.15) holds also for $\text{mes}\, \widetilde{\mathcal{N}}_{j,\delta,t}^{\varepsilon,1}$.

Finally, similar to (20.14) and (20.16),

$$\text{mes}\, \widetilde{\mathcal{N}}_{j,\delta,t}^{\varepsilon,2} \leqslant C(\varepsilon) h^{-n+\delta} \int\limits_{|x|^{k_\varepsilon}<C\lambda^{1/2}} dx\, \lambda^{n/2} \langle x \rangle^{-\delta'(\varepsilon)} \int\limits_{|\tilde{f}(\theta)|>h^\delta \langle x \rangle^{-\delta'(\varepsilon)}} |\tilde{f}(\theta)|^{-1} d\theta$$

$$\leqslant C(\varepsilon, \varepsilon_1) h^{-n+\delta+\varepsilon_1} \lambda^{n/2} \int\limits_{|x|^{k_\varepsilon}<C\lambda^{1/2}} dx \langle x \rangle^{-\delta'(\varepsilon)+\varepsilon_1},$$

and $\text{mes}\, \widetilde{\mathcal{N}}_{j,\delta,t}^{\varepsilon,2}$ is also estimated by (20.15).

Since $\delta \in (0, \frac{2}{3})$ is arbitrary, we can assume that

$$(20.17) \qquad \text{mes}\, \widetilde{\mathcal{N}}_{j,\delta,t}^{\varepsilon} \leqslant C_\varepsilon h^{-n+\delta} \lambda^{n\mathcal{X}-\delta\mathcal{X}_1}.$$

By defining $\widetilde{\widetilde{\mathcal{N}}}_{j,\delta,t}^{\varepsilon,s}$ similarly to $\widetilde{\mathcal{N}}_{j,\delta,t}^{\varepsilon,s}$ and by using the same arguments [here we do not need to change to the spherical coordinate system], we obtain

$$(20.18) \qquad \text{mes}\, \widetilde{\widetilde{\mathcal{N}}}_{j,\delta,t}^{\varepsilon} \leqslant C_\varepsilon h^{-n+\delta} \lambda^{n/2},$$

but (20.17) and (20.18) imply (20.6) for any $\delta \in (0, \frac{2}{3})$, due to condition (20.5).

§21 Asymptotics of a Discrete Spectrum of Schrödinger Operators and Dirac Operators with Decreasing Potentials

21.1. In this section we shall consider the operator $A_{\Omega,h}$, introduced in the previous section, with a potential satisfying the following condition: for any $\varepsilon > 0$,

$$(21.1) \qquad \|q^{(\alpha)}(x)\| \leqslant C_{\alpha,\varepsilon} \langle x \rangle^{-2k+\varepsilon-|\alpha|}, \quad |\alpha| = 0, 1, \ldots,$$

where $k > 0$; the Schrödinger operator, perturbed by first-order differential operators with decreasing coefficients; and the Dirac operator

$$A_h = \mu\beta + \langle \alpha, -ih\nabla \rangle + q(x),$$

where the potential q satisfies condition (21.1) and where $\mu > 0$ is a mass, $\alpha = (\alpha_1, \alpha_2, \alpha_3)$ is a "vector,"

$$\beta = \begin{bmatrix} I_2 & 0 \\ 0 & -I_2 \end{bmatrix}, \qquad \alpha_j = \begin{bmatrix} 0 & \sigma_j \\ \sigma_j & 0 \end{bmatrix},$$

and σ_j are Pauli matrices. The Dirac operator is considered an unbounded operator in $L_2(R^3)^4$ with the domain $H^1(R^3)^4$.

Lemma 21.1. *For any $h > 0$,*

$$\sigma_{\text{ess}} A_{\Omega,h} = [0, \infty), \qquad \sigma_{\text{ess}} A_h = (-\infty, -\mu] \cup [\mu, \infty).$$

Proof. We shall show that $(0 \leqslant)\lambda \in \sigma_{\text{ess}} A_{\Omega,h}$. For this purpose, we fix $\theta \in C_0^\infty(R^1)$, $0 \leqslant \theta \leqslant 1$, $\theta\big|_{|t|<1/2} = 1$, $\theta\big|_{|t|>1} = 0$, and the points $x^j \in \Omega$ satisfying the conditions $\text{dist}(x^j, \partial\Omega) > j$, $\text{dist}(x^j, x^i) > \max\{i, j\}$ $\forall i \neq j$; and we denote

$$v_j(x) = \theta(j^{-1}|x - x^j|)\{1; 0; \ldots; 0\},$$
$$u_j(x) = \exp(ih^{-1}\sqrt{\lambda}\,x_1)\|v_j(x)\|_{L_2}^{-1} v_j(x).$$

Then $\langle u_j, u_i \rangle = 0, \forall i \neq j, \|u_j\| = 1$ and $|\langle (A_{\Omega,h}-\lambda)u_j, u_j \rangle| \leqslant C_h j^{-\min\{1,2k\}}$. The Weyl criterion implies that $\lambda \in \sigma_{\text{ess}} A_{\Omega,h}$ and $\sigma_{\text{ess}} A_{\Omega,h} \supset [0,\infty)$; but next, in proving the theorem on asymptotics of a discrete spectrum, we shall show that $\sigma_{\text{ess}} A_{\Omega,h} \cap (-\infty, 0) = \emptyset$. Therefore, the first statement of the lemma is proved.

The second statement follows from the obvious equality $\sigma_{\text{ess}}(A_h - q(x)) = (-\infty, -\mu] \cup [\mu, \infty)$, from the compactness of the operator $q(x)(A_h - q(x))^{-1}$, and from the Weyl theorem on the essential spectrum. ∎

In what follows, we shall see that if $k < 1$, then the discrete spectrum can accumulate at the boundary of the essential spectrum, and it makes sense to study the combined asymptotics of the spectrum with respect to spectral and small parameters. On the other hand, if $k > 1$, then for any $h > 0$, the number of points of discrete spectrum is finite, and it only makes sense to study the asymptotics as $h \to +0$.

At the end of this section we shall analyze the asymptotics with respect to spectral parameters only for the Schrödinger operator perturbed by first-order differential operators and for the Dirac operator with a matrix potential. Using the fundamental theorems in §7, we can also analyze the asymptotics with respect to a small parameter; but the calculations become more difficult, and, in general, we only succeed in obtaining a worse estimate for the remainder.

21.2. As in §20, we assume that $n \geqslant 2$.

Theorem 21.1. *In* (21.1) *let* $k \in (0,1)$. *Then*
   a) *For any* $\delta \in (0, \tfrac{1}{3})$, $\tau > 0$,

$$(21.2) \qquad N(-\tau, A_{\Omega,h}) = h^{-n}\widehat{V}(-\tau, q, \Omega) + O((h\tau^{\mathcal{X}})^{\delta-n}),$$

uniformly with respect to $h \in (0,1]$, *where* $\mathcal{X} = (1-k)/(2k)$.
   b) *If* q *satisfies the conditions in part* b) *of Theorem 20.1, then estimate* (21.2) *holds for any* $\delta \in (0, \tfrac{1}{2})$.
   c) *If the conditions in part* c) *of Theorem 20.1 hold, then estimate* (21.2) *holds for any* $\delta \in (0, \tfrac{2}{3})$.

Remark. If there exists an open cone $\widetilde{\Omega} \subset \Omega$ such that for any $\varepsilon > 0$,

$$\min_j \lambda_j(x) \leqslant -C(\varepsilon)\langle x \rangle^{-2k-\varepsilon} \quad \forall x \in \widetilde{\Omega},$$

where λ_j are eigenvalues of the matrix q, then (21.2) implies the estimate

$$(21.3) \qquad N(-\tau, A_{\Omega,h}) = h^{-n}\widehat{V}(-\tau, q, \Omega)(1 + O((h\tau^{\mathcal{X}})^{\delta})).$$

Proof of Theorem 21.1. We fix a small $\varepsilon > 0$ and denote

$$k_\varepsilon = k(1-\varepsilon), \qquad \Lambda_h^2(x,\xi) = |h\xi|^2 + \langle x \rangle^{-2k_\varepsilon},$$
$$\overset{\circ}{q}_t = (\Lambda_h^2 + \tau)^{1/2}, \qquad q_t = \overset{\circ}{q}_t{}^{-1}, \qquad d_t = h_t,$$
$$g_{t,x,\xi}(y,\eta) = g_{h,x,\xi}(y,\eta) = h^2\Lambda_h(x,\xi)^{-2}|\eta|^2 + \langle x \rangle^{-2}|y|^2,$$
$$\overset{\circ}{a}_{t,w} = A_h^0 + \tau, \qquad a_t(x,\xi) = q_t(x,\xi)^2(|h\xi|^2 + q(x) + \tau).$$

As in §20, the parameter $t \in \{(h,\tau) \mid 0 < h < 1, \tau > 0\}$ is not scalar.

Choosing $\varepsilon > 0$ so that $k_\varepsilon < 1$, we have $h_t(X) \leqslant C\langle X \rangle^{-c}h^c$, where $c > 0$; therefore, Lemmas 1.2, 1.3, 3.2, and 3.3 imply that, uniformly with respect to t, g_t is a σ-temperate metric and $\overset{\circ}{q}_t \in O(C^1, g_t)$, $A_h^0 + \tau \in \mathcal{L}(g_t, \overset{\circ}{q}_t^2)$. In addition, $A_h^0 + \tau \in \mathcal{L}I^+(g_t, \overset{\circ}{q}_t^2)$ for any fixed t; therefore, all conditions of Theorem 7.1 hold, and the principal term of the asymptotics of the function $N(-\tau, A_{\Omega, h})$ is

$$V(\Omega \times R^n, a_t) = h^{-n} \widehat{V}(-\tau, q, \Omega).$$

Now (21.1) yields

(21.4)

$$\mathcal{J}_c^\delta(a_t, d_t) \leqslant C(\varepsilon) \iint\limits_{|h\xi|^2 - C(\varepsilon)\langle x \rangle^{-2k_\varepsilon} + \tau < 0} \langle x \rangle^{-\delta} h^\delta \Lambda_h(x, \xi)^{-\delta} dx\, d\xi$$

$$\leqslant C'(\varepsilon) h^{-n+\delta} \int\limits_{|x|^{k_\varepsilon} < C(\varepsilon)\tau^{-1/2}} \langle x \rangle^{-nk_\varepsilon - \delta + \delta k_\varepsilon} dx \leqslant C(\varepsilon, \varepsilon') h^{-n+\delta} \tau^{-(n-\delta)\mathcal{X} - \varepsilon'},$$

where $\varepsilon' \to 0$ as $\varepsilon \to 0$; and since Ω is a Lipschitz domain conical at infinity, then similarly to (20.10) and (21.4), we have

(21.5) $V((\partial \Omega \times R^n)(c, h^{-2\delta}g); a_t - c(d_t + h_t^\delta)) \leqslant c(\varepsilon, \varepsilon') h^{-n+\delta} \tau^{-(n-\delta)\mathcal{X} - \varepsilon'}.$

Furthermore,

(21.6) $$W_c(\Omega \times R^n, a_t, d_t + h_t^\delta) \leqslant \sum_{1 \leqslant j \leqslant m} \operatorname{mes} \mathcal{M}_{j, \delta, t}^\varepsilon,$$

where

$$\mathcal{M}_{j, \delta, t}^\varepsilon = \{(x, \xi) \in \Omega \times R^n \mid |x|^{k_\varepsilon} < C\tau^{-1/2}, |h\xi| < c\langle x \rangle^{-k_\varepsilon},$$

$$\big||h\xi|^2 + \lambda_j(x) + \tau\big| \leqslant C(\varepsilon) h^\delta \langle x \rangle^{-\delta(1-k_\varepsilon) - 2k_\varepsilon}\}.$$

Let

$$\mathcal{M}_{j, \delta, t}^{1, \varepsilon} = \{(x, \xi) \in \mathcal{M}_{j, \delta, t}^\varepsilon \mid |\tau + \lambda_j(x)| \leqslant C_1 h^\delta \langle x \rangle^{-\delta(1-k_\varepsilon) - 2k_\varepsilon}\},$$
$$\mathcal{M}_{j, \delta, t}^{2, \varepsilon} = \mathcal{M}_{j, \delta, t}^\varepsilon \backslash \mathcal{M}_{j, \delta, t}^{1, \varepsilon}.$$

Then

(21.7)
$$\operatorname{mes} \mathcal{M}_{j, \delta, t}^{1, \varepsilon} \leqslant C(\varepsilon) h^{-n+n\delta/2} \int\limits_{|x|^{k_\varepsilon} < C\tau^{-1/2}} dx \langle x \rangle^{-n[\delta(1-k_\varepsilon) + 2k_r]/2}$$

$$\leqslant C'(\varepsilon') h^{-n+\delta} \tau^{-\sigma - \varepsilon'},$$

where $\varepsilon' \to 0$ as $\varepsilon \to 0$, and $\sigma = [n - n(\delta(1-k) + 2k)/2](2k)^{-1} \leqslant (n-\delta)\mathcal{X}$, since $n/2 \geqslant 1$.

If C_1 is large and $(x, \xi) \in \mathcal{M}^{2,\varepsilon}_{j,\delta,t}$, then

$$|\tau + \lambda_j(x)|(1 \pm Ch^\delta \langle x \rangle^{-\delta(1-k_\varepsilon)-2k_\varepsilon} |\tau + \lambda_j(x)|^{-1})^{n/2}$$
$$= |\tau + \lambda_j(x)|^{n/2} + O(|\tau + \lambda_j(x)|^{-1+n/2} h^\delta \langle x \rangle^{-\delta(1-k_\varepsilon)-2k_\varepsilon});$$

therefore,

$$\text{mes } \mathcal{M}^{2,\varepsilon}_{j,\delta,t} \leqslant C(\varepsilon) h^{-n+\delta} \int\limits_{|x|^{k_\varepsilon} < C\tau^{-1/2}} dx |\tau + \lambda_j(x)|^{n/2-1} \langle x \rangle^{-\delta(1-k_\varepsilon)-2k_\varepsilon}$$

$$(21.8) \qquad \leqslant C'(\varepsilon) h^{-n+\delta} \int\limits_{|x|^{k_\varepsilon} < C\tau^{-1/2}} dx \langle x \rangle^{-\delta(1-k_\varepsilon)-2k_\varepsilon} [\tau^{n/2-1} + \langle x \rangle^{-k_\varepsilon(n-2)}]$$

$$\leqslant C''(\varepsilon) h^{-n+\delta} \tau^{-(n-\delta)\mathcal{X}-\varepsilon'},$$

where $\varepsilon' \to 0$ as $\varepsilon \to 0$.

Then (21.4)–(21.8) yield estimate (21.2) under conditions in part a) of Theorem 21.1.

By noting that proof of estimates (21.4)–(21.8) is similar to that of (20.9)–(20.11), (20.13), and (20.14), respectively, and by modifying the proof of parts b) and c) of Theorem 20.1 in the same fashion, we obtain the proofs of parts b) and c) of Theorem 21.1. ∎

21.3. Now we shall consider a Dirac operator A_h with a scalar potential q satisfying condition (21.1) with $k \in (0, 1)$.

Theorem 21.2.   *For any $\delta \in (0, \frac{2}{3})$ and $\mu_2 \in (-\mu, \mu)$,*

$$N((\mu_2, \mu - \tau); A_h) = h^{-3} \widehat{V}((\mu_2, \mu - \tau), a, R^3) + O((h\tau^{\mathcal{X}})^{\delta-3}),$$

uniformly with respect to $h \in (0, 1]$ and $\tau \in (0, (\mu - \mu_2)/2)$, where $\mathcal{X} = (1-k)/(2k)$ and

$$\widehat{V}((\mu_2, \mu_1), a, \Omega) = (2\pi)^{-3} \iint\limits_{\Omega \times R^3} N((\mu_2, \mu_1), a(x, \xi)) dx \, d\xi,$$

$$a(x, \xi) = \mu\beta + \langle \alpha, \xi \rangle + q(x).$$

Proof.   We let

$$E(\xi) = (|\xi|^2 + \mu^2)^{1/2}, \qquad \chi(\xi) = E(\xi) + \mu, \qquad \zeta(\xi) = \langle \sigma, \xi \rangle,$$

$$\omega(\xi) = (|\xi|^2 + \chi^2(\xi))^{-1/2}, \qquad T(\xi) = \omega(\xi) \begin{bmatrix} \chi(\xi) I_2 & \zeta(\xi) \\ \zeta(\xi) & -\chi(\xi) I_2 \end{bmatrix}$$

and note that the unitary matrix $T(\xi)$ reduces $H(\xi) = \mu\beta + \langle\alpha,\xi\rangle$ to the diagonal form $H(\xi) = T^*(\xi)E(\xi)\beta T(\xi)$.

Furthermore, let $\mu_1 = \mu - \tau$, $\tilde{a}_t(x,\xi) = H(h\xi) + q(x)$, $f(x) = (x - (\mu_1 + \mu_2)/2)^2 - (\mu_1 - \mu_2)^2/4$, $\overset{\circ}{a}_{t,w} = f(\tilde{a}_{t,w})$. Then

$$N((\mu_2, \mu - \tau), A_h) = \mathcal{N}(\|(A_h - (\mu_1 + \mu_2)/2) \cdot \|^2_{L_2} - (\mu_1 - \mu_2)^2 \| \cdot \|^2/4, D(A_h))$$
$$= \mathcal{N}_0(\overset{\circ}{a}_{t,w}, R^3),$$

since, due to Theorem 6.5, A_h is the closure of the operator $\tilde{a}_{t,w} : C_0^\infty(R^3)^4 \to L_2(R^3)^4$ [indeed, $\tilde{a}_t \in SI(g_t, \Lambda_h)$, where g_t is the same as in §21.2, and $\Lambda_h(x,\xi) = \langle h\xi\rangle$].

Hence, we can apply Theorem 7.4 with the same metric g_t as in §21.2 and with $\Omega_t = R^3$, $d_t = d_{t,j} = h_t$, and

$$\overset{\circ}{q}_t = [\delta_{ij}\overset{\circ}{q}_{t,i}I_2]_{i,j=1,2}, \qquad q_t(x,\xi) = T^*(h\xi)\overset{\circ}{q}_t(x,\xi)^{-1},$$
$$a_t = [\delta_{ij}\overset{\circ}{q}_{t,i}^{-2}a_{t,i}I_2]_{i,j=1,2},$$

where

$$\overset{\circ}{q}_{t,1}(x,\xi)^2 = |h\xi|^2 + \langle x\rangle^{-2k_\epsilon} + \tau, \qquad \overset{\circ}{q}_{t,2}(x,\xi)^2 = |h\xi|^2 + 1,$$
$$a_{t,1}(x,\xi) = ((|h\xi|^2 + \mu^2)^{1/2} + q(x) - (\mu_1 + \mu_2)/2)^2 - (\mu_2 - \mu_1)^2/4$$
$$= [(|h\xi|^2 + \mu^2)^{1/2} - \mu + \tau + q(x)][(|h\xi|^2 + \mu^2)^{1/2} - \mu_2 + q(x)],$$
$$\text{(21.9)}$$
$$a_{t,2}(x,\xi) = [(|h\xi|^2 + \mu^2)^{1/2} + \mu - \tau - q(x)][(|h\xi|^2 + \mu^2)^{1/2} + \mu_2 - q(x)].$$

Since $\Omega_t = R^3$ and the matrix a_t is diagonal, we can estimate the remainders in formula (7.14) as in Theorem 7.3, with $\varphi_{t,1} = \varphi_{t,2} = 1$, and can write formula (7.14) as

$$N((\mu_2, \mu - \tau), A_h) = V(R^3 \times R^3, a_t) + O(W_c(R^3 \times R^3, a_t, h_t^\delta)).$$

The principal term of the asymptotics is

$$h^{-3}\widehat{V}((-\infty, 0), f(a), R^3) = h^{-3}\widehat{V}((\mu_2, \mu - \tau), a, R^3);$$

and due to (21.9), the remainder is estimated by

$$2\sum_{j=1,2} \text{mes}\{(x,\xi) \mid |a_{t,j}(x,\xi)| \leqslant C_\varepsilon \overset{\circ}{q}_{t,j}(x,\xi)h^\delta\langle x\rangle^{-\delta(1-k_\epsilon)}\}$$
$$= 2h^{-3}\sum_{j=1,2} \text{mes}\{(x,\xi) \mid |a_{t_1,j}(x,\xi)| \leqslant C_\varepsilon \overset{\circ}{q}_{t_1,j}(x,\xi)h^\delta\langle x\rangle^{-\delta(1-k_\epsilon)}\}$$
$$= 2h^{-3}\sum_{j=1,2} \text{mes}\, W_{j,t_1},$$

where $t_1 = (\tau, 1)$. Noting that there exists $C_1 = C_1(\varepsilon)$ such that

1) if $|\xi| > C_1$, then $(x, \xi) \notin W_{1,t_1} \cup W_{1,t_2}$;

2) if $|\xi| > C_1$ and $|x| > C_1$, then $(x, \xi) \notin W_{2,t_1}$;

3) if $|\xi|^2 + \tau > C_1 \langle x \rangle^{-2k_\varepsilon}$, then $(x, \xi) \notin W_{1,t_1}$,

we see that for all $(x, \xi) \in W_{1,t_1} \cup W_{2,t_2}$,

$$\left| (|\xi|^2 + \mu^2)^{1/2} \pm (\mu_j - q(x)) \right| \leqslant C;$$

and, therefore,

$$W_{1,t_1} \cup W_{2,t_2} \subset \left\{ (x, \xi) \,\Big|\, \left||\xi|^2 + \mu^2 - (\mu_1 - q(x))^2\right| \left||\xi|^2 + \mu^2 - (\mu_2 - q(x))^2\right| \right.$$
$$\left. \leqslant C'_\varepsilon h^\delta \langle x \rangle^{-\delta(1-k_\varepsilon)-2k_\varepsilon} \right\}.$$

Thus,
(21.10)
$$W_c(R^3 \times R^3, a_t, h_t^\delta)$$
$$\leqslant C h^{-3} \sum_{j=1,2} \mathrm{mes}\left\{ (x, \xi) \,\Big|\, \left||\xi|^2 + \mu^2 - (q(x) - \mu_j)^2\right| \leqslant C'_\varepsilon h^\delta \langle x \rangle^{-\delta(1-k_\varepsilon)-2k_\varepsilon} \right\}.$$

By considering, along with its complement, a subset on which

$$\left| \mu^2 - (q(x) - \mu_j)^2 \right| \leqslant C_1 h^\delta \langle x \rangle^{-\delta(1-k_\varepsilon)-2k_\varepsilon}, \quad \text{for } j = 1, 2$$

and by estimating their measures as in (21.7) and (21.8), we see that the remainder is $O((h\tau^\mathcal{X})^{-3+\delta})$.   ∎

21.4. Now let the Schrödinger operator $A_{\Omega,h}$ be the same as in §21.2, but let condition (21.1) hold for $k > 1$.

Theorem 21.3. *As $h \to 0$,*

$$(21.11) \qquad N(\mu, A_{\Omega,h}) = h^{-n} \widehat{V}(\mu, q, \Omega) + O(h^{-n+\delta})$$

uniformly with respect to $\mu \leqslant 0$, where δ is defined by the potential q and by the boundary of the domain as in Theorem 21.1.

Proof. We can assume that condition (21.1) holds with $\varepsilon = 0$. Let us fix $\rho < (k-1)^{-1}$, and let

$$\Omega_{c,h} = \{x \in \Omega \mid |x| > ch^{-\rho}\}, \qquad \Omega_{-c,h} = \Omega \backslash \Omega_{c,h}.$$

We then construct a function $\tilde{q}_h$ that is equal to q on $\Omega_{-2,h}$, $h^{-2k\rho} I_m$ on $\Omega_{4,h}$ and that satisfies estimate (21.1) uniformly with respect to $h \in (0,1]$. Obviously,

$$N(\mu, A_{\Omega,h}) \geqslant \mathcal{N}_0(-h^2 \Delta I_m + \tilde{q}_h - \mu, \Omega_{-2,h}),$$

and the asymptotics of the last function can be analyzed with the help of the theorems in §7. Let

$$\overset{\circ}{q}_t(x,\xi)^2 = |h\xi|^2 + \langle x \rangle^{-2k} + h^{2k\rho} - \mu, \qquad q_t = \overset{\circ}{q}_t^{-1},$$

$$g_{t,x,\xi}(y\eta) = h^2\overset{\circ}{q}_t(x,\xi)^{-2}|\eta|^2 + \langle x \rangle^{-2}|y|^2, \qquad d_t = h_{g_t},$$

$$a_t(x,\xi) = q_t(x,\xi)^2(|h\xi|^2 + \tilde{q}_h(x) - \mu), \qquad \Omega_t = \Omega_{-2,h}.$$

Then due to the choice of ρ, $\quad h_{g_t}(X) \leqslant C(\langle x \rangle + h^{-1})^{-c}$, where $c > 0$; therefore, Lemmas 1.2, 1.3, 3.2, and 3.3 imply that, uniformly with respect to $t \in \{(\mu,h) \mid \mu \leqslant 0, \ h \in (0,1]\}$, g_t is a σ-temperate metric and $\overset{\circ}{q}_t \in O(C^1; g_t)$, $-h^2\Delta + \tilde{q}_h - \mu \in \mathcal{L}(g_t, \overset{\circ}{q}_t^2)$. In addition, $a_t(x,\xi) \geqslant c > 0$ for $|x| + |\xi| > C_h$; therefore, all the conditions of Theorem 7.1 hold.

The principal term of the asymptotics is

$$(21.12) \qquad\qquad V(\Omega \times R^n, a_t) = h^{-n}\widehat{V}(\mu, q, \Omega_{-2,h});$$

and since (21.1) and the condition $k > 1$ yield
(21.13)

$$\widehat{V}(\mu, q, R^n \backslash (R^n)_{-c,h}) \leqslant C \int\limits_{|x| > ch^{-\rho}} \langle x \rangle^{-nk} dx \leqslant C_1 h^{n(k-1)\rho} = C_1 h^{n-\varepsilon},$$

where $\varepsilon \to 0$ as $\rho \to (k-1)^{-1}$, then in (21.12) we can replace $\Omega_{-2,h}$ with Ω after adding the term $O(h^{-\varepsilon})$.

Since $h_{g_t}(X) \leqslant Ch\langle x \rangle^{k-1}$, we have

$$
\begin{aligned}
(21.14) \qquad \mathcal{J}_c^\delta(a_t, d_t) &\leqslant C \int\limits_{\Omega_{-4,h}} dx \int\limits_{|h\xi| < C\langle x \rangle^{-k}} h^\delta \langle x \rangle^{\delta(k-1)} dx \\
&\leqslant C_1 h^{-n+\delta} \int\limits_{|x| < 4h^{-\rho}} \langle x \rangle^{-nk+\delta(k-1)} dx \leqslant C_2 h^{-n+\delta}.
\end{aligned}
$$

Furthermore,
(21.15)
$$
\begin{aligned}
V((\partial\Omega_t \times R^n)&(c, h_{g_t}^{-2\delta} g_t), a_t - c(d_t + h_{g_t}^\delta)) \\
&\leqslant V((\partial\Omega \times R^n)(c, h_{g_t}^{-2\delta} g_t) \cap ((R^n)_{-4,h} \times R^n), a_t - c(d_t + h_{g_t}^\delta)) \\
&\qquad\qquad + V((R^n)_{1,h} \times R^n, a_t - c(d_t + h_{g_t}^\delta)) \\
&\overset{\text{def.}}{=} \widetilde{V}_{1,t} + \widetilde{V}_{2,t}.
\end{aligned}
$$

Then (21.13) yields $\widetilde{V}_{2,t} < Ch^{-\varepsilon}$, and $\widetilde{V}_{1,t}$ admits estimate (21.14), since Ω is a Lipschitz domain conical at infinity [compare the proof of estimate (20.10)].

Finally, the same arguments as in (21.6)–(21.8) imply

$$(21.16) \qquad W_c(\Omega_{-2,h} \times R^n, a_t, d_t + h_{g_t}^\delta) \leqslant Ch^{-n+\delta},$$

and (21.12)–(21.16) yield the estimate

$$(21.17) \qquad N(\mu, A_h) \geqslant h^{-n}\widehat{V}(\mu, q, \Omega) - Ch^{-n+\delta}$$

for any $\delta \in (0, \frac{1}{3})$.

If q is locally smoothly diagonalizable, in the sense of part b) of Theorem 20.1, then in (21.15), by replacing $\partial\Omega$ with $\Gamma\Omega$, where $\Gamma\Omega$ is set conical at infinity, by dividing domains with different diagonalizations of the potential q, we obtain (21.17) for any $\delta \in (0, \frac{1}{2})$ [using Theorem 7.2 instead of Theorem 7.1].

Finally, if $\partial\Omega$ is piecewise smooth, in the sense of part c) of Theorem 20.1, then we can apply Theorem 7.3 by constructing a function $f(x)$, such that:

1) locally, f is a product of coordinate functions [in suitable local coordinate systems];

2) for $C < |x| < h^{-\rho}$, f does not depend on radial variable;

3) f satisfies the estimates

$$|\partial^\alpha f(x)| \leqslant C_\alpha \langle x \rangle^{-|\alpha|};$$

4) on the connected components of the set $\Omega \backslash \{x \mid f(x) = 0\}$, the potential q is smoothly diagonalizable.

The term $\mathcal{J}_c^\delta(a_t, d_t)$ in formula (7.12) admits estimate (21.14) for any $\delta \in (0, \frac{2}{3})$, and the second term is estimated by measures of sets that depend on the eigenvalues λ_j of the potential q:

$$\mathcal{N}_{j,t} = \{(x, \xi) \mid |h\xi| < C\langle x \rangle^{-k},$$
$$|f(x)|\big||h\xi|^2 + \lambda_j(x) - \mu\big| < Ch^\delta\langle x \rangle^{\delta(k-1)}(|h\xi|^2 + \langle x \rangle^{-2k} + h^{2k\rho} - \mu),$$
$$|x| < 4h^{-\rho}\}.$$

As in (21.13),

$$\operatorname{mes}\mathcal{N}_{j,t} \cap \{(x, \xi) \mid |x| > h^{-\rho}\} \leqslant Ch^{-\varepsilon};$$

and since for $C < |x| < h^{-\rho}$, f coincides with a function homogeneous of degree zero, then the measure of the set $\mathcal{N}_{j,t} \cap \{(x, \xi) \mid |x| < h^{-\rho}\}$ is estimated in exactly the same way as that of $\mathcal{N}_{j,\delta,t}^\varepsilon$ in the proof of Theorem 20.1, and we obtain an estimate for the last summand in the remainder of Theorem 7.3 through $Ch^{-n+\delta}$, where $\delta \in (0, \frac{2}{3})$. This proves estimate (21.17) for any $\delta \in (0, \frac{2}{3})$.

21.5. In order to obtain the estimate from above, we fix $\theta \in C_0^\infty(R^1)$, $0 \leqslant \theta \leqslant 1$, $\theta|_{|t|<1} = 1$, $\theta|_{|t|>2} = 0$ and let

$$\psi_{-c,h}(x) = \theta(c^{-1}|x|h^\rho), \qquad \psi_{c,h} = 1 - \psi_{-c,h},$$
$$\tilde{\psi}_{\pm c,h} = \psi_{\pm c,h}(\psi_{c,h}^2 + \psi_{-c,h}^2)^{-1/2}, \qquad u_{\pm c,h} = \tilde{\psi}_{\pm c,h}u.$$

Then for any $u \in C_0^\infty(\Omega)^m$,

$$\langle(-h^2\Delta I_m + q(x) - \mu)u, u\rangle$$
$$= \sum_{j=\pm 1} \langle(-h^2\Delta I_m + q(x) - \mu)u_{jh}, u_{j,h}\rangle + h^2(\langle \nabla u, k_h^1 u\rangle + \langle k_h^2 u, u\rangle),$$

where $\|k_h^j(x)\| \leqslant Ch^{j\rho}$ and supp $k_h^j \subset \overline{\Omega_{-2,h}\backslash\Omega_{1,h}}$.

For any $\omega > 0$, the last group of terms is estimated by

$$\langle(-h^{2+2\omega}\Delta + Ch^{2-2\omega+2\rho})u, u\rangle;$$

therefore, if we denote

$$A'_{h,\omega} = h^2\psi_{-2,h}\psi_{1/2,h}(-h^{2\omega}\Delta + Ch^{-2\omega+2\rho})\psi_{-2,h}\psi_{1/2,h},$$

then we obtain

$$\langle(-h^2\Delta + q(x) - \mu)u, u\rangle \geqslant \sum_{j=\pm 1} \langle(-h^2\Delta + q(x) - \mu - A'_{h,\omega})u_{j,h}, u_{j,h}\rangle$$
$$\overset{\text{def.}}{=} \sum_{j=\pm 1} \langle A''_{h,\omega} u_{j,h}, u_{j,h}\rangle.$$

Hence,

$$(21.18) \qquad N(\mu, A_h) \leqslant \mathcal{N}_0(A''_{h,\omega}, \Omega_{-2,h}) + \mathcal{N}_0(A''_{h,\omega}, \Omega_{1,h}).$$

On the support of the coefficients of the operator $A'_{h,\omega}$, we have $ch^{-\rho} < |x| < Ch^{-\rho}$; therefore, by choosing ω so that $2 - 2\omega + 2\rho > 2k\rho$, we have

$$A''_{h,\omega} \geqslant (-\tfrac{1}{2}h^2\Delta - C|x|^{-2k})I_m = A'''_h,$$

and

$$\mathcal{N}_0(A''_{h,\omega}, \Omega_{1,h}) \leqslant \mathcal{N}_0(A'''_h, (R^n)_{1,h}).$$

After the substitution $x = h^{-\rho}y$, we obtain

$$\mathcal{N}_0(A'''_h, (R^n)_{1,h}) \leqslant \mathcal{N}_0(-\Delta - 2Ch^{-\chi}\langle x\rangle^{-2k}, R^n),$$

where $\mathcal{X} = 2(1 - \rho(k - 1)) > 0$ can be made arbitrarily small by choice of ρ. Since $k > 1$, the last function is estimated through $C_1 h^{-n\mathcal{X}/2}$ [see §28]; therefore, we can assume that

(21.19)
$$\mathcal{N}_0(A''_{h,\omega}, \Omega_{1,h}) \leqslant Ch^{-n+1}.$$

Estimate (21.13) and the equality

$$\sigma^w(A''_{h,\omega})(x, \xi) = h^2|\xi|^2 + q(x) - \mu \quad \forall (x, \xi) \in \Omega_{-1/4,h} \times R^n$$

imply that the first term in (21.18) is estimated in exactly the same way as

$$\mathcal{N}_0(-h^2\Delta + q(x) - \mu, \Omega_{-2,h})$$

above; therefore, (21.18) and (21.19) yield the estimate

(21.20)
$$N(\mu, A_h) \leqslant h^{-n}\widehat{V}(\mu, q, \Omega) + Ch^{-n+\delta},$$

and (21.17) and (21.20) imply (21.11).   ∎

21.6. Now we shall consider the Dirac operator with a scalar potential satisfying estimate (21.1) with $k > 1$ and $\varepsilon = 0$.

Theorem 21.4.   *For any $\delta \in (0, \frac{2}{3})$,*

(21.21)
$$N((\mu_2, \mu_1), A_h) = h^{-3}\widehat{V}((\mu_2, \mu_1), a, R^3) + O(h^{-3+\delta})$$

uniformly with respect to $h \in (0, 1]$ and $-\mu \leqslant \mu_2 < \mu_1 \leqslant \mu$.

Proof.   We define $\overset{\circ}{a}_{t,w}$ as in Section 21.3. Then

(21.22)
$$N((\mu_2, \mu_1), A_h) \geqslant \mathcal{N}_0(\overset{\circ}{a}_{t,w}, (R^3)_{-2,h}),$$

and in the definition of the operator $\overset{\circ}{a}_{t,w}$, we can replace the function q with the function $\tilde{q}_h$ constructed in the previous section. Then we apply Theorem 7.4 with the metric g_t constructed in §21.4, with the set $\Omega_t = \Omega_{-2,h}$, and with the functions $\varphi_{t,1}(x)\varphi_{t,2}(x) = \theta(|x|h^\rho)$,

$$d_t = d_{t,j} = h_t, \qquad \overset{\circ}{q}_t = [\delta_{ij}\overset{\circ}{q}_{t,i}I_2]_{i,j=1,2},$$
$$q_t(x, \xi) = T^*(h\xi)\overset{\circ}{q}_t(x, \xi)^{-1}, \qquad a_t = [\delta_{ij}\overset{\circ}{q}_{t,i}^{-2}a_{t,i}I_2]_{i,j=1,2},$$

where

$$\overset{\circ}{q}_{t,i}(x, \xi) = |h\xi|^2 + \langle x \rangle^{-2k} + h^{2k\rho} + \mu + (-1)^i\mu_i,$$
$$a_{t,i}(x, \xi) = [(-1)^{i+1}(|h\xi|^2 + \mu^2)^{1/2} + \tilde{q}_h(x) - \mu_1]$$
$$\times [(-1)^{i+1}(|h\xi|^2 + \mu^2)^{1/2} + \tilde{q}_h(x) - \mu_2].$$

The principal term of the asymptotics is equal to

(21.23) $V(\Omega_{-2,h} \times R^3, a_t) = h^{-3}\widehat{V}((\mu_2, \mu_1), a, (R^3)_{-2,h});$

but, as in (21.13),

(21.24) $\widehat{V}((\mu_2, \mu_1), a, (R^3)_{c,h}) \leqslant Ch^{3-\varepsilon},$

where $\varepsilon \to 0$ as $\rho \to (k-1)^{-1}$. Therefore, in (21.23) we can replace $(R^3)_{-2,h}$ with R^3 by adding the term $O(h^{-\varepsilon})$.

Since in this case, the function $\varphi_{t,1}\varphi_{t,2}$ is equal to 1 for $|x| < h^{-\rho}$ and the function a_t is diagonal, then the remainder of formula (7.14) can be estimated through [compare the arguments in §21.3]
(21.25)
$C \operatorname{mes}\{(x,\xi) \mid |h\xi| < C\langle x \rangle^{-k}, \ |x| > h^{-\rho}\}$

$+ Ch^{-3} \sum_{j=1,2} \operatorname{mes}\{(x,\xi) \mid |\|\xi\|^2 + \mu^2 - (q(x) - \mu_j)^2| \leqslant Ch^\delta \langle x \rangle^{-\delta(1-k)-2k}\};$

but the first term in (21.25) is $O(h^{-\varepsilon})$ [compare (21.24)], and the second term, by the same arguments as in §21.3, is estimated through $Ch^{-3+\delta}$. Thus, (21.22) yields the estimate

$$N((\mu_2, \mu_1), A_h) \geqslant h^{-3}\widehat{V}((\mu_2, \mu_1), a, R^3) - Ch^{-3+\delta}.$$

To obtain a similar estimate from above, we note that for some $c > 0$,

$$(\mu\beta + \langle \alpha, \xi \rangle - (\mu_1 + \mu_2)/2)^2 - (\mu_1 - \mu_2)^2/4 \geqslant c|\xi|^2 I_4;$$

therefore, there exist $c_1, C_1 > 0$ such that

$$\overset{\circ}{a}_{t,w} \geqslant (-c_1 h^2 \Delta - C_1 \langle x \rangle^{-2k}) I_4.$$

Using the last estimate, we can obtain the estimate from above

$$N((\mu_2, \mu_1), A_h) \leqslant h^{-3}\widehat{V}((\mu_2, \mu_1), a, R^3) + Ch^{-3+\delta}$$

in exactly the same way as in Section 21.5.
This proves Theorem 21.4.   ∎

21.7. Now we shall analyze the asymptotics of a discrete spectrum of the operator A_h for $h = 1$, i.e., of the operator A_1, but with a matrix potential satisfying condition (21.1) with $k \in (0,1)$. To be more specific, we analyze the asymptotics of the function $N((0, \mu - \tau), A_1)$ as $\tau \to +0$.

We write q in the block form $q = [q_{ij}]_{i,j=1,2}$, where q_{ij} are 2×2 matrices, and let

$$\mathcal{X}_1 = 3(1-k)/(2k), \qquad \delta_*(\delta) = \min\{\delta(1-k), k\}/2k.$$

Theorem 21.5. *As $\tau \to +0$, $\forall \delta \in (0, \frac{1}{3})$,*

$$(21.26) \qquad N((0, \mu - \tau), A_1) = (2\mu)^{3/2} \widehat{V}(-\tau, q_{11}, R^3) + O(\tau^{-\mathcal{X}_1 + \delta_*(\delta)});$$

in addition, if the matrix q_{11} is locally smoothly diagonalizable for $|x| > C$, in the sense of part b) of Theorem 20.1, then estimate (21.26) holds for any $\delta \in (0, \frac{2}{3})$.

Proof. Assume that $\mu_1 \in (0, \mu)$ does not belong to the spectrum of the operator A_1. Then, due to discreteness of the spectrum on $(-\mu, \mu)$, we have

$$(21.27) \quad N((0, \mu - \tau), A_1) = O(1) + N(O, (\mu - \mu_1)^{-1} - (A_1 - \mu_1)^{-1} + \tau_1),$$

where

$$\tau_1 = (\mu - \mu_1 - \tau)^{-1} - (\mu - \mu_1)^{-1} = (\mu - \mu_1)^{-2} \tau + O(\tau^2).$$

Since $T(D) : L_2(R^3)^4 \to L_2(R^3)^4$ is an isomorphism, then in (21.27) we can replace A_1 with $A_1' = T(D) A_1 T(D)^*$.

Let

$$\Lambda(x, \xi)^2 = |\xi|^2 + \langle x \rangle^{-2k_\epsilon}, \qquad M_1(x, \xi) = \langle \xi \rangle,$$
$$M_2(x, \xi) = \langle x \rangle, \qquad g_{x,\xi}(y, \eta) = \Lambda(x, \xi)^{-2} |\eta|^2 + \langle x \rangle^{-2} |y|^2.$$

We write $A_1' - \mu_1$ in the block form $A_1' - \mu_1 = [A_{ij}]_{i,j=1,2}$, where

(21.28)
$$A_{11} = (E(D) - \mu_1)I_2 + (\omega\chi)(D)q_{11}(x)(\omega\chi)(D) + (\omega\chi)(D)q_{12}(x)(\zeta^*\omega)(D)$$
$$+ (\zeta\omega)(D)q_{21}(\omega\chi)(D) \quad + (\omega\zeta)(D)q_{22}(x)(\omega\zeta^*)(D),$$

$$(21.29) \qquad A_{22} = -(\mu_1 + E(D))I_2 + B_{22} \overset{\text{def.}}{=} A_{22}^0 + B_{22},$$

$$(21.30) \qquad A_{12}, A_{21}, B_{22} \in \mathcal{L}(g, M_2^{-2k_\epsilon}).$$

Obviously, $\zeta \in S(g, \Lambda)$, $\omega \in S(g, M_1^{-1})$, $\chi \in S(g, M_1)$, and $\omega\chi - I \in S(g, \Lambda^2 M_1^{-2})$, since $(\omega\chi)(\xi) - I = O(|\xi|^2)$ as $|\xi| \to 0$. Therefore, (21.28) and (21.30) yield

$$(21.31) \qquad A_{11} = (E(D) - \mu_1)I_2 + q_{11}(x) + B_{11} \overset{\text{def.}}{=} A_{11}^0 + B_{11},$$

where $B_{11} \in \mathcal{L}(g, \Lambda M_1^{-1} M_2^{-2k_\epsilon})$; and

$$(21.32) \qquad A' - \mu_1 = [A_{ij}^0 + B_{ij}]_{i,j=1,2},$$

where
$$A_{12}^0 = A_{21}^0 = 0, \qquad B_{12}, B_{21}, B_{22} \in \mathcal{L}(g, M_2^{-2k_\epsilon}).$$

Now (21.29) and (21.30) imply $A_{22} \in \mathcal{LI}(g, M_1)$; therefore, there exists a paramatrix $R_{22} = R_{22}^*$ of the operator A_{22}. Let

$$P = \begin{bmatrix} I_2 & 0 \\ -R_{22}B_{12}^* & I_2 \end{bmatrix}, \qquad P_1 = P^{-1} = \begin{bmatrix} I_2 & 0 \\ R_{22}B_{12}^* & I_2 \end{bmatrix}.$$

Then (21.32) yields

$$P^*(A_1' - \mu_1)P = [\delta_{ij}(A_{ii}^0 + B_{ii}')]_{i,j=1,2} + T,$$

and

$$A_1' - \mu_1 = P_1^*[\delta_{ij}(A_{ii}^0 + B_{ii}')]_{i,j=1,2}P_1 + T_1,$$

where $T, T_1 \in \mathcal{L}^{-\infty}(g, 1)$ and where $B_{22}' = B_{22}$, and $B_{11}' = B_{11} - B_{12}R_{22}B_{12}^* \in \mathcal{L}(g, \Lambda M_1^{-1}M_2^{-2k_\epsilon} + M_1^{-1}M_2^{-4k_\epsilon}) \subset \mathcal{L}(g, \Lambda M_1^{-1}M_2^{-2k_\epsilon})$.

Since (21.31) and (21.28) yield $A_{jj}^0 + B_{jj}' \in \mathcal{LI}(g, M_1)$, let $R_{jj} = R_{jj}^* \in \mathcal{L}(g, M_1^{-1})$ be their paramatrices. Then

$$R = P[\delta_{ij}R_{jj}]_{i,j=1,2} \times P^*$$

is a paramatrix of the operator $A_1' - \mu_1$. As shown in Lemma 6.3,

$$T_2 = (A_1' - \mu_1)^{-1} - R \in \mathcal{L}^{-\infty}(g, 1).$$

Furthermore,

$$((\mu - \mu_1)^{-1} + \tau_1)I_4 = P((\mu - \mu_1)^{-1} + \tau_1)I_4 P^* + K_\tau,$$

where

$$K_\tau = \begin{bmatrix} 0 & K_{1\tau}^* \\ K_{1\tau} & 0 \end{bmatrix}, \qquad K_{1\tau} \in \mathcal{L}(g, M_1^{-1}M_2^{-2k_\epsilon}).$$

Obviously,

$$P^{-1}(K_\tau + T_2)P^{*-1} = K_\tau' = [K_{\tau,ij}']_{i,j=1,2},$$

where $K_{\tau,11}' \in \mathcal{L}^{-\infty}(g, 1)$ and $K_{\tau,ij}' \in \mathcal{L}(g, M_1^{-1}M_2^{-2k_\epsilon})$ for $i+j > 2$; therefore, by choosing $\omega \in (0, (\mu - \mu_1)^{-1})$ and by noticing that for $u = (u_1, u_2) \in L_2(R^3)^2 \oplus L_2(R^3)^2$,

$$|\langle K_\tau' u, u \rangle| \leqslant \langle (\omega + F_{2,t})u_2, u_2 \rangle + \langle F_{1,\tau}u_1, u_1 \rangle,$$

where $F_{j,\tau} \in \mathcal{L}(g, M_1^{-2}M_2^{-4k_\epsilon})$, we obtain
(21.33)
$$P[\delta_{ij}Q_{j,\tau}^+]_{i,j=1,2} \times P^* \leqslant (\mu - \mu_1)^{-1} - (A_1' - \mu_1)^{-1} + \tau_1 \leqslant P[\delta_{ij}Q_{j,\tau}^-]_{i,j=1,2}P^*,$$

where

$$Q_{1,\tau}^{\pm} = (\mu - \mu_1)^{-1} - R_{11} + \tau_1 \mp F_1, \qquad Q_{2,\tau}^{\pm} = (\mu - \mu_1)^{-1} \mp \omega - R_{22} + \tau_1 \mp F_2.$$

Since $P : L_2(R^3)^4 \to L_2(R^3)^4$ are isomorphisms, (21.33) yields the estimate

$$(21.34) \qquad \sum_{j=1,2} \mathcal{N}_0(Q_{j,\tau}^-, R^3) \leqslant N((0, \mu - \tau), A_1) \leqslant \sum_{j=1,2} \mathcal{N}_0(Q_{j,\tau}^+, R^3).$$

Then (21.29) and (21.32), and the choice of ω, imply that $Q_{2,\tau}^{\pm} \in \mathcal{L}I^+(g, 1)$ uniformly with respect to τ; therefore, Lemma 8.2 yields

$$(21.35) \qquad\qquad \mathcal{N}_0(Q_{2,\tau}^{\pm}, R^3) \leqslant C.$$

Let $R_{11}^0 = R_{11}^{0*} \in \mathcal{L}(g, M_1^{-1})$ be a paramatrix of the operator A_{11}^0. Then

$$R_{11}^0 A_{11} = I + K, \quad \text{where } K \in \mathcal{L}(g, \Lambda M_1^{-2} M_2^{-2k_{\epsilon}});$$

and, by multiplying this equality by R_{11} on the left, we obtain

$$R_{11}^0 - R_{11} \in \mathcal{L}(g, M_1^{-3} M_2^{-2k_{\epsilon}}).$$

But

$$R_{11}^0 = (E - \mu_1)^{-1}(D) - (E - \mu_1)^{-1}(D) q_{11}(x)(E - \mu_1)^{-1}(D) + K,$$

where $K \in \mathcal{L}(g, M_1^{-3} M_2^{-4k_{\epsilon}})$; and since as $|\xi| \to 0$,

$$(21.36) \qquad (E - \mu_1)^{-1}(\xi) = (\mu - \mu_1)^{-1} - (\mu - \mu_1)^{-2}(2\mu)^{-1}|\xi|^2 + O(|\xi|^4),$$

then

$$(\mu - \mu_1)^{-1} - (E - \mu_1)^{-1}(D) \in \mathcal{L}(g, \Lambda^2 M_1^{-2}),$$

and

$$R_{11} = (E - \mu_1)^{-1}(D) - (\mu - \mu_1)^{-2} q_{11}(x) + K,$$

where $K \in \mathcal{L}(g, \Lambda^2 M_1^{-2} M_2^{-2k_{\epsilon}} + \Lambda M_1^{-3} M_2^{-2k_{\epsilon}})$.

Hence,

$$\begin{aligned}
Q_{1,\tau}^{\pm} &= (\mu - \mu_1)^{-1} - (E - \mu_1)^{-1}(D) + (\mu - \mu_1)^{-2} q_{11}(x) + \tau_1 \mp K_{\pm}' \\
&= (\mu - \mu_1)^{-2}[\mu - \mu_1 - (\mu - \mu_1)^2(E - \mu_1)^{-1}(D) + q_{11}(x) + \tau_2 \mp K_{\pm}''] \\
&\overset{\text{def}}{=} (\mu - \mu_1)^{-2}(\overset{\circ}{a}_w + \tau_2 \mp K_{\pm}''),
\end{aligned}$$

where $\tau_2 = (\mu - \mu_1)^2 \tau_1 = \tau + O(\tau^2)$ and

$$\begin{aligned}
K_{\pm}', K_{\pm}'' &\in \mathcal{L}(g, \Lambda^2 M_1^{-2} M_2^{-2k_{\epsilon}} + \Lambda M_1^{-3} M_2^{-2k_{\epsilon}} + M_1^{-2} M_2^{-4k_{\epsilon}}) \\
&\subset \mathcal{L}(g, \Lambda^2 M_1^{-2} M_2^{-k_{\epsilon}}).
\end{aligned}$$

Let $\widehat{K} = C_0(\Lambda^2 M_1^{-2} M_2^{-k_\epsilon})_w$. If C_0 is sufficiently large, then Lemma 8.1 yields

$$|\langle K''_\pm u, u \rangle| \leqslant \langle (\widehat{K} + T)u, u \rangle \quad \forall u \in L_2(R^3)^2,$$

where $T \in \mathcal{L}^{-\infty}(g, I_2)$; hence, by denoting

$$\widehat{Q}_\tau^\pm = \overset{\circ}{a}_w + \tau_2 \mp \widehat{K} \mp T,$$

we have

$$(21.37) \qquad \mathcal{N}_0(Q_{1,\tau}^-, R^3) \geqslant \mathcal{N}_0(\widehat{Q}_\tau^-, R^3), \qquad \mathcal{N}_0(Q_{1,\tau}^+, R^3) \leqslant \mathcal{N}_0(\widehat{Q}_\tau^+, R^3).$$

The functions $\mathcal{N}_0(\widehat{A}_\tau^\pm, R^3)$ can be estimated using the theorems in §7. Let

$$t = h^{-1}, \qquad \overset{\circ}{q}_t = M_1^{-2}\Lambda^2 + t^{-1}, \qquad q_t = \overset{\circ}{q}_t^{-1},$$
$$d_t = h, \qquad a_t = q_t^2(\overset{\circ}{a} + \tau_2 \mp \sigma^w(\widehat{K})).$$

Then (21.36) implies that there exists C_ϵ such that all the sets whose measures appear in the asymptotic formulas in §7, are subsets of the set

$$\{(x, \xi) \mid |\xi| < C_\epsilon \langle x \rangle^{-k_\epsilon}\};$$

but on it,

$$a_t(x, \xi) - q_t^2(x, \xi)(\tfrac{1}{2\mu}|\xi|^2 + q_{11}(x) + \tau) = O(\tau^2 + \langle x \rangle^{-\delta^* + \epsilon'}),$$

where $\delta^* = \min\{(1 - k), k\}$, and $\epsilon' \to 0$ as $\epsilon \to 0$. Hence, the arguments in Section 21.2 yield

$$\mathcal{N}_0(\widehat{Q}_\tau^\pm, R^3) = (2\mu)^{3/2}\widehat{V}(-\tau, q_{11}, R^3) + O(\tau^{-\mathcal{X}_1 + \delta_*(\delta)}),$$

where $\mathcal{X}_1$, $\delta_*(\delta)$ are the same as in (21.26).

Now (21.26) follows from (21.37), (21.34), and (21.35); and Theorem 21.5 is proved. ∎

21.8. We now consider a Schrödinger operator perturbed by first-order differential operators

$$A = -\Delta I_m - i\langle Q(x), \nabla \rangle - i\langle \nabla, Q(x) \rangle + q_0(x),$$

where $Q = (q_1, \dots, q_n)$ is a "vector" and q_i are $m \times m$ matrix functions. In the scalar case, the operator A describes a quantum particle in electric and magnetic fields; and in the matrix case, it allows us to take into account a spin orbital interaction, etc. (see Birman and Yafaev [1, 2]).

For the sake of simplicity, we shall consider the operator A in R^n, assuming that the functions q_i are homogeneous at infinity, i.e., for $|x| \geqslant 1$,

(21.1') $$q_i(x) = q_i^1(x/|x|)|x|^{-k} + \hat{q}_i(x) \quad i \geqslant 1,$$
(21.2') $$q_0(x) = q_0^1(x/|x|)|x|^{-2k} + \hat{q}_0(x),$$

where $k \in (0,1)$; and for any α,

(21.3') $$\|\hat{q}_i^{(\alpha)}(x)\| + \|\hat{q}_0^{(\alpha)}(x)\|\langle x\rangle^k \leqslant C_\alpha \langle x\rangle^{-k-k_1-|\alpha|},$$

where $k_1 > 0$.

The same arguments as in Lemma 21.1 show that under conditions (21.1')–(21.3'), $\sigma_{\mathrm{ess}} A = [0, \infty)$ [A is considered as an operator in $L_2(R^n)^m$ with the domain $\{u \mid Au \in L_2(R^n)^m\}$]. In order to formulate the theorem on asymptotics of a discrete spectrum, we let

$$\delta^*(\delta) = \min\{\delta(1-k), k_1\},$$
$$\rho(\delta^*(\delta)) = \max\{(n(1-k) - \delta^*(\delta))/2k, \ n(1-k)/(2k + \delta^*(\delta))\},$$
$$\tilde{a}(x,\xi) = |\xi|^2 + 2 \sum_{1 \leqslant i \leqslant n} q_i^1(x/|x|)|x|^{-k}\xi_i + q_0^1(x/|x|)|x|^{-2k},$$
$$c_0 = (2\pi)^{-n} \sum_{1 \leqslant j \leqslant m} \mathrm{mes}\{(x,\xi) \mid \tilde{\lambda}_j(x,\xi) + 1 < 0\},$$

where $\tilde{\lambda}_j$ are eigenvalues of the matrix $\tilde{a}$.

Theorem 21.6. *As $\tau \to +0$, $\forall \delta \in (0, \frac{1}{3})$,*

(21.4') $$N(-\tau, A) = c_0 \tau^{-n(1-k)/(2k)} + O(\tau^{-\rho(\delta^*(\delta))});$$

if $m = 1$, then (21.4') holds for any $\delta \in (0, \frac{2}{3})$.

Proof. Let $M_1(x,\xi) = \langle \xi \rangle$, $\Lambda(x,\xi)^2 = |\xi|^2 + \langle x \rangle^{-2k}$, and $g_{x,\xi}(y,\eta) = \Lambda(x,\xi)^{-2}|\eta|^2 + \langle x \rangle^{-2}|y|^2$. (21.1')–(21.3') imply that $A + I \in \mathcal{L}I(g, M_1^2)$; therefore, Theorem 6.5 implies that A is a closure of the operator $A : C_0^\infty(R^n)^m \to L_2(R^n)^m$, and hence, $N(-\tau, A) = \mathcal{N}_0(A + \tau, R^n)$.

The asymptotics of the last function can be analyzed with the help of the theorems in §7, by denoting $t = \tau^{-1}$, $\overset{o}{q}_t = (\Lambda^2 + \tau)^{1/2}$, $q_t = \overset{o}{q}_t^{-1}$ and by constructing the function a_t so that for $|x| > 1$,

$$a_t(x,\xi) = q_t(x,\xi)^2(\tilde{a}(x,\xi) + \tau);$$

then $d_t = h_t$.

Quasi-homogeneity of the function $\tilde{a}$ implies that

$$V(R^n \times R^n, a_t) = O(1) + c_0 \tau^{-n(1-k)/(2k)};$$

and since $h_t(x,\xi) \leqslant C\langle x\rangle^{k-1}$, then

$$
J_c^\delta(a_t, d_t) \leqslant C \int\limits_{C|x|^{-2k}>\tau} dx \langle x\rangle^{\delta(k-1)} \int\limits_{|\xi|<C\langle x\rangle^{-k}} d\xi
$$

(21.5')

$$
\leqslant C_1 \int\limits_{|x|<C_1\tau^{-1/(2k)}} \langle x\rangle^{-nk+\delta(k-1)} dx \leqslant C_2 \tau^{(n-\delta)(k-1)/(2k)}.
$$

By changing to the spherical coordinate system (r,θ) in R_x^n and by making the substitution $\xi = r^{-k}\eta$, we obtain

$$
W_c^\delta(a_t, d_t) = O(1) + C \sum_{1\leqslant j\leqslant m} \int\limits_{S_{n-1}} d\theta \int\limits_{|\eta|\leqslant C} d\eta \int\limits_{\mathcal{M}_{j,r}(\theta,\eta)} r^{-nk+n-1} dr,
$$

where

$$
\mathcal{M}_{j,\tau}(\theta,\eta) = \{r > 1 \mid |1 + t\tilde{a}(\theta,\eta)r^{-2k}| \leqslant Cr^{-\delta^*(\delta)}(1 + tr^{-2k})\},
$$
$$
\tilde{a}(\theta,\eta) = |\eta|^2 + 2 \sum_{1\leqslant j\leqslant n} q_j^1(\theta)\eta_j + q_0^1(\theta).
$$

By applying Lemma 13.2 with $q = n(1-k)$, $p = 2k$, and $\Sigma = \{(\theta,\eta) \mid \theta \in S_{n-1}, \ |\eta| \leqslant C\}$, we obtain

(21.6')
$$
W_c^\delta(a_t, d_t) \leqslant C\tau^{-\rho(\delta^*(\delta))}.
$$

Theorem 7.1 and (21.5') and (21.6') yield (21.4') for any $\delta \in (0, \frac{1}{3})$.

If $m = 1$, then instead of Theorem 7.1, one can apply Theorem 7.3 with $\varphi_{1,t} = \varphi_{2,t} = 1$. For such $\varphi_{j,t}$, the remainder of Theorem 7.3 has the same form as the remainder of Theorem 7.1 [but with $\delta \in (0, \frac{2}{3})$]; therefore, (21.5') and (21.6') imply (21.4') for any $\delta \in (0, \frac{2}{3})$.

CHAPTER 5

ASYMPTOTICS OF THE SPECTRUM OF
PSEUDODIFFERENTIAL OPERATORS
WITH OPERATOR-VALUED
SYMBOLS AND SOME APPLICATIONS

§22 Pseudodifferential Operators with Operator-Valued Symbols

22.1. Let $t \geqslant 1$ be a parameter, let $H_{t,0} \subset H_t \subset H_{t,0}^*$ be Hilbert spaces, and let $\langle \cdot, \cdot \rangle$ be a pairing between $H_{t,0}^*$ and $H_{t,0}$. Let $a_t \in C^\infty(R^{2n}; \operatorname{Hom}(H_{t,0}; H_{t,0}^*))$ satisfy the condition

$$\langle a_t(X)u, v \rangle = \overline{\langle a_t(X)v, u \rangle} \quad \forall u, v \in H_{t,0}, \ X \in R^{2n},$$

and let $\Omega_t \subset R^n$ be an open set. We shall study the asymptotics of the function

$$\mathcal{N}(a_{t,w}, C_0^\infty(\Omega_t, H_{t,0})) = \mathcal{N}_0(a_{t,w}, \Omega),$$

as $t \to \infty$, assuming that the following conditions are satisfied:

A. There exist a σ-temperate splitting metric g_t and an operator function $\overset{\circ}{q}_t : R^{2n} \to \operatorname{Hom}(H_{t,0}, H_t)$ such that

(22.1) $\quad h_t(X) = h_{g_t}(X) \to 0 \quad \text{as} \quad t \to \infty, \quad$ uniformly with respect to X;

and there exists an isomorphism $J_t : H_t \to H_{t,0}$ such that

$$\overset{\circ}{q}'_t = \overset{\circ}{q}_t J_t \in O_r(H_t, g_t), \qquad a'_t = J_t^* a_t J_t \in S(g_t, \overset{\circ}{q}'^*_t, \overset{\circ}{q}'_t).$$

In order to formulate the remaining conditions, we denote

$$\widehat{Q}_{t,X}[u] = \|\overset{\circ}{q}_t(X)u\|_{H_t}^2, \qquad \widehat{\mathcal{A}}^\varepsilon_{t,X}[u] = \langle a_t(X)u, u \rangle - \varepsilon \widehat{Q}_{t,X}[u],$$

$$U_\varepsilon = U(\varepsilon^2, g_t), \qquad U_{-\varepsilon} = U \backslash (\partial U)_\varepsilon$$

$$\mathcal{M}^\varepsilon_t = \{ X \in (\Omega_t \times R^n)_\varepsilon \mid \exists u \in H_{t,0} : \widehat{\mathcal{A}}^\varepsilon_{t,X}[u] < 0 \},$$

$$V(\mathcal{A}, H, U) = (2\pi)^{-n} \iint_U \mathcal{N}(\mathcal{A}_X, H) dX.$$

Now we can formulate the second set of conditions.

B. There exist a σ-temperate metric g_t' and constants $\varepsilon, c, C > 0$ such that

$$(22.2) \qquad cg_{t,X}(\cdot) \leqslant g_{t,X}^1(\cdot) \leqslant Cg_{t,X}(\cdot) \quad \forall X \in \mathcal{M}_t^\varepsilon;$$

$$(22.3) \qquad c(t + |X|)^{-C} \leqslant h_{g_t^1}(X) \leqslant C(t + |X|)^{-\varepsilon};$$

$$(22.4) \qquad g_{t,X}^1(\cdot) \leqslant C(t + |X|)^C |\cdot|^2;$$

$$(22.5) \qquad \mathcal{N}(\widehat{A}_{t,X}^\varepsilon, H_{t,0}) = 0 \quad \forall X \in (\Omega_t \times R^n)_\varepsilon, \ |X| > t^C;$$

$$(22.6) \qquad \mathcal{N}(\widehat{A}_{t,X}^\varepsilon, H_{t,0}) < Ct^C \quad \forall X \in (\Omega_t \times R^n)_\varepsilon.$$

Theorem 22.1. *If conditions A and B are satisfied uniformly with respect to $t \geqslant 1$, then there exist C_i such that as $t \to \infty$,*
(22.7)
$$V(\widehat{A}_t^{-\varepsilon_1(t)}; H_{t,0}; (\Omega_t \times R^n)_{-\varepsilon(t)}) - \varepsilon_1(t)V(\widehat{A}_t^{\varepsilon_1(t)}; H_{t,0}; (\Omega_t \times R^n)_{-\varepsilon(t)}) - \varepsilon_1(t)$$
$$\leqslant \mathcal{N}_0(a_{t,w}, \Omega_t) \leqslant \varepsilon_1(t) + V(\widehat{A}_t^{\varepsilon_1(t)}; H_{t,0}; (\Omega_t \times R^n)_{\varepsilon_1(t)})(1 + \varepsilon_1(t)),$$

where $\varepsilon_i(t) = C_i\nu(t)$ and $\nu(t) = \max_X h_t(X)^{1/7}$.

22.2.

Proof. As in Chapter 2, we shall drop the index t. Furthermore, we fix an isomorphism $J : H \to H_0$ and denote $a' = J^*aJ$, $\overset{\circ}{q}{}' = \overset{\circ}{q}J$. Then $\mathcal{N}(a_w; C_0^\infty(\Omega; H_0)) = \mathcal{N}(a_w'; C_0^\infty(\Omega; H))$; and if, similar to the form $\widehat{A}_X^\varepsilon$, we define the form $\widehat{A}_X'^\varepsilon$ in terms of the functions a' and q', then we shall see that $\mathcal{N}(\widehat{A}_X'^\varepsilon; H) = \mathcal{N}(\widehat{A}_X^\varepsilon; H_0)$ and that a', $\overset{\circ}{q}{}'$, $\widehat{A}'^\varepsilon$ satisfy conditions A and B with $H_0 = H$.

Therefore, we can assume that

$$(22.8) \qquad a \in S(g, \overset{\circ}{q}{}^*, q), \qquad \overset{\circ}{q} \in O_r(H; g), \qquad H_0 = H.$$

Unlike in Chapter 2, here we must apply some hybrid of the ASP method and the classical variational method [with localization in the space R_x^{2n} instead of R_X^n]. Therefore, we need the following lemma:

Lemma 22.1. *For any sufficiently small $\nu > 0$, there exist points $X_j \in R^{2n}$ and sets U_j such that*

(22.9)
$$\bigcup_j U_j = R^{2n}, \quad (X_j)_{\nu/C} \subset U_j \subset (X_j)_\nu, \qquad U_j \cap U_k = \emptyset \quad \forall j \neq k;$$

(22.10)
multiplicity of the covering $\{(U_j)_\nu\}$ is uniformly bounded with respect to ν;

$$(22.11) \qquad \forall \nu_1 \in (0, \nu), \quad \mathrm{mes}(\partial U_j)_{\nu_1} \leqslant C\nu_1\nu^{-1} \mathrm{mes}\, U_j.$$

Constant C in (22.9) and (22.11) does not depend on j, ν, ν_1.

Proof. If $\nu > 0$ is sufficiently small, then due to Lemma 1.4, there exist points X_j such that the balls $V_{j,\nu} = (X_j)_\nu$ cover R^{2n}. The multiplicity of the covering $\{V_{j,2\nu}\}$ is bounded uniformly with respect to ν, j; and for $\varepsilon = \nu/C$ for some C, the balls $V_{j,\varepsilon}$ do not intersect. Therefore, we can assume that

$$U_j = V_{j,\nu} \backslash (\bigcup_{1 \leqslant i \leqslant j-1} U_i \cup \bigcup_{i>j} V_{i,\varepsilon}).$$

Indeed, then $U_j \supset V_{j,\varepsilon}$ and conditions (22.9) and (22.10) are satisfied. Condition (22.11) is also satisfied, since

1) ∂U_j is a union of a finite number of ellipsoids $\partial V_{k,\nu}$, $\partial V_{i,\varepsilon}$, and this number is bounded uniformly with respect to j, ν;

2) for all such k, i, due to σ-temperateness of the metric g, $\operatorname{mes} U_j \asymp \operatorname{mes} V_{k,\nu} \asymp \operatorname{mes} V_{i,\varepsilon}$; and

3) $V_{k,\nu}$, $V_{i,\varepsilon}$ satisfies condition (22.11). ∎

Let $\nu = \max h(X)^{1/7}$. Lemma 5.3 implies that $\nu^{-4}g$ is a σ-temperate metric. Starting with ν and using Lemma 22.1, we shall construct the sets U_j and fix a small c_0; and, using the partition of unity associated with $\nu^{-4}g$, c_0, and $c_0/2$, we shall construct the functions $f_{ij,c}$ ($i = 0,1,2$, $c = \frac{1}{2},1,2,4$, $j = 1,2,\dots$) such that, uniformly with respect to j,

(22.12) $f_{ij,c} \in S(\nu^{-4}g, 1)$, $0 \leqslant f_{ij,c} \leqslant 1$, $\sum_i f_{0j,c}^2 = 1$,

(22.13) $\operatorname{supp} f_{0j,c}, \operatorname{supp} f_{2j,c} \subset (U_j)_{\nu^2}$, $\operatorname{supp} f_{1j,c} \subset U_j$,

(22.14) $f_{0j,c}|_{(U_j)_{-\nu^2}} = f_{1j,c}|_{(U_j)_{-\nu^2}} = 1$, $f_{2j,c}|_{U_j} = 1$,

(22.15) $f_{0j,c}f_{2j,2c} = f_{0j,c}$, $f_{2j,c}f_{2j,2c} = f_{2j,c}$,

(22.16) $f_{1j,c}f_{1j,2c} = f_{1j,2c}$.

22.3. We fix $X_j \in U_j$, $C_0 > 0$ and denote $\varepsilon_0 = C_0\nu$,

$$\overset{\circ}{q}_j = \overset{\circ}{q}(X_j), \qquad a_j = a(X_j), \qquad \widehat{A}_j = \widehat{A}_{X_j}, \qquad \widehat{Q}_j = \widehat{Q}_{X_j},$$
$$a_j^{\pm\varepsilon_0} = \overset{\circ}{q}_j^{*-1} a_j \overset{\circ}{q}_j^{-1} \mp 4\varepsilon_0 I, \qquad \widehat{A}_j^{\pm\varepsilon_0} = \widehat{A}_j \mp \varepsilon_0 \widehat{Q}_j.$$

Lemma 22.2. a) *If $X \in (U_j)_\nu$, then*

(22.17) $\|\overset{\circ}{q}_j^{*-1}(a_j - a(X))\overset{\circ}{q}_j^{-1}\|_{H \to H} < \varepsilon$, $|\widehat{A}_j[u] - \widehat{A}_X[u]| \leqslant \varepsilon\widehat{Q}_j[u]$.

b) *Let a_j, $\overset{\circ}{q}_j$ be PDOs with constant symbols a_j, $\overset{\circ}{q}_j$. Then*

(22.18) $\|\overset{\circ}{q}_j^{*-1}(a_j - a_w)\overset{\circ}{q}_j^{-1} f_{ij,c,w}\|_{L_2 \to L_2} < \varepsilon$

(from here on, ε denotes functions (depending on t) of the type $C\nu$).

Proof. a) $(22.17)_1$ is proved in exactly the same way as Lemma 12.1. Since $\overset{\circ}{q}$ is a right g-continuous function, then $\widehat{Q}_X[u] \leqslant C\widehat{Q}_j[u]$, and $(22.17)_2$ follows from $(22.17)_1$.

b) Let d_w denote the operator in (22.18). $(22.17)_1$ and $(22.12)_1$ yield, for $k = 0$,

$$(22.19) \qquad n_k(\nu^{-6}g, 1, d) \leqslant C_k\nu.$$

But $(22.8)_1$ and $(22.12)_1$ imply that $d \in S(\nu^{-4}g, 1)$; and since $\nu^3\nu^{-2} = \nu$, then (22.19) holds for all $k \geqslant 1$. Applying Theorem 4.2, we obtain (22.18). ∎

22.4. Let χ be a characteristic function of the interval $(-\infty, 0)$, let $p_j^{\pm\varepsilon_0} = \chi(a_j^{\pm\varepsilon_0})$ be a spectral projection, let $\|\cdot\|_1$ be a trace norm, and let

$$\mathcal{J}_- = \{j \mid U_j \subset \Omega \times R^n\}, \qquad \mathcal{J}'_+ = \{j \mid (U_j)_{\nu^2} \cap \Omega \times R^n \neq \emptyset\},$$
$$\mathcal{J}_+ = \{j \mid \exists i \in \mathcal{J}'_+ : (U_j)_{\nu^2} \cap (U_i)_{\nu^2} \neq \emptyset\}.$$

Then (22.1) implies that for $t \geqslant t_0$, $(U_j)_{\nu^2} \subset \mathcal{M}^\varepsilon$ $\forall j \in \mathcal{J}_+$; therefore, due to condition (22.6), for the same j, we have,

$$(22.20) \qquad \|p_j^{\pm\varepsilon_0}\|_1 = \operatorname{Tr} p_j^{\pm\varepsilon_0} = \mathcal{N}(\widehat{A}_j^{\ \pm\varepsilon_0}, H) < Ct^C.$$

Now $(22.17)_2$ and $(22.8)_2$ imply that for any ε, there exists ε' such that

$$\mathcal{N}(\widehat{A}_j^{-\varepsilon'}, H) \leqslant \mathcal{N}(\widehat{A}_X^{\pm\varepsilon}, H) \leqslant \mathcal{N}(\widehat{A}_j^{\varepsilon'}, H) \quad \forall X \in (U_j)_{\nu^2};$$

therefore, (22.11) yields the estimates

$$V(\widehat{A}^{-\varepsilon}; H; U_j) - \varepsilon_1 V(\widehat{A}^\varepsilon, H, U_j) \leqslant (2\pi)^{-n} \operatorname{mes}(U_j)_{\pm\nu^2} \operatorname{Tr} p_j^{\pm\varepsilon_0}$$
$$\leqslant V(\widehat{A}^\varepsilon; H; U_j)(1 + \varepsilon_1),$$
$$(22.21) \qquad \operatorname{mes}(\partial U_j)_{\nu^2} \operatorname{Tr} p_j^{\pm\varepsilon_0} \leqslant C\nu V(\widehat{A}^\varepsilon, H, U_j).$$

Let

$$(22.22) \qquad \mathcal{E}^{-j} = f_{1j,2,w} p_j^{-\varepsilon_0}, \qquad \mathcal{E} = \sum_{j \in \mathcal{J}_-} \mathcal{E}^{-j}, \qquad \mathcal{E}_1 = \mathcal{E}^2(3 - 2\mathcal{E}).$$

Then $(22.9)_2$ and (22.2)–(22.4) imply that each set U_j, on which $\mathcal{E}^{-j} \neq 0$, contains a ball of the radius $c(|X_j| + t)^{-C}$; and due to condition (22.5), all these sets are contained in a ball of the radius Ct^C. Hence,

$$(22.23) \qquad \operatorname{card}\{j \mid \mathcal{E}^{-j} \neq 0\} \leqslant C_1 t^{C_1}.$$

Since $f_{ij,2} \in C_0^\infty(R^{2n})$ and the operators $p_j^{-\varepsilon_0}$ are finite-dimensional, then (22.23) implies that $\sigma^w(\mathcal{E}) \in \mathcal{S}(R^{2n}; \mathrm{Hom}(H', H'))$, where H' is a finite-dimensional subspace of H. Therefore, its kernel has the same property, and $\mathcal{E}$ is an operator of the trace class. Hence, $\mathcal{E}_1$ is of the trace class. Let ν_i, ν_{1i} be nonzero eigenvalues of the operators $\mathcal{E}$, $\mathcal{E}_{1i}$, numbered in any order [counting multiplicity]. Since $(22.12)_1$, $(22.12)_2$, $(22.13)_3$, and $(22.9)_3$ imply that $cI + \mathcal{E} \in \mathcal{L}I^+(\nu^{-4}g, 1)$ and $(1 + c)I - \mathcal{E} \in \mathcal{L}I^+(\nu^{-4}g, 1)$ for any $c > 0$, then Lemma 8.4′ implies that $\frac{1}{4}I + \mathcal{E} > 0$ and $\frac{5}{4}I - \mathcal{E} > 0$ for $t \geqslant t_0$. Hence, $\nu_i \in (-\frac{1}{4}, \frac{5}{4})$, and [see the proof of estimate (9.6)]

$$(22.24) \qquad \nu_{1i} \in [0, 1] \quad \forall i.$$

Since the kernel of the operator $\mathcal{E}_1$ belongs to $\mathcal{S}(R^{2n}; \mathrm{Hom}(H', H'))$, Lemma 8.5′ yields

$$(22.25) \qquad \mathrm{Tr}\, \mathcal{E}_1 = (2\pi)^{-n} \iint \mathrm{Tr}\, \sigma^w(\mathcal{E}_1)(X)dX.$$

Applying Theorem 3.1 and taking into account (22.2), $(22.12)_1$, $(22.13)_3$, and $(22.9)_3$, we obtain

$$(22.26)$$
$$\sigma^w(\mathcal{E}_1) = \sum_{j \in J_-} (e_j + f_j p_j^{-\varepsilon_0}) + \sum_{j_1, j_2, j_3 \in J_-} r_{j_1 j_2 j_3} p_{j_1}^{-\varepsilon_0} p_{j_2}^{-\varepsilon_0} p_{j_3}^{-\varepsilon_0} \overset{\text{def.}}{=} E_1 + E_2,$$

where

$$(22.27) \qquad e_j = f_{1j,2}^2 p_j^{-\varepsilon_0}(3 - 2f_{1j,2}p_j^{-\varepsilon_0}), \qquad r_{j_1 j_2 j_3} \in \mathcal{L}^{-\infty}(g^1, 1),$$
$$f_j \in S(\nu^{-4}g, \nu), \qquad \mathrm{supp}\, f_j \subset (\partial U_j)_{\nu^2}.$$

Now (22.3), (22.23), (22.20), and two well-known properties of the trace norm

$$\|AB\|_1 \leqslant \|A\|_1 \|B\|, \qquad \|A + B\|_1 \leqslant \|A\|_1 + \|B\|_1$$

yield

$$(22.28) \qquad \int \|E_2(X)\|_1 dX \leqslant C_N t^{-N} \quad \forall N,$$

$$(22.29) \qquad \|E_1(X)\|_1 \leqslant C\|p_j^{-\varepsilon_0}\|_1 = C\, \mathrm{Tr}\, p_j^{-\varepsilon_0} \quad \forall X \in U_j.$$

Furthermore, if $X \in U_j \backslash \mathrm{supp}(1 - f_{1j,2})$, then $e_j(X) + f_j(X)p_j^{-\varepsilon_0} = p_j^{-\varepsilon_0}$; thus, (22.21), (22.25), (22.26), (22.28), and (22.29) yield the estimate

$$(22.30) \quad \mathrm{Tr}\, \mathcal{E}_1 = V(\hat{A}^{-\varepsilon_1}; H; (\Omega \times R^n)_{-\varepsilon_2}) + O(\nu)(1 + V(\hat{A}^{\varepsilon_2}; H; (\Omega \times R^n)_{-\varepsilon_2})).$$

Similarly, we can verify the same estimate for $\operatorname{Tr} \mathcal{E}_1^2$; and (22.24) and (22.30) yield [see §9] the estimate
(22.31)

$$\operatorname{card}\{i \mid \nu_{1i} \geqslant \tfrac{1}{2}\}$$

$$= V(\widehat{A}^{-\varepsilon_1}; H; (\Omega \times R^n)_{-\varepsilon_2}) + O(\nu)(1 + V(\widehat{A}^{\varepsilon_2}; H; (\Omega \times R^n)_{-\varepsilon_2})).$$

22.5. Now we can prove the left inequality in (22.7). Lemma 1.5 allows us to assume that $\overset{\circ}{q} \in S(g, \overset{\circ}{q})$ and $q = \overset{\circ}{q}{}^{-1} \in S(g, \overset{\circ}{q}{}^{-1})$. Let $\widehat{L} = \operatorname{span}\{u \mid \mathcal{E}_1 u = \nu_{1i} u, \nu_{1i} \geqslant \tfrac{1}{2}\}$, $\psi = \sum_{j \in \mathcal{J}_-} f_{1j,1}$, $L = \psi_\ell q_w \widehat{L}$. Condition $(22.13)_3$ implies that $\psi \in C_0^\infty(\Omega \times R^n)$; thus, $L \subset C_0^\infty(\Omega; H)$. Let $v \in \widehat{L}$ and $u = \psi_\ell q_w \mathcal{E}_1 v \in L$. Then (22.16) and Theorems 3.2 and 3.1 imply

$$\overset{\circ}{q}_w u = \mathcal{E}_1 v + Tv, \quad \text{where } T \in \mathcal{L}(\nu^{-4} g, \nu).$$

But

(22.32)
$$\|\mathcal{E}_1 v\| \geqslant \tfrac{1}{2}\|v\|,$$

and $\|T\| \leqslant C\nu$, according to Theorem 4.2; therefore [for $t \geqslant t_0$],

(22.33)
$$\dim L = \dim \widehat{L} = \operatorname{card}\{i \mid \nu_{1i} \geqslant \tfrac{1}{2}\}.$$

Now (22.31) and (22.33) imply that left estimate in (22.7) would follow from the estimate

(22.34)
$$\langle a_w u, u \rangle < 0 \quad \forall (0 \neq) u \in L.$$

Then (22.16) yields $\psi_\ell q_w \mathcal{E}_1 - q_w \mathcal{E}_1 \in \mathcal{L}^{-\infty}(g, q, I)$; therefore, by using (22.18), (22.19), $(22.13)_3$, and $(22.9)_3$, we obtain
(22.35)

$$\langle a_w u, u \rangle = \langle (a_w + 4\varepsilon_0 \overset{\circ}{q}{}_w^* \overset{\circ}{q}_w) u, u \rangle - 4\varepsilon_0 \|\overset{\circ}{q}_w u\|^2$$

$$= \sum_{j \in \mathcal{J}_-} \langle \mathcal{E}^{-j}(3 - 2\mathcal{E}^{-j}) f_{1j,2,w} p_j^{-\varepsilon_0} a_j^{-\varepsilon_0} p_j^{-\varepsilon_0} f_{1j,2,w}(3 - 2\mathcal{E}^{-j}) \mathcal{E}^{-j} v, v \rangle$$

$$- 4\varepsilon_0 \|\mathcal{E}_1 v\|^2 + \langle Tv, v \rangle,$$

where $T \in \mathcal{L}(\nu^{-4} g, \nu)$. But $p_j^{-\varepsilon_0} a_j^{-\varepsilon_0} p_j^{-\varepsilon_0} \leqslant 0$; and hence, the first term in (22.35) is nonpositive. Applying Theorem 4.2 to the latter and taking (22.32) into account, we obtain $\langle a_w u, u \rangle \leqslant (-\varepsilon_0 + C_1 \nu)\|v\|^2$, where $\varepsilon_0 = C_0 \nu$ and where C_1 does not depend on C_0. By choosing $C_0 > C_1$, we obtain (22.34).

22.6. Thus, the left estimate in (22.7) is proved. In order to prove the right estimate, we let

$$\widetilde{Q} = \overset{\circ}{q}{}_w^* \overset{\circ}{q}_w, \qquad A^{\varepsilon_0} = a_w - 2\varepsilon_0 \widetilde{Q}, \qquad \psi = \sum_{j \in \mathcal{J}_+} f_{0j,1/2}^2,$$

and notice that similar to (22.19),

$$(22.19') \qquad f_{0j,1/2,w}(\widetilde{Q} - \overset{\circ}{q}{}_j^*\overset{\circ}{q}_j)f_{0j,1/2,w} \in \mathcal{L}(\nu^{-6}g, \nu\overset{\circ}{q}{}_j^*, \overset{\circ}{q}_j)$$

and that due to $(22.12)_3$, $\psi(X) = 1 \quad \forall X \in \Omega \times R^n$; therefore, $\psi_r u = u \quad \forall u \in C_0^\infty(\Omega; H)$. In addition,

$$\psi_r - \sum_{j \in \mathcal{J}_+} f_{0j,1/2,w} f_{0j,1/2,w} \in \mathcal{L}(\nu^{-4}g, \nu);$$

hence, by applying (22.12), $(22.15)_1$, (22.19), and (22.19)', we obtain
(22.36)

$$
\begin{aligned}
\langle a_w u, u \rangle &= \sum_{j \in \mathcal{J}_+} \langle A^{\varepsilon_0} f_{0j,1/2,w} u, f_{0j,1/2,w} u \rangle + \langle T_1 u, u \rangle + 2\varepsilon_0 \langle \widetilde{Q} u, u \rangle \\
&= \sum_{j \in \mathcal{J}_+} \langle \overset{\circ}{q}{}_j^* [f_{2j,1,w} a_j^{\varepsilon_0} f_{2j,1,w} + 2\varepsilon_0] \overset{\circ}{q}_j f_{0j,1/2,w} u, f_{0j,1/2,w} u \rangle \\
&\qquad\qquad\qquad\qquad\qquad + \langle (2\varepsilon_0 \widetilde{Q} + T_2) u, u \rangle,
\end{aligned}
$$

where $T_k \in \mathcal{L}(\nu^{-6}g, \nu\overset{\circ}{q}{}^*, \overset{\circ}{q})$. But $Q \in \mathcal{L}I^+(g, \overset{\circ}{q}{}^*, \overset{\circ}{q})$; therefore $2\varepsilon_0 \widetilde{Q} + T_2 \in \mathcal{L}I^+(\nu^{-6}g, \nu\overset{\circ}{q}{}^*, \overset{\circ}{q})$, too, for sufficiently large C_0 in the definition of ε_0. Lemma 8.4' implies that the last term in (22.36) is nonnegative [for $t \geqslant t_0$]. Therefore, $\forall u \in C_0^\infty(\Omega; H)$,

$$(22.37) \qquad\qquad \langle a_w u, u \rangle \geqslant \sum_{j \in \mathcal{J}_+} \langle \overset{\circ}{q}{}_j^* M_j^{\varepsilon_0} \overset{\circ}{q}_j u_j, u_j \rangle,$$

where

$$u_j = f_{0j,1/2,\ell} u, \qquad M_j^{\varepsilon_0} = f_{2j,1,w} a_j^{\varepsilon_0} f_{2j,1,w} + 2\varepsilon_0 I.$$

But

$$\sum_{j \in \mathcal{J}_+} \|u_j\|^2 = \langle (I + T) u, u \rangle, \quad \text{where } T \in \mathcal{L}(\nu^{-4}g, \nu);$$

therefore, by applying Theorem 4.2 to T, we see that the mapping

$$C_0^\infty(\Omega; H) \ni u \to (u_1, \ldots, u_N) \in \mathcal{S}(R^n, H)^N,$$

where $N = \operatorname{card} \mathcal{J}_+$, is injective for $t \geqslant t_0$; and (22.37) yields the estimate

$$(22.38) \qquad\qquad \mathcal{N}_0(a_w, \Omega) \leqslant \sum_{j \in \mathcal{J}_+} \mathcal{N}(M_j^{\varepsilon_0}; \mathcal{S}(R^n; H)).$$

Let $\mathcal{E}_j = f_{2j,2,w} p_j^{\varepsilon_0}$, $\mathcal{E}_{1j} = \mathcal{E}_j^2(3 - 2\mathcal{E}_j)$, $\mathcal{E}_{1j}' = I - (p_j^{\varepsilon_0})^2(3 - 2p_j^{\varepsilon_0}) = I - p_j^{\varepsilon_0}$. The same arguments as in §22.4 show that $\mathcal{E}_{1j}$ are operators of the trace class and

$$(22.39) \qquad \operatorname{card}\{i \mid \nu_{1ji} \geqslant \tfrac{1}{2}\} = V(\widehat{A}^\varepsilon; H; U_j)(1 + \varepsilon_1) + \varepsilon_2,$$

where ν_{1ji} are nonzero eigenvalues of the operator $\mathcal{E}_{1j}$. Let

$$L_j = \mathrm{span}\{u \mid \mathcal{E}_{1j}u = \nu_{1ji}u, \nu_{1ji} \geqslant \tfrac{1}{2}\}.$$

If $v \in {}^{\perp}L_j$ and $u = (I - \mathcal{E}_{1j})v$, then $(22.15)_2$ yields

$$\langle M_j^{\varepsilon_0}u, u \rangle = \langle f_{2j,1,w}\mathcal{E}_{1j}'a_j^{\varepsilon_0}\mathcal{E}_{1j}'f_{2j,1,w}u, u \rangle + 2\varepsilon_0\|u\|^2 + \langle Tv, v\rangle.$$

Where $T \in \mathcal{L}(\nu^{-4}g, \nu)$, by construction $\mathcal{E}_{1j}'a_j^{\varepsilon_0}\mathcal{E}_{1j}' \geqslant 0$ and $\|u\| \geqslant \|v\|/2$, and Theorem 4.2 yields $\|T\| \leqslant C\nu$. Hence, for $t \geqslant t_0$,

$$\langle M_j^{\varepsilon_0}u, u \rangle \geqslant 0 \quad \forall u \in {}^{\perp}L_j,$$

and

$$\mathcal{N}(M_j^{\varepsilon_0}, \mathcal{S}(R^n; H)) \leqslant \dim L_j = \mathrm{card}\{i \mid \nu_{1ji} \geqslant \tfrac{1}{2}\}.$$

Now the right estimate in (22.7) follows from (22.38) and (22.39), and Theorem 22.1 is proved.

§23 Boundary Value Problems in Strongly Anisotropic Domains

23.1. In anisotropic solid media physics and in mechanics, there frequently appear partial differential equations of the type

$$(23.1) \qquad L_1(x, hD', D'')u = \lambda L_2(x, hD', D'')u,$$

where $x = (x', x'') \in \Omega^{(1)} \times \Omega^{(2)}$ and where $\Omega^{(i)}$ are bounded domains and $h > 0$ is a typical small parameter. Among other problems that reduce to this type are problems for PDOs with constant symbols with respect to x' (i.e., not depending on x') in the domains of the type $\Omega_h^{(1)} \times \Omega^{(2)}$, where $\Omega_h^{(1)} = \{x' \mid hx' \in \Omega^{(1)}\}$—it suffices to make the substitution $x' = h^{-1}y'$.

For equations of the type in (23.1), one can formulate natural boundary value problems, including spectral ones. When $h > 0$ is fixed, as a rule one can find distribution functions for positive and negative spectra; their asymptotics as $h \to +0$ are of great interest in applications.

In order to calculate such asymptotics, it is natural to consider the operators in equation (23.1) as operators in $R_{x'}^{n'}$-space with operator-valued symbols, i.e., to study equations of the type:

$$(23.2) \qquad \hat{L}_1(x', hD')u = \lambda \hat{L}_2(x', hD')u,$$

where $\forall (x', \xi')$, $\hat{L}_j(x', \xi')$ are operators in $L_2(\Omega^{(2)})$. If $\hat{L}_1$ is positive and $\hat{L}_2$ is subordinate to $\hat{L}_1$, then the number of positive (negative) eigenvalues of problem (23.2) is equal to the number of negative eigenvalues of the operator $\hat{L}_1 - \lambda \hat{L}_2$ ($\hat{L}_1 + \lambda \hat{L}_2$); thus, it suffices to study the case of $\hat{L}_2 = I$.

In §23.2, we shall study the Dirichlet problem for PDOs; in §23.3, $\hat{L}_1$ is defined by an integro-differential form with continuous coefficients; in §23.4, we consider anisotropic resonators (whose problem reduces to an operator that is not even pseudodifferential, but this obstacle is overcome similarly to the one in §15 and §16); and in §23.5, we analyze the asymptotics $N(\lambda, \hat{L}_1(hD'))$ [as $h \to +0$] and λ approaching the lower bound of the spectrum of the symbol $\hat{L}_1(\xi')$. From the latter asymptotics, we derive asymptotics for the eigenvalues $\lambda_{N,h}$ of the operator $\hat{L}_1$ for $N \gg 1$, as $h \to +0$.

23.2. Let $m \geqslant 0$; let H_0, H be Hilbert spaces, where $H_0 \subset H$ continuously; let p be a self-adjoint, positive-definite operator in H with domain H_0, which we shall consider as an operator $p : H_0 \to H$; and let the function $a \in C^\infty(R^{2n}; \mathrm{Hom}(H_0; H_0^*))$ satisfy the conditions

$$(23.3) \qquad \|(|\xi|^m + p)^{*-1} a^{(\alpha)}_{(\beta)}(x, \xi)(|\xi|^m + p)^{-1}\| \leqslant C_{\alpha\beta} \quad \forall \alpha, \beta,$$

$$(23.4) \qquad \langle a(x, \xi)u, v \rangle = \overline{\langle a(x, \xi)v, u \rangle} \quad \forall u, v \in H_0;$$

$$(23.5) \qquad \langle a(x, \xi)u, u \rangle \geqslant c\|(|\xi|^m + p)u\|^2_{H_0} - C\|u\|^2_H \quad \forall u \in H_0,$$

uniformly with respect to (x, ξ). Let $a_h(x, \xi) = a(x, h\xi)$, $\mathcal{A}_h[u] = \int \langle a_{h,w} u(x), u(x) \rangle dx$, $u \in \mathcal{S}(R^n; H_0)$. In proving Theorem 23.1, we shall show that for $h \leqslant h_0$, the form $\mathcal{A}_h$ is semibounded from below on $\mathcal{S}(R^n; H_0) \subset L_2(R^n; H)$. Let $\Omega \subset R^n$ be a bounded open domain and let $A_{\Omega,h}$ and $\hat{a}(x, \xi)$ be operators associated with the variational triples $\mathcal{A}_h$, $C_0^\infty(\Omega; H_0)$, $L_2(\Omega; H)$ and $\langle a(x, \xi)\cdot, \cdot \rangle$, H_0, H, respectively.

Let

$$\widehat{V}(\hat{a} - \lambda, \Omega) = (2\pi)^{-n} \iint\limits_{\Omega \times R^n} N(\lambda, \hat{a}(x, \xi)) dx \, d\xi.$$

Theorem 23.1. *Let conditions (23.3)–(23.5) be satisfied and*
  1) *let* $\mathrm{mes}_n \, \partial\Omega = 0$;
  2) *let* $\lambda \in R$ *and let* $c, M > 0$ *exist such that for all* $\xi \in R^n$, $x \in \Omega_c = \{x \mid \mathrm{dist}(x, \Omega) < c\}$

$$(23.6) \qquad\qquad N(\lambda + c, \hat{a}(x, \xi)) < M,$$

and for all $x \in \Omega_c$ *and* $|\xi| > M$,

$$(23.7) \qquad\qquad N(\lambda + c, \hat{a}(x, \xi)) = 0;$$

  3) *As* $\lambda' \to \lambda$, *let*

$$(23.8) \qquad\qquad \widehat{V}(\hat{a} - \lambda', \Omega) \to \widehat{V}(\hat{a} - \lambda, \Omega).$$

Then as $h \to 0$,

$$(23.9) \qquad N(\lambda, A_{\Omega,h}) = h^{-n} \widehat{V}(\hat{a} - \lambda, \Omega) + o(h^{-n}).$$

Remark 23.1.

a) If $m > 0$, then (23.7) follows from (23.5).

b) If $m > 0$ and $H_0 \subset H$ is compact, then (23.6) follows from (23.5).

c) (23.6) and (23.7) imply that $\widehat{V}(\hat{a} - \lambda', \Omega) < \infty$ for $\lambda' < \lambda + c$.

d) Since $F(\lambda) = \widehat{V}(\hat{a} - \lambda, \Omega)$ is a monotone function, condition (23.8) holds for almost all λ satisfying conditions (23.6) and (23.7).

e) If $a(x, \xi)$ is an analytic function of one of the ξ_i (for example, a polynomial), then (23.8) follows from (23.6) and (23.7). Indeed, for all x and $\xi' = (\xi_j)_{j \neq i}$, the eigenvalues $\mu_j(x, \xi', \cdot)$ of the operator $\hat{a}(x, \xi', \cdot)$ are analytic functions of ξ_i and, hence, can take a constant value on a set of nonzero measure only if they do not depend on ξ_i. (23.7) implies that all such $\mu_j(x, \xi', \cdot) \geqslant \lambda + c$. Therefore, $N(\lambda', \hat{a}(x, \xi)) \to N(\lambda, \hat{a}(x, \xi))$ almost everywhere; and the theorem on approaching the limit under the sign of the Lebesgue integral yields (23.8). ∎

Proof of Theorem 23.1. We apply Theorem 22.1 with $t = h^{-1}$,

$$H_{t,0} = H_0, \qquad H_t = H, \qquad \Omega_t = \Omega, \qquad \overset{\circ}{q}_t(x, \xi) = |h\xi|^m + p,$$
$$a_t(x, \xi) = a_h(x, \xi) - \lambda, \qquad g_{t,x,\xi}(y, \eta) = h^2 |\eta|^2 + |y|^2,$$
$$g_{t,x,\xi}^1(y, \eta) = (h^{-2} + |\xi|^2)^{-1} |\eta|^2 + \langle x \rangle^{-2} |y|^2.$$

Condition A in Section 22.1 holds, due to (23.3) and (23.4); conditions (22.3) and (22.4) are obvious; condition (22.6) follows from (23.6); condition (22.5) follows from (23.7) and the boundedness of Ω; and condition (22.2) holds, since for some $\varepsilon, C > 0$,

$$\mathcal{M}_t^\varepsilon \subset \{(x, \xi) \mid |x| < C, \ |\xi| < Ch^{-1}\}.$$

In order to obtain formula (23.9) from (22.7), we notice that due to (23.5), for any ε, there exist $\varepsilon', \varepsilon''$ such that

$$(23.10) \qquad \widehat{\mathcal{A}}_{t,X}{}^{\varepsilon}[u] = \langle a_h(X)u, u \rangle - \varepsilon \|(|\xi|^m + p)u\|_H^2 - \lambda \|u\|_H^2$$
$$\geqslant \langle a_h(X)u, u \rangle (1 - \varepsilon') - (\lambda + \varepsilon'') \|u\|_H^2,$$

$$(23.11) \qquad \widehat{\mathcal{A}}_{t,X}{}^{-\varepsilon}[u] \leqslant \langle a_h(X)u, u \rangle (1 + \varepsilon') - (\lambda - \varepsilon'') \|u\|_H^2,$$

for all $X = (x, \xi)$ (from here on, $\varepsilon, \varepsilon'$, etc., denote functions (of t) such that $\varepsilon \to 0$ as $h \to +0$).

Then (23.10), (23.11), and (22.7) yield the estimate
(23.12)
$$|N(\lambda, A_{\Omega,h}) - \widehat{V}(\hat{a}_h - \lambda, \Omega)|$$
$$\leqslant C(\widehat{V}(\hat{a}_h - \lambda - \varepsilon_1, \Omega) - \widehat{V}(\hat{a}_h - \lambda + \varepsilon_1, \Omega)) + C\widehat{V}(\hat{a}_h - \lambda - \varepsilon_1, (\partial\Omega)_{\varepsilon_1}).$$

But $\widehat{V}(\hat{a}_h - \lambda, \Omega) = h^{-n}\widehat{V}(\hat{a} - \lambda, \Omega)$; therefore, condition (23.8) implies that the first term on the right-hand side of (23.12) is $o(h^{-n})$, and the second term is $o(h^{-n})$, since $\mathrm{mes}(\partial\Omega)_{\varepsilon_1} = \varepsilon_2$.

This proves (23.9). ∎

23.3. Let $m > 0$ be an integer, let $\Omega \subset R^n$ be a bounded Lipschitz domain, and let $H_h^m(\Omega, H_0) \subset L_2(\Omega, H)$ be a subspace of functions with the finite norm

$$\|u\|_{h,m,\Omega}^2 = \sum_{|\alpha| \leqslant m} h^{2|\alpha|} \|p^{1-|\alpha|/m} D^\alpha u\|_{L_2(\Omega;h)}^2.$$

We define on $H_h^m(\Omega; H_0)$ the form

$$\mathcal{A}_h[u] = \sum_{|\alpha|=|\beta|\leqslant m} h^{2|\alpha|} \int_\Omega \langle a_{\alpha\beta}(x)D^\alpha u(x), D^\beta u(x)\rangle dx,$$

assuming that

$$a_{\alpha\beta}(x) = a_{\beta\alpha}^*(x) \in \mathrm{Hom}(H_0, H_0^*) \quad \forall x \in \overline{\Omega};$$
$$a'_{\alpha\beta} = (p^{-1+|\beta|/m})^* a_{\alpha\beta} p^{-1+|\alpha|/m} \in C(\overline{\Omega}; \mathrm{End}\, H);$$
$$\exists c > 0 \; \Big| \; \forall j = 0, 1, \ldots, m, \; \forall u_\alpha \in H_0, \; \forall x \in \overline{\Omega},$$

(23.13)
$$\sum_{|\alpha|=|\beta|=j} \langle a_{\alpha\beta}(x)u_\alpha, u_\beta\rangle \geqslant c \sum_{|\alpha|=j} \|p^{1-j/m}u_\alpha\|_H^2.$$

Then (23.13) yields the estimate

(23.14)
$$\mathcal{A}_h[u] \geqslant c\|u\|_{m,h,\Omega}^2 \quad \forall u \in H_h^m(\Omega; H_0).$$

Let

(23.15)
$$C_0^\infty(\Omega; H_0) \subset V_h \subset H_h^m(\Omega, H_0),$$

and let $A_{\Omega,h}$ and $\hat{a}(x,\xi)$ be operators associated with the variational triples $\mathcal{A}_h$, V_h, $L_2(\Omega, H)$ and $\langle a(x,\xi)\cdot, \cdot\rangle$, H_0, H, respectively, where

$$a(x,\xi) = \sum_{|\alpha|=|\beta|\leqslant m} a_{\alpha\beta}(x)\xi^{\alpha+\beta}.$$

Theorem 23.2.   *Let $H_0 \subset H$ be compact embedding and let conditions* (23.13) *and* (23.15) *be satisfied. Then formula* (23.9) *is valid for any λ.*

Proof.   Let $\omega = h^{1/2}$. We construct a partition of unity $\sum_i \varphi_i^\omega(x) = 1$ on R^n, associated with $\omega^{-2}|\cdot|^2$, $\frac{1}{2}$, $\frac{1}{4}$, and extend the coefficients $a_{\alpha\beta}$ on R^n, preserving their properties. We fix $x_i \in \operatorname{supp} \varphi_i^\omega$, denote

$$a_{\alpha\beta}^\omega(x) = \sum_i \varphi_i^\omega(x) a_{\alpha\beta}(x_i),$$

and define, through the functions $a_{\alpha\beta}^\omega$, the form $\mathcal{A}_h^\omega$ and the functions a^ω and $\hat{a}^\omega$, in the same way that $\mathcal{A}_h$, a, $\hat{a}$ were defined through the functions $a_{\alpha\beta}$.

Since $a_{\alpha\beta}' \in C(\overline{\Omega}; \operatorname{End} H)$ and $\operatorname{diam} \operatorname{supp} \varphi_i^\omega \leqslant \omega/2$, we have

$$|\langle (a_{\alpha\beta}^\omega(x) - a_{\alpha\beta}(x))u, v\rangle| \leqslant \varepsilon \|p^{1-|\alpha|/m} u\|_H \|p^{1-|\beta|/m} v\|_H;$$

therefore,

$$|\mathcal{A}_h^\omega[u] - \mathcal{A}_h[u]| \leqslant \varepsilon_1 \|u\|_{m,h,\Omega}^2 \quad \forall u \in H_h^m(\Omega, H_0),$$
$$|\langle (a^\omega(x,\xi) - a(x,\xi))u, u\rangle| \leqslant \varepsilon_1 \|(|\xi|^m + p)u\|_H^2 \quad \forall u \in H_0.$$

These estimates and (23.13)–(23.15) yield

(23.16) $\mathcal{N}(\mathcal{A}_h^\omega - (\lambda - \varepsilon')\|\cdot\|_{L_2}^2; C_0^\infty(\Omega; H_0))$
$$\leqslant N(\lambda, A_{\Omega,h}) \leqslant \mathcal{N}(\mathcal{A}_h^\omega - (\lambda + \varepsilon')\|\cdot\|_{L_2}^2; H_h^m(\Omega; H_0)),$$

(23.17) $\widehat{V}(\hat{a} - \lambda + \varepsilon', \Omega) \leqslant \widehat{V}(\hat{a}^\omega - \lambda, \Omega) \leqslant \widehat{V}(\hat{a} - \lambda - \varepsilon', \Omega).$

Since $|\varphi_i^{(\alpha)}| \leqslant C_\alpha \omega^{-|\alpha|}$ $\forall \alpha$, then $\forall u \in C_0^\infty(\Omega, H_0)$,

$$\mathcal{A}_h^\omega[u] = \langle a_{h,w}^\omega u, u\rangle + O(\omega \|u\|_{h,m,\Omega}^2);$$

hence, by using (23.14) again, we obtain

$$\mathcal{N}_0(\mathcal{A}_h^\omega - (\lambda - \varepsilon')\|\cdot\|_{L_2}^2; \Omega) \geqslant \mathcal{N}_0(a_{h,w}^\omega - \lambda + \varepsilon'', \Omega).$$

But the asymptotics of the last function can be calculated with the help of Theorem 22.1, by choosing t, $H_{t,0}$, H_t, Ω_t, $\overset{\circ}{q}_t$ in exactly the same way as in §23.2 and by assuming that

$$a_t = a_h^\omega - \lambda + \varepsilon'', \qquad g_{t,x,\xi}(y,\eta) = h^2|\eta|^2 + h^{-1}|y|^2,$$
$$g_{t,x,\xi}^1(y,\eta) = (h^{-1} + |\xi|)^{-2}|\eta|^2 + h^{-1}\langle x\rangle^{-2}|y|^2.$$

Since $H_0 \subset H$ is compact and $m > 0$, conditions (23.6) and (23.7) hold; hence, the conditions of Theorem 22.1 also hold (see Remark 23.1 and the verification of these conditions in §23.2), and we obtain, as in §23.2,

$$|\mathcal{N}_0(a^\omega_{h,w} - \lambda + \varepsilon'', \Omega) - \widehat{V}(\hat{a}^\omega_h - \lambda, \Omega)|$$
$$\leqslant C(\widehat{V}(\hat{a}^\omega_h - \lambda - \varepsilon_1, \Omega) - \widehat{V}(\hat{a}^\omega_h - \lambda + \varepsilon_1, \Omega)) + C\widehat{V}(\hat{a}^\omega_h - \lambda - \varepsilon_1, (\partial\Omega)_{\varepsilon_1}).$$

Now (23.17) implies that on the right-hand side, we can replace $\hat{a}^\omega_h$ with $\hat{a}_h$ (then ε_1 will also change); and since, due to part (e) of Remark 23.1, condition (23.8) holds, then the arguments in §23.2 give

$$\mathcal{N}_0(\mathcal{A}^\omega_h - (\lambda - \varepsilon')\| \cdot \|^2_{L_2}, \Omega) \geqslant h^{-n}\widehat{V}(\hat{a} - \lambda, \Omega) - o(h^{-n}).$$

Returning to (23.16), we obtain the estimate from below for $N(\lambda, A_{\Omega,h})$.

In order to obtain the estimate from above, we use the partition of unity $\sum_i \varphi^\omega_i(x) = 1$ to construct the functions $\psi_\omega, \psi^\pm_\omega \in C^\infty(R^n)$ such that

$$\operatorname{supp} \psi^-_\omega \subset \Omega_{-\omega}, \qquad \psi^-_\omega|_{\Omega_{-2\omega}} = 1, \qquad \psi_\omega|_{\Omega_\omega} = 1,$$
$$\operatorname{supp} \psi_\omega \subset \Omega_{2\omega}, \qquad (\psi^+_\omega)^2 + (\psi^-_\omega)^2 = 1,$$
(23.18)$$\qquad |\psi^{(\alpha)}_\omega| + |(\psi^\pm_\omega)^{(\alpha)}| \leqslant C_\alpha \omega^{-|\alpha|}, \qquad |\alpha| = 0, 1, \ldots,$$

and let $u^\pm_\omega = \psi^\pm_\omega u$. By using (23.18) and (23.14), we obtain
(23.19)
$$\mathcal{A}^\omega_h[u] - (\lambda + \varepsilon')\|u\|^2_{L_2}$$
$$\geqslant \{\mathcal{A}^\omega_h[u^-_\omega](1 - C\omega) - (\lambda + \varepsilon'')\|u^-_\omega\|^2_{L_2}\} + \{c\|u^+_\omega\|^2_{h,m,\Omega} - C\|u^+_\omega\|^2_{L_2}\},$$

where $c, C > 0$ do not depend on $h \in (0, 1)$. Since Ω is a bounded Lipschitz domain, there exists a linear extension operator $\ell : H^m_h(\Omega; H_0) \to H^m_h(R^n; H_0)$ bounded uniformly with respect to $h \in (0, 1]$ (the existence of ℓ can easily be proved, using Stein's extension theorem [1]). Taking (23.18) into account and denoting $u'_\omega = \psi_\omega \ell u^+_\omega$, we have

$$c\|u^+_\omega\|^2_{h,m,\Omega} - C\|u^+_\omega\|^2_{L_2} \geqslant c_1\|u'_\omega\|^2_{h,m,R^n} - C_1\|u'_\omega\|^2_{L_2};$$

and (23.19) yields the estimate

$$N(\lambda, A_{\Omega,h}) \leqslant \mathcal{N}(\mathcal{A}^\omega_h - (\lambda + \varepsilon_1)\| \cdot \|^2_{L_2}, C^\infty_0(\Omega, H_0))$$
$$+ \mathcal{N}(\| \cdot \|^2_{h,m,R^n} - C_2\| \cdot \|^2_{L_2}, C^\infty_0((\partial\Omega)_{4\omega}, H_0)).$$

Here we used the fact that the functions from $H^m_h(\Omega, H_0)$ with supports in Ω are approximated, with respect to the norm of $H^m_h(\Omega, H_0)$, by functions of the class $C^\infty_0(\Omega; H_0)$.

But $\text{mes}(\partial\Omega)_{4\omega} \leqslant Ch^{1/2}$; therefore, Theorem 22.1 yields an estimate for the second term through $O(h^{-n+1/2})$, and we calculate the asymptotics of the first term in exactly the same way as the asymptotics of the function

$$\mathcal{N}_0(\mathcal{A}_h^\omega - (\lambda - \varepsilon'')\| \cdot \|_{L_2}^2, \Omega).$$

∎

23.4. We shall analyze the asymptotics of frequencies ω_n of an electromagnetic resonator $\Omega \subset R^3$ of the form $\Omega = \Omega^{(1)} \times \Omega^{(2)}$, where $\Omega^{(2)}$ is a finite interval and $\Omega^{(1)}$ is a bounded Lipschitz domain with a piecewise smooth boundary. We assume that the dielectric and magnetic permeabilities have the forms

$$(23.20) \qquad \varepsilon_h = E_h \hat{\varepsilon} E_h, \qquad \mu_h = E_h \hat{\mu} E_h,$$

where

$$E_h = [h^{-\delta_{i+j,6}} \delta_{ij}]_{i,j=1,2,3}, \qquad \hat{\varepsilon}, \hat{\mu} \in C^\infty(\overline{\Omega}; \text{End}\, C^3)$$

(the last condition can be weakened as in §23.3), and $\hat{\varepsilon} \geqslant cI_3$, $\hat{\mu} \geqslant cI_3$.

Then, as shown in §13, the distribution functions for squares of eigenfrequencies of the resonator are given by the formula

$$(23.21) \qquad N_h(\lambda) = \mathcal{N}(\mathcal{A}_h[\cdot] - \lambda\langle\varepsilon_h P_h\cdot, \cdot\rangle_0, V_1),$$

where $P_h = I - \nabla F_h^{-1}\nabla^*\varepsilon_h$, F_h is an operator of the Dirichlet problem in Ω for the operator $-\nabla^*\varepsilon_h\nabla$, and

$$\mathcal{A}_h[u,v] = \langle\mu_h^{-1}\operatorname{rot} u, \operatorname{rot} v\rangle_0 + h^2\langle\nabla^*\varepsilon_h u, \nabla^*\varepsilon_h v\rangle_0$$

(i.e., given the notations of §16, we assume that $\rho = h^2$).

Let

$$\operatorname{rot}_h u = \begin{vmatrix} e_1 & e_2 & e_3 \\ h\partial_1 & h\partial_2 & \partial_3 \\ u_1 & u_2 & u_3 \end{vmatrix}, \qquad \nabla_h u = \begin{bmatrix} h\partial_1 u \\ h\partial_2 u \\ \partial_3 u \end{bmatrix}.$$

By $\overset{o}{H}{}_h^1(\Omega)$, we denote a space $\overset{o}{H}{}^1(\Omega)$ with the norm

$$\|u\|_{1,h}^2 = h^2\|\partial_1 u\|_0^2 + h^2\|\partial_2 u\|_0^2 + \|\partial_3 u\|_0^2;$$

and by $F_h' : \overset{o}{H}{}_h^1(\Omega) \to H_h^1(\Omega)^*$, we denote an operator of the Dirichlet problem in Ω for the operator $\nabla_h^*\hat{\varepsilon}\nabla_h$. We let

$$P_h' = I - \nabla_h F_h'^{-1}\nabla_h^*\hat{\varepsilon},$$
$$\mathcal{A}_h'[u,v] = \langle\hat{\mu}^{-1}\operatorname{rot}_h u, \operatorname{rot}_h v\rangle_0 + \langle\nabla_h^*\hat{\varepsilon}u, \nabla_h^*\hat{\varepsilon}v\rangle_0;$$

and in (23.21), we make the substitution $u \to E_h^{-1} u$. Then, we have

(23.22) $$N_h(\lambda) = \mathcal{N}(\mathcal{A}'_h[\cdot] - \lambda \langle \hat{\varepsilon} P'_h \cdot, \cdot \rangle, V_1).$$

Formula (23.22) is also valid for a "long" resonator with ε, μ not depending on x_1, x_2—it suffices to make the substitution $x_i \to h^{-1} x_i$, $i = 1, 2$ in (16.5).

The form in (23.22) defines an operator that can be realized as an operator of the type $A(x', hD')$ with an operator-valued symbol and can be studied as in §§23.2 and 23.3 [here $x' = (x_1, x_2)$]. But $A(x', hD')$ is not even a pseudodifferential operator, since it contains an inverse of the operator of the boundary value problem. This difficulty is overcome the same way as in §15. We use the estimate

(23.23) $$\mathcal{A}'_h[u, u] \geqslant c\|u\|_{1,h}^2 - C\|u\|_0^2 \quad \forall u \in V_1,$$

where $C, c > 0$ do not depend on h, which can be easily obtained by integrating by parts [see §16].

In order to formulate the theorem on the asymptotics of the function $N_h(\lambda)$, we define, for all $(x', \xi') \in R^4$, the operators on a straight line

$$\mathrm{rot}_{h,\xi'} u = \begin{vmatrix} e_1 & e_2 & e_3 \\ ih\xi_1 & ih\xi_2 & \partial_3 \\ u_1 & u_2 & u_3 \end{vmatrix}, \qquad \nabla_{h,\xi'} v = \begin{bmatrix} ih\xi_1 v \\ ih\xi_2 v \\ \partial_3 v \end{bmatrix},$$

and the forms on $H^1(\Omega^{(2)})^3$ and on $\overset{\circ}{H}{}^1(\Omega^{(2)})$

$$\widehat{A}_{h,x',\xi'}[u] = \langle \hat{\mu}^{-1}(x', \cdot)\, \mathrm{rot}_{h,\xi'} u, \mathrm{rot}_{h,\xi'} u \rangle_0 + \langle \nabla^*_{h,\xi'} \hat{\varepsilon}(x', \cdot)u, \nabla^*_{h,\xi'} \hat{\varepsilon}(x', \cdot)u \rangle_0,$$
$$\mathcal{F}_{h,x',\xi'}[v_1, v_2] = \langle \hat{\varepsilon}(x', \cdot)\nabla_{h,\xi'} v_1, \nabla_{h,\xi'} v_2 \rangle_0;$$

and we introduce in $H^1(\Omega^{(2)})$ the norm

$$\|v\|_{h,\xi',1}^2 = (h^2|\xi'|^2 + 1)\|v\|_0^2 + \|\partial_3 v\|_0^2.$$

The forms $\mathcal{F}_{h,x',\xi'}$ define the operators $\widehat{F}_{h,x',\xi'} : \overset{\circ}{H}{}^1_{h,\xi'}(\Omega^{(2)}) \to \overset{\circ}{H}{}^1_{h,\xi'}(\Omega^{(2)})^*$ that are bounded and invertible, uniformly with respect to h, x', ξ.

Let

$$\widehat{P}_{h,x',\xi'} = I - \nabla_{h,\xi'} \widehat{F}^{-1}_{h,x',\xi'} \nabla^*_{h,\xi'} \hat{\varepsilon}(x', \cdot),$$
$$\widehat{A}^{\lambda,\nu}_{h,x',\xi'}[u] = \widehat{A}_{h,x',\xi}[u](1 - \nu) - \nu\|u\|_0^2 - \lambda \langle \hat{\varepsilon}(x', \cdot)\widehat{P}_{h,x',\xi'} u, u \rangle_0,$$
$$H_1 = \{u \in H^1(\Omega^{(2)})^3 \mid u_1\big|_{\partial\Omega^{(2)}} = u_2\big|_{\partial\Omega^{(2)}} = \partial_3 u_3\big|_{\partial\Omega^{(2)}} = 0\}.$$

Theorem 23.3. *For fixed λ, as $h \to +0$,*

(23.24) $$N_h(\lambda) = h^{-2} V(\mathcal{A}_1^{\lambda,0}, H_1, \Omega^{(1)} \times R^n) + o(h^{-2}).$$

Proof. One can easily show that $A^\lambda_{1,x',\xi'}$—an operator associated with the variational triple $\widehat{A}^{\lambda,0}_{1,x',\xi'}$, H_1, $L_2(\Omega^{(2)})^3$—is an analytic function of ξ_2 in some neighborhood of the real axis; therefore, the arguments in part e) of Remark 23.1 and the proof of Theorem 23.1 show that it suffices to prove the estimate

$$(23.25) \quad \begin{aligned} o(h^{-2}) + h^{-2}V(\widehat{A}_1{}^{\lambda,-\nu}; H_1; \Omega^{(1)} \times R^2) &\leqslant N_h(\lambda) \\ &\leqslant h^{-2}V(\widehat{A}_1{}^{\lambda,\nu}; H_1; \Omega^{(1)} \times R^2) + o(h^{-2}), \end{aligned}$$

where $\nu \to 0$ as $h \to +0$.

Let us continue $\hat{\varepsilon}$, $\hat{\mu}$ in R^3 to positive-definite matrix functions of the class C^∞, constant outside some compact set. Let $\Omega_0 = R^2 \times \Omega^{(2)}$, and let $F_h^0 : \overset{\circ}{H}{}^1_h(\Omega_0) \to \overset{\circ}{H}{}^1_h(\Omega_0)^*$ be an operator of the Dirichlet problem in Ω_0 for $\nabla_h^* \hat{\varepsilon} \nabla_h$. Furthermore, let $H = L_2(\Omega^{(2)})$ and $H_0 = \overset{\circ}{H}{}^1(\Omega^{(2)})$. By $\hat{p}$ we denote a Friedrichs extension of the operator $-\partial^2 : C_0^\infty(\Omega^{(2)}) \to H$. Then $p = \hat{p}^{1/2} : H_0 \to H$ and $p^* : H \to H_0^*$ are isomorphisms. We identify p, p^* with $I \otimes p$, $I \otimes p^*$ and let $\widetilde{\nabla}_h^* = p^{*-1}\nabla_h^*$, $\widetilde{\nabla}_h = \nabla_h p^{-1}$, $\widetilde{F}_h' = p^{*-1}F_h'p^{-1}$, $\widetilde{F}_h^0 = p^{*-1}F_h^0 p^{-1}$, $b_h(x',\xi') = h|\xi|p^{-1}+1$, $g_{h,x',\xi'}(y',\eta') = h^2|\eta'|^2 + h^{-1}|y'|^2$. Then $P_h' = I - \widetilde{\nabla}_h\widetilde{F}_h'^{-1}\widetilde{\nabla}_h^*\hat{\varepsilon}$, and

$$(23.26) \quad \widetilde{\nabla}_h^* \in \mathcal{L}(g_h, b_h^*, I_{H^3}), \qquad \widetilde{\nabla}_h \in \mathcal{L}(g_h, I_{H^3}, b_h),$$

$$(23.27) \quad \widetilde{F}_h^0 \in \mathcal{L}(g_h, b_h^*, b_h),$$

$$(23.28) \quad \sigma^w(\widetilde{F}_h^0) - p^{*-1}\widehat{F}_{h,\cdot,\cdot}p^{-1} \in \mathcal{L}(g_h, hb_h^*, b_h).$$

Let

$$(23.29) \quad R_h = (p(\widehat{F}_{h,\cdot,\cdot})^{-1}p^*)_w \in \mathcal{L}(g_h, b_h^{-1}, b_h^{*-1}).$$

Applying Theorem 3.1 and taking (23.27)–(23.29) into account, we obtain

$$(23.30) \quad \widetilde{F}_h^0 R_h = I + K_h, \quad K_h \in \mathcal{L}(g_h, hb_h^*, b_h^{*-1}).$$

Let $\omega = h^{1/2}$. Starting with the set $\Omega^{(1)}$, we construct the functions ψ_ω, $\psi_\omega^\pm$ as in §23.3. Then

$$(23.31) \quad \psi_\omega, \psi_\omega^\pm \in S(g_h, 1);$$

and due to (23.30) and equality $\psi_\omega^- \psi_{\omega/2}^- = \psi_\omega^-$, we have

$$(23.32) \quad \psi_\omega^- = \widetilde{F}_h^0 \psi_\omega^- R_h + \psi_{\omega/2}^- K_h', \quad K_h' \in \mathcal{L}(g_h, hb_h^*, b_h^*)^{-1}.$$

We identify the functions $f \in \overset{\circ}{H}{}^1(\Omega_0)^*$, for which $\operatorname{supp} f \subset \Omega$, with the function from $\overset{\circ}{H}{}^1(\Omega)^*$ and notice that u, for which $\operatorname{supp} u \subset \Omega$, satisfy the equality $F'_h u = f^0_h u$. Therefore, in (23.32) we can replace $\widetilde{F}^0_h$ with $\widetilde{F}'_h$. But $\psi^-_{2\omega} u = u$ $\forall u \in C^\infty_0(\Omega^{(1)}_{-4\omega}, H)$; hence, using (23.32), we obtain

$$(23.33) \qquad P'_h u = (I - \widetilde{\nabla}_h R_h \widetilde{\nabla}^*_h \hat{\varepsilon})u + \widetilde{\nabla}_h \widetilde{F}'^{-1}_h \psi^-_{\omega/2} K'_h \widetilde{\nabla}^*_h u + \psi^-_{\Omega/2} T_h u,$$

where $T_h \in \mathcal{L}^{-\infty}(g_h; 1)$; Theorem 4.2 yields $\|\psi^-_{\omega/2} T_h u\|_0 \leqslant Ch\|u\|_0$.

In order to obtain the same estimate for the second term, we let

$$\overset{\circ}{\widetilde{H}}_h{}^1(\Omega) = \{u \mid \|p^{-1}u\|_{\overset{\circ}{H}^1_h(\Omega)} < \infty\}.$$

Since $\nabla_h F'^{-1}_h : \overset{\circ}{H}^1_h(\Omega)^* \to L_2(\Omega)^3$ is uniformly bounded, so is $\widetilde{\nabla}_h \widetilde{F}'^{-1}_h :$
$\overset{\circ}{\widetilde{H}}_h{}^1(\Omega)^* \to L_2(\Omega)^3$, but

$$\|u\|_{\overset{\circ}{\widetilde{H}}_h{}^1(\Omega)^\bullet} \leqslant \|u\|_{\overset{\circ}{\widetilde{H}}_h{}^1(\Omega_0)^\bullet} \qquad \forall u \in C^\infty_0(\Omega^{(1)}, H);$$

hence, it suffices to prove the estimate

$$(23.34) \qquad \|\psi^-_{\omega/2} K'_h \widetilde{\nabla}^*_h u\|_{\overset{\circ}{\widetilde{H}}_h{}^1(\Omega_0)^\bullet} \leqslant Ch\|u\|_0.$$

Due to Lemma 5.6, $\overset{\circ}{\widetilde{H}}_h{}^1(\Omega_0)$ is identified with $\mathcal{H}(b_h, H)$; and due to Theorem 5.2, $\mathcal{H}(b_h^{-1}, H)$ is identified with $\mathcal{H}(b_h, H)^*$. Therefore, on the left-hand side we can replace $\overset{\circ}{\widetilde{H}}_h{}^1(\Omega)^*$ with $\mathcal{H}(b_h^{-1}, H)$.

Now (23.26) and (23.32) imply that $\psi^-_{\omega/2} K'_h \widetilde{\nabla}^*_h \in \mathcal{L}(g_h, hb_h, I)$. Hence, (23.34) follows from Theorem 5.2; and (23.33) and (23.34) yield

$$\|P'_h u - \widetilde{P}_h u\|_0 \leqslant \nu\|u\|_0, \quad \text{where } \widetilde{P}_h = I - \widetilde{\nabla}_h R_h \widetilde{\nabla}^*_h \hat{\varepsilon}.$$

By identifying $\widetilde{V}_1 = C^\infty_0(\Omega_{-4\omega}, H_1)$ with a subspace in V_1 and taking (23.23) into account, we obtain

$$(23.35) \qquad N_h(\lambda) \geqslant \mathcal{N}(\mathcal{A}'_h - \lambda\langle \hat{\varepsilon}\widetilde{P}_h \cdot, \cdot\rangle + \nu\| \cdot \|^2_0, \widetilde{V}_1).$$

Let $\hat{p}_1$ be an operator associated with the variational triple $\| \cdot \|^2_1, H_1, H^3$ and with $p_1 = \hat{p}_1^{1/2}$. The form in (23.35) can be written as $\langle \widetilde{A}_{h,\lambda,-\nu} \cdot, \cdot\rangle$, where $\widetilde{A}_{h,\lambda,-\nu}$ is a PDO in R^2 with an operator-valued symbol satisfying the conditions

$$(23.36)$$

$$\|(|h\xi'| + p_1)^{*-1}(\tilde{a}_{h,\lambda,-\nu})^{(\alpha)}_{(\beta)}(x', \xi')(|h\xi'| + p_1)^{-1}\| \leqslant C_{\alpha\beta}h^{|\alpha|}, \quad \forall \alpha, \beta,$$

$$\langle \tilde{a}_{h,\lambda,-\nu}(x', \xi')u, u\rangle \geqslant c\|(|h\xi'| + p_1)u\|^2_0 - C\|u\|^2_0,$$

$$\langle \tilde{a}_{h,\lambda,-\nu}(x', \xi')\cdot, \cdot\rangle = \widetilde{\mathcal{A}}^{\lambda,-\nu}_{h,x',\xi'}[\cdot] + \langle d_h(x', \xi')\cdot, \cdot\rangle,$$

where d_h satisfies conditions (23.36) with exponents $|\alpha| + \frac{1}{2}$ instead of $|\alpha|$. Therefore, by applying Theorem 22.1 to the operator $\widetilde{A}_{h,\lambda,-\nu}$ as in §23.2, we obtain the left estimate in (23.25).

In order to obtain the estimate from above, we use the functions $\psi_\omega^\pm$ in the same way as in §23.3. Here an additional difficulty appears as we estimate the commutator

(23.37)
$$[P_h', \psi_\omega^\pm] = P_h' \psi_\omega^\pm - \psi_\omega^\pm P_h'$$
$$= -\widetilde{\nabla}_h \widetilde{F}_h'^{-1}[\widetilde{\nabla}_h^* \hat{\varepsilon}, \psi_\omega^\pm] - \widetilde{\nabla}_h \widetilde{F}_h'^{-1}[\psi_\omega^\pm, \widetilde{F}_h'] \widetilde{F}_h'^{-1} \widetilde{\nabla}_h^* \hat{\varepsilon} - [\widetilde{\nabla}_h, \psi_\omega^\pm] \widetilde{F}'^{-1} \widetilde{\nabla}_h^* \hat{\varepsilon}.$$

But $[\psi_\omega^\pm, \widetilde{F}_h'] = \psi_{\omega/2}^-[\psi_\omega^\pm, \widetilde{F}_h^0]$; therefore, the same arguments used to estimate the second term in (23.33) show that the norms of all operators in (23.37) are small, and our proof is completed as in §23.3.

This proves Theorem 23.3. ∎

23.5. Assume that the symbol a does not depend on x and that conditions (23.3)–(23.5) are satisfied. Let $\lambda_0 = \inf \sigma(\hat{a}(\xi))$ for only a single point ξ_0, where λ_0 is an isolated point of the spectrum of the operator $\hat{a}(\xi_0)$, and let $r = \dim \operatorname{Ker}(a(\xi) - \lambda_0 I_{H_0}) < \infty$.

Let there also exist $c, c_1 > 0$ such that

(23.38)
$$\hat{a}(\xi) > (\lambda_0 + c_1) I_H \quad \forall \xi \notin \{\xi_0\}_c.$$

By $\pi_2 : H \to H$ we denote an orthoprojection onto $L_2 = \operatorname{Ker}(a(\xi_0) - \lambda_0 I_{H_0})$. Let $\pi_1 = I - \pi_2$, $L_1' = \pi_1 H$, $L_1 = L_1' \cap H_0$ and let the operators $a_{ij}(\xi) : L_j \to L_i^*$ be defined by the equalities $a_{ij}(\xi) u = a(\xi) u|_{L_i}$.

Now let $\hat{a}_{11}(\xi)$ be an operator that is associated with the variational triple $\langle a_{11}(\xi)\cdot, \cdot \rangle$, L_1, L_1'. Then the choice of λ_0 and condition (23.38) yield

(23.39)
$$(\hat{a}_{11}(\xi) - \lambda_0) \geqslant c_2 I_{L_1'} \quad \forall \xi,$$

where $c_2 > 0$; and (23.5) implies that

(23.40)
$$\langle (\hat{a}_{11}(\xi) - \lambda_0) u, u \rangle \geqslant c(|\xi|^{2m} \|u\|_{L_1'}^2 + \|u\|_{L_1}^2) - C\|u\|_{L_1}^2 \quad \forall u \in D(\hat{a}_{11}(\xi)),$$

uniformly with respect to ξ.

Then (39) implies that there exists C_1 such that

$$C_1(\hat{a}_{11}(\xi) - \lambda_0) \geqslant \hat{a}_{11}(\xi) - \lambda_0 + C;$$

thus, (23.40) holds with $C = 0$ (and with different $c > 0$); and since $D(\hat{a}_{11}(\xi))$ is dense in L_1, then $\forall u \in L_1$, $\forall \xi$, we have

$$\langle (a_{11}(\xi) - \lambda_0) u, u \rangle \geqslant c(|\xi|^{2m} \|u\|_{L_1'}^2 + \|u\|_{L_1}^2).$$

This estimate implies invertibility of the operator $a_{11}(\xi) - \lambda_0$. Moreover, if p_1 is a self-adjoint positive-definite operator in L_1' with domain L_1, which we consider as an operator $p_1 : L_1 \to L_1'$, then (23.3) yields the estimate

$$(23.41) \qquad \|(|\xi|^m + p_1)((a_{11}(\xi) - \lambda_0)^{-1})^{(\alpha)}(|\xi|^m + p_1)^*\| \leqslant C_\alpha$$

for all α, ξ.

Suppose that for $\xi \in \{\xi_o\}_{c_1}$,

$$(23.42) \qquad \|a_{12}(\xi)\|_{L_2 \to L_1^*} \leqslant C|\xi - \xi_0|^k, \qquad \|a_{22}(\xi) - \lambda_0\| \leqslant C|\xi - \xi_0|^{2k},$$

where $k > 0$ is an integer. (23.41) implies that the matrix

$$a_0(\xi) = a_{22}(\xi) - \lambda_0 I_{L_2} - (a_{21}(a_{11} - \lambda_0 I_{L_1'})^{-1} a_{12})(\xi)$$

is defined; here a_0 is uniformly bounded, as are all of its derivatives.

By expanding a_0 into a Taylor series in a neighborhood of the point ξ_0 and taking (23.42) into account, we obtain, as $\xi \to \xi_0$,

$$(23.43) \qquad a_0(\xi) = a^0(\xi - \xi_0) + O(|\xi - \xi_0|^{2k+1}),$$

where a^0 is a homogeneous polynomial of degree $2k$.

Theorem 23.4. *Let conditions (23.3)–(23.5), (23.38), and (23.42) hold, and let*

$$(23.44) \qquad\qquad a^0(\eta) \geqslant c|\eta|^{2k} I_r, \quad c > 0,$$
$$(23.45) \qquad\qquad \mathrm{mes}(\partial\Omega)_\varepsilon \leqslant C\varepsilon, \quad \forall \varepsilon > 0.$$

Then for $1 \ll t \ll h^{-1}$, $\lambda = \lambda_0 + (th)^{2k}$, and $\delta \in (0, \tfrac{1}{3})$,

$$(23.46) \qquad\qquad N(\lambda, A_{\Omega,h}) = V_0 t^n (1 + O(t^{-\delta}) + O(th)),$$

where

$$V_0 = (2\pi)^{-n} \,\mathrm{mes}\,\Omega \sum_{1 \leqslant j \leqslant r} \mathrm{mes}\{\eta \mid \lambda_j^0(\eta) < 1\},$$

and λ_j^0 are eigenvalues of the matrix a^0.

If the matrix a^0 is locally smoothly diagonalizable on $R^n \backslash O$ (for example, if $r = 1$), then estimate (23.46) holds for all $\delta \in (0, \tfrac{1}{2})$.

Proof. Without loss of generality, we can assume that $\lambda_0 = 0$, $\xi_0 = 0$. We fix an isomorphism $\hat{p} : H \to H_0$ such that $\hat{p}|_{L_2} = I_{L_2}$, $\hat{p}L_1' = L_1$, $\hat{p}^*|_H = \hat{p}$; and we denote $a' = \hat{p}^* a \hat{p}$, $a'_{ij} = \pi_i a'|_{\pi_j H}$, $\hat{p}_1 = \pi_1 \hat{p}|_{L_1'}$, $a'_0 = a'_{22} - \lambda_0 - a'_{21}(a'_{11} - \lambda_0)^{-1} a'_{12}$. Then $a'_{22} = a_{22}$, $a'_0 = a_0$ and

$$(23.47) \quad N(\lambda, A_{\Omega,h}) = \mathcal{N}(a_{h,w} - \lambda; C_0^\infty(\Omega; H_0)) = \mathcal{N}(a'_{h,w} - \lambda\hat{p}^2; C_0^\infty(\Omega; H)).$$

We note that due to (23.39),

$$(23.48) \qquad a'_{11}(\xi) \geqslant c\hat{p}_1^2, \qquad \hat{a}_{11}(\xi) - \lambda\hat{p}_1^2 \geqslant c_1\hat{p}_1^2$$

uniformly with respect to ξ (the last inequality is valid for small th). Let $t^{-\delta} = \omega$. We construct the functions $\psi_\omega^\pm$, ψ_ω as in §23.3 and denote $\varphi_\omega^- = \psi_\omega^-$, $\varphi_\omega^+ = \psi_\omega$,

$$Q_{h,\omega}^\pm = \begin{bmatrix} I_{L_1'} & -\varphi_{2\omega}^\pm a'^{-1}_{11,h,w} a'_{12,h,w} \\ 0 & I_{L_2} \end{bmatrix},$$

$$g_{t,x,\xi}(y,\eta) = t^{-2}|\eta|^2 + t^{2\delta}|y|^2, \qquad b_{t,h}^1 = (a'_{11,h})^{1/2},$$

$$b_{t,h}^2(x,\xi) = \langle h\xi\rangle^{m-k}(|h\xi|^{2k} + (th)^{2k})^{1/2} I_{L_2},$$

$$b_{t,h} = [b_{t,h}^i \delta_{ij}]_{i,j=1,2}.$$

One can easily verify that, uniformly with respect to $1 \ll t \ll h^{-1}$, g_t is a σ-temperate metric and $b_{t,h}^i$, $b_{t,h}$ are σ,g_t-temperate functions; and due to (23.3), (23.41)–(23.43), and (23.48),

$$(23.49) \qquad a'_{11,h} \in SI^+(g_t, b_{t,h}^{1*}, b_{t,h}^1), \qquad a'_{12,h} \in S(g_t, b_{t,h}^{1*}, b_{t,h}^2),$$

$$(23.50) \qquad \begin{aligned} \lambda\hat{p}_1^2 &\in S(g_t, \lambda b_{t,h}^{1*}, b_{t,h}^1), \\ a_{22,h}, a_{0,h}, \lambda &\in S(g_t, b_{t,h}^2, b_{t,h}^2), \\ Q_{h,\omega}^\pm &\in \mathcal{L}(g_t, b_{t,h}^{-1}, b_{t,h}). \end{aligned}$$

By construction, $\operatorname{supp}\varphi_{2\omega}^- \subset \Omega_{-2\omega}$ and $\operatorname{supp}\varphi_{2\omega}^+ \subset \Omega_{4\omega}$; therefore,

$$Q_{h,\omega}^+ : C_0^\infty(\Omega_{4\omega}, H) \to C_0^\infty(\Omega_{4\omega}, H),$$
$$Q_{h,\omega}^- : C_0^\infty(\Omega_{-2\omega}, H) \to C_0^\infty(\Omega_{-2\omega}, H),$$

are isomorphisms.

Furthermore, $\varphi_{2\omega}^- \varphi_{4\omega}^- = \varphi_{4\omega}^-$, $\varphi_{2\omega}^+ \varphi_\omega^+ = \varphi_\omega^+$, and

$$\varphi_{4\omega}^- u = u \quad \forall u \in C_0^\infty(\Omega_{-8\omega}, H),$$
$$\varphi_\omega^+ u' = u' \quad \forall u \in C_0^\infty(\Omega, H);$$

therefore, by denoting

$$\tilde{Q}_h = \begin{bmatrix} I_{L_1'} & -a'^{-1}_{11,h,w} a'_{12,h,w} \\ 0 & I_{L_2} \end{bmatrix},$$

we have

(23.51) $$Q^+_{h,\omega}u' = \tilde{Q}_h u' + T^+_{h,\omega}u', \qquad Q^-_{h,\omega}u = \tilde{Q}_h u + T^-_{h,\omega}u,$$

where $T^{\pm}_{h,\omega} \in \mathcal{L}^{-\infty}(g_t, b^{-1}_{t,h}, b_{t,h})$.

Assuming that $u_j = \pi_j(Q^-_{h,\omega})^{-1}u$ for $u \in C^\infty_0(\Omega_{-8\omega}, H)$ and using (23.49)–(23.51), we obtain

(23.52)
$$\langle(a'_{h,w} - \lambda\hat{p}^2)u, u\rangle = \langle(a'_{11,h,w} - \lambda\hat{p}^2_1)u_1, u_1\rangle + \langle(a_{0,h,w} - \lambda I_{L_2})u_2, u_2\rangle + \langle K_{h,\omega}u, u\rangle,$$

where

$$K_{h,\omega} \in \mathcal{L}(g_t, d_t b^*_{t,h}, b_{t,h}), \qquad d_t = t^{\delta-1} + (th)^{2k}.$$

For sufficiently large C_0, $Q^{-*}_{h,\omega}K_h Q^-_{h,\omega} + C_0 d_t(b^*_{t,h}b_{t,h})_w \in \mathcal{LI}^+(g_t, d_t b^*_{t,h}, b_{t,h})$; but as $t \to \infty$, $h_{g_t} = t^{\delta-1} \to 0$. Therefore, by applying Lemma 8.4', we see that for $t \geqslant t_0$,

$$|\langle K_{h,\omega}u, u\rangle| \leqslant C_0 d_t(\langle b^2_{t,h,w}u_2, u_2\rangle + \langle a'_{11,h,w}u_1, u_1\rangle);$$

and hence,

(23.53) $$\langle(a'_{h,w} - \lambda\hat{p}^2)u, u\rangle \leqslant \sum_{j=1,2}\langle A^{-j}_{t,h}u_j, u_j\rangle,$$

where

$$A^{\pm 2}_{t,h} = a_{0,h,w} - \lambda I \mp C_0 d_t b^2_{t,h,w},$$
$$A^{\pm 1}_{t,h} = a'_{11,h,w} - \lambda\hat{p}^2_1 \mp C_0 d_t a''_{11,h,w}.$$

Denoting $u_j = \pi_j(Q^+_{h,\omega})^{-1}u$ for $u \in C^\infty_0(\Omega; H)$ and repeating the same arguments, we obtain

(23.54) $$\langle(a'_{h,w} - \lambda\hat{p}^2)u, u\rangle \geqslant \sum_{j=1,2}\langle A^j_{t,h}u_j, u_j\rangle.$$

But due to (23.48) and (23.49), $A^1_{t,h} \in \mathcal{LI}^+(g_t, b^{1*}_{t,h}, b^1_{t,h})$ for small t, th; and hence, it is positive-definite [Lemma 8.4']. Therefore, (23.47), (23.53), and (23.54) yield the estimate

(23.55) $$\mathcal{N}_0(A^{-2}_{t,h}, \Omega_{-8\omega}) \leqslant N(\lambda, A_{\Omega,h}) \leqslant \mathcal{N}_0(A^2_{t,h}, \Omega_{8\omega}).$$

The asymptotics of the functions $\mathcal{N}_0(A^{\pm 2}_{t,h}, \Omega_{\pm 8\omega})$ can be analyzed with the help of Theorems 7.1 and 7.2. Condition (23.44) implies that on sets whose measures appear in asymptotic formulas of §7, we have $|h\xi| \leqslant Cth$; hence, on these sets [due to (23.43) and (23.44)],

$$
\begin{aligned}
(23.56) \qquad \sigma^w(A_{t,h}^{\pm2})(x,\xi) &= a^0(h\xi) - (th)^{2k}I + O((t^{\delta-1} + th)(th^{2k})I \\
&= h^{2k}[a^0(\xi) - t^{2k}I + O(t^{\delta-1} + th)t^{2k}].
\end{aligned}
$$

Using (23.56) and taking into account the fact that due to (23.45),

$$
\operatorname{mes} \Omega_{\pm 8\omega} = \operatorname{mes} \Omega + O(\omega),
$$

we can repeat the arguments from §13 to obtain

$$
(23.57) \qquad \mathcal{N}_0(A_{t,h}^{\pm2}, \Omega_{\pm 8\omega}) = V_0 t^n (1 + O(t^{-\delta} + th)).
$$

Then (23.55) and (23.57) yield (23.46), thus proving Theorem 23.4. ∎

Corollary 23.1. *Let* $\lambda_{N,h}$ *be the* N*th eigenvalue of the operator* $A_{\Omega,h}$, $1 \ll N \ll h^{-n}$, *and let the conditions of Theorem 23.4 hold. Then*

$$
(23.58) \qquad \lambda_{N,h} = \lambda_0 + (V_0^{-1}N)^{2k/n} h^{2k}(1 + O(N^{-\delta/n} + N^{1/n}h)),
$$

where δ *is defined as in Theorem 23.4.*

Proof. In (23.46) let $\lambda = \lambda_{N,h}$. Then

$$
t = (\lambda_{N,h} - \lambda_0)^{1/2k} h^{-1}, \quad 1 \ll t \ll h^{-1}.
$$

Therefore, the conditions $\lambda = \lambda_0 + (th)^{2k}$, $1 \ll t \ll h^{-1}$ of Theorem 23.4 hold, and

$$
N = V_0(\lambda_{N,h} - \lambda_0)^{n/2k} h^{-n}(1 + O(N^{-\delta/n} + N^{1/n}h)).
$$

This yields (23.58).

CHAPTER 6

DEGENERATE DIFFERENTIAL OPERATORS

§24 General Analysis of Degenerate Operators and Generalizations of the Weyl Formula

24.1. In §13, §14, and §§19–21, we studied operators that either satisfied the conditions of the theorems in §7, or reduced to such operators after changing their symbols outside $\Omega \times R^n$; but even the simplest degenerate operator

$$(24.1) \qquad A = \sum_{|\alpha|=|\beta|\leqslant m} D^\beta a'_{\alpha\beta}(x)\rho(x)^{2k}D^\alpha,$$

where ρ is a regularized distance to a submanifold $\Gamma \subset \overline{\Omega}$, is not so easy to study. A function ρ is called a regularized distance to a set Γ, if $\rho(\cdot) \asymp \operatorname{dist}(\cdot,\Gamma)$ and

$$|\rho^{(\alpha)}| \leqslant C_\alpha \rho^{1-|\alpha|} \quad \forall\alpha.$$

Indeed, one can easily see that a σ-temperate metric g_t and a function $\overset{\circ}{q}_t \in O_r(C^m; g_t)$ such that $A_t = A - t \in \mathcal{L}I^+(g_t, \overset{\circ}{q}{}^*_t, q_t)$ can be found only in the case where k is an integer, $k \geqslant m$; but if $k \geqslant m$, then the spectrum of the operator A is not discrete (to be shown below).

At the same time, in each subdomain $\Omega_1 \subset \Omega$ such that $\overline{\Omega}_1 \subset \Omega$, the operator $A_t = a_{t,w}$ is sufficiently good, since its coefficients can be changed outside Ω so that the conditions of the theorems in §7 hold; on the other hand, along the tangent to Γ directions, the operator A does not degenerate; therefore, in thin strip adjacent to Γ, the operator A can be realized as a nondegenerate operator on Γ with an operator-valued symbol. These remarks suggest the following way of studying degenerate operators.

1. We shall find the scalar functions $\Psi_{t,j}$, $\psi_{t,j}$ $(j = 1, \ldots, n)$ and a matrix function $\overset{\circ}{q}_t$ such that on $\Omega \times R^n$

$$(24.2) \qquad 1) \qquad \|\overset{\circ}{q}{}^{-1*}_t a^{(\alpha)}_{t(\beta)}\overset{\circ}{q}{}^{-1}_t\| \leqslant c_{\alpha\beta} \Psi_t^{-\alpha}\psi_t^{-\beta}$$

for all α, β, where $\psi_t^\beta = \psi_{t,1}^{\beta_1} \cdots \psi_{t,n}^{\beta_n}$;

(24.3)   2) $a_t(x,\xi) \geqslant c \overset{\circ}{q}_t(x,\xi)^* \overset{\circ}{q}_t(x,\xi)$ for $|x| + |\xi| \geqslant C(t)$.

Condition (24.3) is one of the fundamental conditions in §7; condition (24.2) can be written as $a_t \in S(g_t, \overset{\circ}{q}_t^*, \overset{\circ}{q}_t, \Omega \times R^n)$, where

$$g_{t,x,\xi}(y,\eta) = \sum_{1 \leqslant j \leqslant n} (\Psi_{t,j}(x,\xi)^{-2}|\eta_j|^2 + \psi_{t,j}(x,\xi)^{-2}|y_j|^2),$$

but in many interesting cases, the metric g_t and the function $\overset{\circ}{q}_t$ will not be σ-temperate and σ, g_t-temperate, respectively.

2. We shall find a sufficiently wide open set $\Omega_t \subset \Omega$ such that

(24.4) $$\max_{\Omega_t \times R^n} h_t(x,\xi) = M^{-1} \to 0 \quad \text{as} \quad t \to \infty;$$

for the given metric

$$h_t(x,\xi) = \max_{1 \leqslant j \leqslant n} (\Psi_{t,j} \psi_{t,j})(x,\xi)^{-1}.$$

Naturally, a maximal Ω_t does not exist but the choice of Ω_t is not essential if M does not increase too rapidly. Unsuccessful choice of Ω_t results in more calculations but also leads to the needed result. We note that in all considered cases, it is sufficient to take $M = \ln \ln t$.

3. Using standard variational arguments (utilized in §14 and §23.3), we must reduce the original problem to problems on Ω_t and on $\Omega_t \backslash \overline{\Omega}_t$. Denoting by $N_1(t)$, $N_2(t)$ distribution functions for spectra of these problems, we have

$$N(0, A_t) \sim N_1(t) + N_2(t).$$

4. If conditions (24.2)–(24.4) hold and the coefficients of the operator do not oscillate, then as a rule, the functions $\Psi_{t,j}$, $\psi_{t,j}$, $\overset{\circ}{q}_t$ and the symbol a_t can be changed outside $\Omega_t \times R^n$ such that g_t becomes a σ-temperate metric, the function $\overset{\circ}{q}_t$ becomes σ, g_t-temperate, and conditions (24.2)–(24.4) hold on $R^n \times R^n$. Then the conditions of Theorem 7.5 will also hold, and using this theorem, we can obtain a classical Weyl formula for $N_1(t)$:

(24.5) $$N_1(t) \sim (2\pi)^{-n} \iint\limits_{\Omega_t \times R^n} N(0, a_t(x,\xi)) dx \, d\xi.$$

5. In order to study the problems in $\Omega \backslash \overline{\Omega}_t$, we further localize and then straighten Γ; we obtain a set of problems in domains of the type $\Pi \times K(\omega)$, where $\Pi \subset R^{n-n_1}$ is the straightened part of Γ, $K(\omega) = \{x'' \in K \mid |x''| < \omega\}$,

and $K \subset R^{n_1}$ is a cone. Along the tangent with respect to the Γ directions, the original operator does not degenerate; thus, if we realize the operator in $\Pi \times K(\omega)$ as an operator with an operator-valued symbol, then we obtain a nondegenerate operator. The asymptotic function of its spectrum is studied with the help of Theorem 22.1, and we obtain the classical formula for an operator with an operator-valued symbol

$$(24.6) \qquad N_2(t) \sim (2\pi)^{n_1-n} \iint\limits_{T^{\bullet}\Gamma} N(0, \tilde{a}_t(\tilde{x}, \tilde{\xi})) d\tilde{x}\, d\tilde{\xi},$$

where $n_1 = \operatorname{codim}\Gamma$, $\tilde{a}_t$ is an auxiliary operator-valued symbol.

6. Calculating asymptotics of integrals in the asymptotic formulae (24.5), (24.6) : Here it is useful to keep in mind that, usually, either $N_1(t) = o(N_2(t))$ or $N_2(t) = o(N_1(t))$; thus, it is sufficient to roughly estimate one of the integrals. In §24.3 and §24.4, we shall give the criteria for distinguishing these two cases and also the formulas *a priori* predicting the asymptotics of $N(0, A_t)$.

24.2. Formal procedure of the definition of the functions $\Psi_{t,j}$, $\psi_{t,j}$, $\overset{\circ}{q}_t$ can be formulated easier for the operator in a divergent form

$$A = \sum_{\alpha,\beta} D^{\beta} a_{\alpha\beta} D^{\alpha}.$$

Let degeneracy of the derivative D^{α} be defined by the function $\rho_{\alpha} = \rho_{\alpha}(x)$ such that the operator

$$A' = \sum_{\alpha,\beta} D^{\beta} a'_{\alpha\beta} D^{\alpha}, \quad a'_{\alpha\beta} = \rho_{\beta}^{-1} a_{\alpha\beta} \rho_{\alpha}^{-1}$$

is not degenerate. Then we must let

$$(24.7) \qquad q_t(x, \xi)^2 = \sum_{\alpha} \rho_{\alpha}(x)^2 \xi^{2\alpha} + t.$$

Furthermore, if degeneracy is isotropic, i.e., degeneracy of the derivatives D^{α} with the same $|\alpha|$ is defined by one function $\rho_{|\alpha|}$, then we can usually take

$$(24.8) \qquad \psi_{t,1}(x, \xi) = \cdots = \psi_{t,n}(x, \xi) = \psi_t(x, \xi) = \rho(x),$$

$$(24.9)$$
$$\Psi_{t,1}(x, \xi) = \cdots = \Psi_{t,n}(x, \xi) = \Psi_t(x, \xi) = \min_{0 < |\alpha|} (t^{1/2} \rho_{|\alpha|}(x)^{-1})^{1/|\alpha|}.$$

For example, for the operator (24.1),

$$\overset{\circ}{q}_t(x, \xi)^2 = \rho(x)^{2k} |\xi|^{2m} + t, \quad \psi_t(x, \xi) = \rho(x), \quad \Psi_t(x, \xi) = t^{1/2m} \rho(x)^{-k/m},$$

and the domain Ω_t is defined by the condition $t^{1/2m}\rho(x)^{1-k/m} \geqslant M$, i.e., $\rho(x) > M_1 t^{-\kappa}$, where $\kappa = [2(m-k)]^{-1}$, $M_1 = M^{2m\kappa}$.

If Ω is a bounded domain, then the metric g_t can be made σ-temperate by replacing $\rho(x)$ with $\max\{\rho(x), M_1 t^{-\kappa}\}$ in (24.8) and (24.9). $\overset{\circ}{q}_t$ changes similarly.

In the case of anisotropic degeneracy, it is more difficult to formalize the choice of $\psi_{t,j}$, $\Psi_{t,j}$, as well as the choice of regularization of g_t, $\overset{\circ}{q}_t$ in the case of an unbounded domain. The author hopes that the reader will be able to get a clear view of these procedures after analysis of concrete problems in this chapter.

24.3. Analysis of degenerate operators becomes more difficult also because in some cases, the classical Weyl formula incorrectly predicts their asymptotics of spectrum; and moreover, it can give $N(t, A) = \infty$ for operators with a discrete spectrum that are semi-bounded from below. In particular, in some cases the asymptotic coefficient is expressed through eigenvalues of auxiliary operators—the so-called case of strong degeneracy; clearly, the Weyl formula cannot predict such asymptotics.

If an asymptotic function is consistent with the classical Weyl formula, then the degeneracy is called weak; degeneracy that is neither weak nor strong is called intermediate.

One can give the following generalization of the Weyl formula for the problem $Au = tBu$ with a positive-definite degenerate operator A and an operator B subordinate to it, which predicts the asymptotics of spectrum in the cases of weak and intermediate degeneracy and the order of asymptotics in the case of strong degeneracy.

Let a, b be Weyl symbols of operators A, B; if A, B are defined by the forms

$$\sum_{\alpha,\beta}\langle a_{\alpha\beta}(x)D^\alpha u, D^\beta u\rangle, \qquad \sum_{\alpha,\beta}\langle b_{\alpha\beta}(x)D^\alpha u, D^\beta u\rangle,$$

then a, b are symbols of the forms:

$$a(x,\xi) = \sum_{\alpha,\beta} a_{\alpha\beta}(x)\xi^{\alpha+\beta}, \qquad b(x,\xi) = \sum_{\alpha,\beta} b_{\alpha\beta}(x)\xi^{\alpha+\beta}.$$

The classical Weyl formula predicts $N_\pm(t) \sim \widehat{V}(a \mp tb, \Omega)$; it turns out that for each kind of degeneracy, there exists a function $d = d(x)$ such that for a sufficiently small constant c, the formula

$$N_+(t) \sim V_c(t) \overset{\text{def.}}{=} \widehat{V}(a + cd \mp tb, \Omega)$$

predicts the principal term of asymptotics in the cases of weak and intermediate degeneracy and, in the case of a strong degeneracy, correctly predicts the order of asymptotics.

In most cases of power degeneracy, d can be defined as a function that is maximum with respect to order and that satisfies the condition $d \leqslant c_1 A_1 + c_2 I$, where A_1 is an operator of the Dirichlet problem on $\Omega \backslash \Gamma$ for the operator A. More precisely, d could be defined, as usual, if Hardy inequality were valid for all exponents (in some cases, one successfully calculates d with the help of the Poincaré inequality or other arguments—see §25).

However, the Hardy inequality is not always correct, and besides, for a nonpower degeneracy of the general type, an analogy for the Hardy inequality is not known; thus, for the general case, d has to be defined in a more complicated way.

We assume that $d(x) = 1$ outside some neighborhood of Γ; we divide this neighborhood into sets Ω_i, each of which in an appropriate local coordinate system can be written in the form $\{y = (y', y'') \mid y' \in \Pi_i, \ y'' \in K, \ |y''| < c\}$, where K is a cone; and we write the operator A in these coordinates as:

$$A = \sum_{\alpha, \beta} D_y^\beta a_{\alpha\beta}(y) D_y^\alpha.$$

If degeneracy of the coefficients $a_{\alpha\beta}$ is defined by the products $\rho_\alpha(y'')\rho_\beta(y'')$, then on the set Ω_i, the function d is defined by the equality

$$d(x) = \max_{\alpha \neq 0} \rho_\alpha(y''(x))^2 |y''(x)|^{-2|\alpha''|},$$

where $\alpha = (\alpha', \alpha'')$ is a decomposition of the multi-index, which corresponds to the decomposition of coordinates into the groups y', y''.

If degeneracy is isotropic (see §24.2) and domain Ω does not have a point in the neighborhood of Γ, then this yields the formula

$$d(x) = \max_{j > 0} \rho_j(x)^2 \rho(x)^{-2j}, \quad \text{where } \rho(x) = \text{dist}(x, \Gamma).$$

24.4. If degeneracy is sufficiently regular, then three cases are possible:

1) for any c, there exists t such that $V_c(t) = \infty$; then the positive spectrum is not discrete;

2) there exists c_0 such that for all $c_1, c_2 > c_0$ and as $t \to \infty$,

$$V_{c_1}(t) \sim V_{c_2}(t), \quad \text{and then} \quad N_+(t) \sim V_{c_1}(t);$$

3) there exist c_0 such that for all $c_1, c_2 > c_0$, $V_{c_1}(t) \asymp V_{c_2}(t)$ but $V_{c_1}(t) \not\sim V_{c_2}(t)$; then $N_+(t) \asymp V_{c_1}(t)$.

In the second case, degeneracy is weak or intermediate; in the third case, it is strong. In order to predict the asymptotic functions in the case of strong degeneracy, we must consider the problem in a small neighborhood Ω' of the set Γ in Ω as a problem in $\Gamma \times \tilde{\Omega}$, must realize the original operators as operators on Γ with operator valued symbols, and must write out the Weyl formula with operator-valued symbols. This formula is valid in all cases where the Weyl formula with principal symbols predicts $N_+(t) = \infty$.

Warning. In some cases the operator-valued symbol is defined on the subset $\mathcal{M} \subset T^*(\Gamma \times \tilde{\Omega}), \mathcal{M} \neq T^*\Gamma$. It is also possible that $\mathcal{M} \times T^*\Gamma = T^*(\Gamma \times \tilde{\Omega})$. This is related to the fact that it is natural to talk not about manifolds of degeneracy, but about directions of degeneracy, or, in other words, about variables with respect to which one cannot localize up to Γ.

In local coordinates, described in §24.3, $\mathcal{M}$ is defined as follows. Let $I \subset \{1, \ldots, n\}$ be a maximal subset such that there exist $\alpha^i = (\delta_{ij} k_i)$, $i \in I$ for which

$$d(x) \asymp \rho_{\alpha^i}(y''(x))^2 |y''(x)|^{-2|(\alpha^i)''|}.$$

Then $\mathcal{M} = T^*\Gamma'$, where $\Gamma' = \{y \mid y_i = 0, i \in I\}$.

Example. If $A = \nabla^{*m} x_n^{2k} \nabla^m$ and $\Omega = \{x = (x', x_n) \mid x_n > 1, |x'| < x_n^\gamma\}$, where $\gamma < 1$, then $\Gamma' = (1, \infty)$.

24.5. In the case of a strong degeneracy, the method of predicting the asymptotics of spectrum, described in §24.4, is not effective; but if degeneracy is defined by power functions, then a set of formulas can be offered that enables us to find the orders of asymptotics and to calculate asymptotic coefficients in all cases. We shall give them for the simple case of $B = I$.

We associate with the operator A:

1) a principal symbol $a^{-\infty}$ (for the Schrödinger operator $a^{-\infty}(x, \xi) = |\xi|^2 + q_0(x)$, where q_0 is a principal homogeneous term of the potential);

2) a function a^∞ on $T^*\widetilde{N\Gamma}$ that is defined in a natural way through the symbol a. Here $\widetilde{N\Gamma}$ is a subbundle of $N\Gamma$ that is defined in a natural way (for example, the bundle of inner normals, if $\Gamma = \partial\Omega$; if $\Gamma \subset \Omega$, then $\widetilde{N\Gamma} = N\Gamma$);

3) an operator-valued symbol $\tilde{a}$ on $T^*\Gamma'$ that is defined in a natural way through the symbol a^∞;

4) a principal operator-valued symbol that is defined below (and does not always exist).

Let I, Γ' be the same as at the end of §24.4, and let $y^1 = (y_i)_{i \notin I}$, $y^2 = (y_i)_{i \in I}$, and $y^3 = (y_i)_{i \in j}$ be a group of variables normal to Γ. If it is possible to find changes of variables $y^3 = t^S \zeta^3$ and variables $\eta_i = t^{S_i} z_i$ dual to y_i, $i \notin I$ such that

$$d(y) = td(z), \qquad \tilde{a}(y^1, \eta^1) = t(\tilde{a}_0(z^1, \zeta^1) + \sum t^{-r_j} \tilde{a}_j(z^1, \zeta^1)),$$

where $r_j > 0$, and $\tilde{a}_0$ is a quasi-elliptic polynomial with respect to (ζ_1, D_{y_2}), then we say that there exists a principal operator-valued symbol $\tilde{a}_0$ (if Γ is conic (cylindrical), then we also have to change the radial (axial) variable).

The following is the complete set of formulas for predicting the asymptotics of spectrum. If there exists a principal operator-valued symbol, if the full operator-valued symbol is coercive in a natural way, and if the integral in the classical Weyl formula with the principal operator-valued symbol converges, then this formula is valid. Due to quasi-homogeneity of $\tilde{a}_0$, we have the following formula:

$$(24.10) \qquad N(t, A) \sim t^{\kappa} (2\pi)^{n'-n} \iint\limits_{T \cdot \Gamma'} N(1, \tilde{a}_0(z^1, \zeta^1)) dz^1 \, d\zeta^1.$$

On the other hand, if the integral in the classical Weyl formula converges, then this formula is valid (this integral and the integral in (24.10) can not converge simultaneously). If both integrals diverge, then

$$(24.11) \qquad N(t, A) \sim (2\pi)^{-n} \iint N(t, a^{\infty}(x, \xi) + cd(x)) dx \, d\xi,$$

where we integrate over a subset in $T^* \widetilde{N\Gamma}$, defined by the condition $|y^3| \leqslant c_1$ (or by the condition $|y^3| \leqslant c_1 r$ if Γ is conical, where r is a radial variable). Here the principal term of the asymptotics does not depend on $c_1 > 0$, $c > c_0$; and if the principal operator-valued symbol does not exist, then we must assume that $c = 0$ (i.e., the usual Weyl formula with "full" symbol holds).

Finally, if the integrals in all the listed formulas diverge, then the spectrum is nondiscrete.

In some cases, (24.11) yields the formula

$$(24.12) \qquad N(t, A) \sim t^{\kappa_1} (2\pi)^{-n} \iint\limits_{T \cdot \widetilde{N\Gamma}} N(1, a^1(x, \xi)) dx \, d\xi;$$

in others, it yields the formula

$$(24.13) \qquad N(t, A) \sim c_2 t^{\kappa_2} \ln t \int\limits_{\mathcal{M}} N(1, a^2(x, \xi)) d\mu,$$

where $\mathcal{M} = \{(\tilde{x}, y^3, \tilde{\xi}, \eta^3) \in T^* \widetilde{N\Gamma} \mid |y^3| = 1\}$, $d\mu$ is a natural measure on $\mathcal{M}$, and a^1, a^2 are some parts of the symbol a^{∞}. The constant c_2 is defined by orders of degeneracy. If the symbol a^{∞} is positively homogeneous with respect to the variables normal to Γ, then (24.11) always implies (24.13) with $a^2 = a^{\infty}$.

Thus, four kinds of formulas are possible [classical with the principal symbol and (24.10), (24.12), and (24.13)], three of which have quite a classical form, i.e., the form

$$N(t, A) \sim t^{\kappa} (2\pi)^{-\tilde{n}} \iint\limits_{T \cdot \widetilde{\Omega}} N(1, \tilde{a}(\tilde{x}, \tilde{\xi})) d\tilde{x} \, d\tilde{\xi},$$

where $\widetilde{\Omega}$ is some set, $\tilde{n} = \dim \widetilde{\Omega}$, and $\tilde{a}$ is some symbol of A defined on $\widetilde{\Omega}$, and the fourth is close to classical (a coefficient, depending on the type of degeneracy, appears); however, if we do not reduce (24.11) to (24.12) and (24.13), then all (three) formulas have a classical form.

Remark. The proposed formal scheme for calculation of the asymptotics of spectrum can give a correct answer for a Dirichlet problem and an incorrect answer for a Neumann problem. An example is a Laplace operator in the domain

$$\Omega = \{(x', x_n) \mid x_n > 1, \ |x'| < x_n^\gamma\}, \quad \gamma < 0$$

(for a Dirichlet problem, the prediction is correct, whereas the spectrum of a Neumann problem, contrary to the prediction, is nondiscrete).

At the same time, for the case of $\Gamma' = \partial\Omega$, the prediction is correct for a broad class of operators—both Dirichlet and Neumann problems.

24.6. One can show that the positive-definiteness of the symbols $a^{-\infty}$, $\tilde{a}_0$ and the coerciveness of the symbol a^∞ [or the positive-definiteness of $a^{-\infty}$, $\tilde{a}_0$, and all the parts of the symbol a^∞ that could appear in (24.12) and (24.13)] are sufficient for the coerciveness of the operator. The system of these symbols can be treated the following way: $T^*\Omega$ is divided into zones, a principal quasi-homogeneous symbol is selected in each, and it defines a classical Weyl formula (with integration over that zone); for a zone adjacent to $T^*\Gamma$, it is an operator-valued symbol defined on $T^*\Gamma'$ (with integration over $T^*\Gamma'$). An integral in one of these formulas increases faster than others and defines the principal term of the asymptotics.

24.7. The scheme given above can be generalized to the case of degeneracy on transversally intersecting manifolds Γ_j; in this case, the operator-valued symbols are connected with all Γ_j and their intersections.

24.8. Ideologies of this section (with some variations) are applicable also to hypoelliptic PDO with multiple characteristics.

§25 Schrödinger Operators with Degenerate Homogeneous Potential

25.1. Let $n \geqslant 2$, let q be a continuous matrix $m \times m$ potential positively homogeneous of degree $a > 0$, let $q(x) = r^a f(\varphi)$, where (r, φ) are spherical coordinates, and

A. outside some smooth submanifold $K \subset S_{n-1}$ of dimension $n_1 - 1$, where $1 \leqslant n_1 \leqslant n - 1$, let $f(\varphi)$ be a positive-definite matrix, and on K, let $f(\varphi) = 0$ (if $n_1 = 1$, then K is a finite set of points);

B. let $(\varphi_1, \varphi_2) \in R^{n_1-1} \times R^{n-n_1}$ be local coordinates in a neighborhood of K such that $K = \{\varphi \mid \varphi_2 = 0\}$ and the surfaces $\varphi_1 = \varphi_1^0$ intersect K at right angles. Then, uniformly with respect to φ_1 and $\varphi_2/|\varphi_2| \neq 0$, there exists

(25.1) $$\lim_{t \to +0} f(\varphi_1, t\varphi_2)t^{-a+a_1} = \hat{f}(\varphi_1, \varphi_2) > 0,$$

where $0 < a_1 < a$.

Let $A = -\Delta I_m + q$ be an operator in R^n. We shall show later that the spectrum of the operator A is discrete; obviously, it is positive. The classical Weyl formula yields that for A,

$$(25.2) \qquad N(t,A) \sim (2\pi)^{-n} \iint\limits_{R^{2n}} N(0, |\xi|^2 + q(x) - t)dx \, d\xi.$$

Integrating first with respect to ξ, then changing to a spherical coordinate system in R_x^n and integrating with respect to r, we obtain that as $t \to \infty$,

$$(25.2') \qquad N(t,A) \sim \gamma_{n,a} t^{s_{n,a}} \int\limits_{S_{n-1}} dS(x) \operatorname{Tr} q(x)^{-n/a},$$

where

$$s_{n,a} = \tfrac{n}{2} + \tfrac{n}{a}, \quad \gamma_{n,a} = a^{-1}(2\sqrt{\pi})^{-n}\Gamma(n/a)/\Gamma(s_{n,a}+1).$$

Due to (25.1), integral (2)' converges if and only if $n_1/a_1 < n/a$. Therefore, we can assume that formula (25.2') holds if and only if $n_1/a_1 < n/a$.

In order to predict the asymptotics in the other cases, we single out the principal quasi-homogeneous part of the symbol in a neighborhood of $\overset{\circ}{K} = \{(r, \varphi) \mid \varphi_2 = 0\}$. We obtain the following function on $T^* N \overset{\circ}{K}$:

$$a^\infty(r, \varphi, \xi_r, \xi_\varphi) = \xi_r^2 + r^{-2}|\xi_\varphi|^2 + r^a \cdot \hat{f}(\varphi_1, \varphi_2),$$

where $(r, \varphi_1) = \tilde{x}$ and (r, φ) are coordinates in $\overset{\circ}{K}$ and $N\overset{\circ}{K}$, respectively. Using functions a^∞, we construct the operator-valued symbol on $T^*\overset{\circ}{K}$:

$$\tilde{a}(\tilde{x}, \tilde{\xi}) = \xi_r^2 + r^{-2}|\xi_{\varphi_1}|^2 - r^{-2}\Delta_{\varphi_2} + r^a \hat{f}(\varphi_1, \cdot).$$

Taking into account the fact that $\hat{f}$ is positively homogeneous with respect to φ of degree $a - a_1$ and making the substitution $\varphi_2 = r^\alpha z$, $\partial_{\varphi_2} = r^{-\alpha}\partial_z$, where $\alpha = (2 + a)(2 + a - a_1)^{-1}$ is a solution of the equation $-2 - 2\alpha = \alpha(a - a_1) + a$, we can easily see that for all $(\tilde{x}, \tilde{\xi})$, the operators $\tilde{a}(\tilde{x}, \tilde{\xi})$ are unitarily equivalent to the operators

$$\tilde{\tilde{a}}(\tilde{x}, \tilde{\xi}) = \xi_r^2 + r^{-2}|\xi_{\varphi_1}|^2 + r^\beta \Lambda(\varphi_1),$$

where $\Lambda(\varphi_1) = -\Delta_{\varphi_2} + \hat{f}(\varphi_1, \cdot)$ are Schrödinger operators in R^{n-n_1} and $\beta = 2a_1/(2 + a - a_1)$. Since $\hat{f}(\varphi_1, \varphi_2) > 0$ for $\varphi_2 \neq 0$ and it is positively homogeneous with respect to φ_2 of degree $a - a_1 > 0$, then

$$(25.3) \qquad \Lambda(\varphi_1) \geqslant cI \quad \forall \varphi_1 \in K;$$

and as $\lambda \to \infty$ (see §20),

(25.4) $N(\lambda, \Lambda(\varphi_1)) \sim c_0 \lambda^\sigma$,   where $\sigma = s_{n-n_1, a-a_1}$.

In the integral

$$I(t) = (2\pi)^{-n_1} \iint_{T^\bullet \mathring{K}} N(t, \tilde{a}(\tilde{x}, \tilde{\xi})) d\tilde{x}\, d\tilde{\xi},$$

we make the substitutions $\xi_r = t^{1/2} \xi_r'$, $\xi_{\varphi_1} = r t^{1/2} \xi_{\varphi_1}'$, $r = t^{1/\beta} r'$ and yield

$$I(t) = (2\pi)^{-n_1} t^{s_{n_1,\beta}} \int_K dK(\varphi_1) \int_0^\infty dr' \cdot r'^{n_1-1} \iint_{R^{n_1}} d\tilde{\xi}' N(1, |\tilde{\xi}'|^2 + r'^\beta \Lambda(\varphi_1)).$$

Integrating with respect to $\tilde{\xi}'$ and denoting by $\lambda_j(\varphi_1)$ eigenvalues of the operator $\Lambda(\varphi_1)$, we see that the inner integral is equal to

$$|v_{n_1}| \sum_j (1 - r'^\beta \lambda_j(\varphi_1))_+^{n_1/2},$$

where $a_+ = \max\{a, 0\}$ and $|v_n|$ is the volume of a unit ball v_n in R^n. Due to (25.3), we can make the substitution $r' = \lambda_j(\varphi_1)^{-1/\beta} dr_1$ in each term of the sum and obtain, using the equality

$$(2\pi)^{-n_1} |v_{n_1}| \int_0^\infty r^{n_1-1} (1 - r^\beta)^{n_1/2} dr = \gamma_{n_1,\beta},$$

the formula

(25.5) $N(t, A) \sim t^{s_{n_1,\beta}} \gamma_{n_1,\beta} \int_K \mathrm{Tr}\, \Lambda(\varphi_1)^{-n_1/\beta} dK(\varphi_1).$

Due to (25.4), $\mathrm{Tr}\, \Lambda(\varphi_1)^{-n_1/\beta} < \infty$ if and only if $n_1/a_1 > n/a$; therefore, the last condition is sufficient and necessary (in accordance with the general method in §24) for the validity of formula (25.5).

In order to predict the asymptotics in the case of $n_1/a_1 = n/a$, we note that due to (25.3), $A \geqslant c|x|^\beta$, and the exponent β cannot be increased. Thus, we can assume that for any $c > 0$, and sufficiently large $c > 0$,

(25.6) $N(t, A) \sim (2\pi)^{-n} \iint_{T^\bullet n \mathring{K}, |\varphi_2| < c} N(t, a^\infty(r, \varphi, \xi) + C r^\beta) dr\, d\varphi\, d\xi.$

Let $M = \ln \ln t$, $\alpha = (2+a)/(2+a-a_1)$. We select the subsets $U_j \subset T^* N \overset{\circ}{K}$ ($j = 1, 2, 3, 4$), using the conditions $M^{-1} < |\varphi_2| < C$, $Mr^{-\alpha} < |\varphi_2| < M^{-1}$, $M^{-1}r^{-\alpha} < |\varphi_2| < Mr^{-\alpha}$, $|\varphi_2| < M^{-1}r^{-\alpha}$, respectively, and let $I_j(t)$ denote integral (25.6) over the set U_j. By taking into account the fact that on U_1, U_2 and on U_4, as $t \to \infty$,

$$r^\beta = o(r^a \hat{f}(\varphi_1, \varphi_2)), \qquad r^a \hat{f}(\varphi_1, \varphi_2) = o(r^\beta),$$

respectively, and by omitting Cr^β from $I_j(t)$ ($j = 1, 2, 3$) (here we estimate $I_3(t)$ only from above), and $r^a \hat{f}(\varphi_1, \varphi_2)$ from $I_4(t)$, we can obtain (provided that $n_1/a_1 = n/a$),

$$I_2(t) \sim \hat{c}_f t^{s_{n,a}} \ln t, \qquad I_j(t) \leqslant C \ln M t^{s_{n,a}}, \quad j \neq 2,$$

where

$$\hat{c}_f = \frac{a+2}{2a_1} \gamma_{n,a} \int_K dK(\varphi_1) \int_{S_{n-n_1-1}} dS(\varphi_2) \operatorname{Tr} \hat{f}(\varphi_1, \varphi_2).$$

Hence, if $n_1/a_1 = n/a$, then

$$(25.7) \qquad\qquad N(t, A) \sim \hat{c}_f t^{s_{n,a}} \ln t.$$

We do not go into details, since formula (25.7) will be strictly proved below.

We note that if $n_1/a_1 > n/a$, then by using the same division of domain of integration, one can show that (25.6) yields the formula

$$(25.7') \qquad\qquad N(t, A) \sim \hat{c}(c, \hat{f}) t^{s_{n_1, \beta}};$$

and if $n_1/a_1 < n/a$, then by adding the term $C|x|^\beta$ to (25.2), we still obtain (25.2') (and (25.7) and (25.7'), if $n_1/a_1 = n/a$ and $n_1/a_1 > n/a$, respectively).

Formulas (25.2'), (25.7), and (25.7') can be obtained from the general Weyl formula by choosing the function d, as in §24.3, to be $d(x) = \rho(x)^{-2}$, where $\rho(x) = \operatorname{dist}(x, \overset{\circ}{K})$. For this, we must repeat the same arguments as above, replacing r^β with $r^{-2}|\varphi_2|^{-2}$.

Theorem 25.1. *As $t \to \infty$ in the cases $n_1/a_1 < n/a$, $n_1/a_1 > n/a$, $n_1/a_1 = n/a$, formulas (25.2'), (25.5), and (25.7) hold, respectively.*

Example. Let us consider the potential

$$q(x) = \sum_{i<j\leqslant k} |x_i \times x_j|^2, \quad x = (x_1, \ldots, x_k) \in R^{3k}$$

(such potentials appear in Young–Mills field theory). The manifold K admits the parametrization $\varphi = (\lambda_1 \mu, \ldots, \lambda_k \mu)$, where $\mu \in S_2$ and $\lambda = (\lambda_1, \ldots, \lambda_k) \in S_{k-1}$. Thus, $n_1 = 2 + k$. The vector $(X_1, \ldots, X_k) \in R^{3k}$ is normal with respect to the cone $\overset{\circ}{K}$ at the point $(x_1, \ldots, x_k) = (q_1\mu, \ldots, q_k\mu)$, $q_i \in R^3$, if and only if $\langle X_i, \mu \rangle = 0$, $i = 1, \ldots, k$, $q_1 X_1 + \cdots + q_k X_k = 0$; thus, simple calculations show that $a_1 = 2$ and $\hat{f}(\varphi) = |\varphi_2|^2$. Hence, if $k \geqslant 5(= 4, < 4)$, then $n_1/a_1 < n/a$ $(n_1/a_1 = n/a, \; n_1/a_1 > n/a)$, and the spectral asymptotics is calculated by formula $(25.2')$ $[(25.7)$ and $(25.5)]$.

In particular, for $k = 3$, the spectrum of the operator $-\Delta_{\varphi_2} + |\varphi_2|^2$ consists of eigenvalues $\lambda_j = 2j$, $j \geqslant 2$ of multiplicity $j(j^2 - 1)/6$; and according to formula (25.5),

$$N(t, A) \sim \gamma_{5,1} \operatorname{mes}_4 K \sum_{j \geqslant 2} \frac{j(j^2 - 1)}{6(2j)^5} t^{15/2}$$

$$= \gamma_{5,1} \frac{1}{2} \operatorname{mes}_2 S_2 \operatorname{mes}_2 S_2 \left(\frac{\pi^2}{6} - \frac{\pi^3}{90} \right) t^{15/2} = \frac{2^3}{15!!} \left(\frac{\pi}{6} - \frac{\pi^3}{90} \right) t^{15/2};$$

and for $k = 4$, according to formula (25.7),

$$N(t, A) \sim \frac{3}{2^{14}\pi^6 g!} \operatorname{mes}_5 K \operatorname{mes}_5 S_5 t^g \ln t$$

$$= \frac{3}{2^{14}\pi^6 g!} \frac{1}{2} \operatorname{mes}_2 S_2 \operatorname{mes}_3 S_3 \operatorname{mes}_5 S_5 t^g \ln t$$

$$= \frac{3}{2^{12} \cdot g!} t^g \ln t.$$

■

25.2.

Beginning of the Proof of Theorem 25.1. Using (25.1), we can construct a positively homogeneous potential q_ε for any $\varepsilon > 0$ such that

$$(25.8) \qquad (1 - \varepsilon_1)q_\varepsilon(x) \leqslant q(x) \leqslant (1 + \varepsilon_1)q_\varepsilon(x),$$

where $\varepsilon_1 \to 0$ as $\varepsilon \to 0$, and for $r = r(x) > 0$, $\varphi_2 = \varphi_2(x) \neq 0$,

$$(25.9) \qquad \pm q_\varepsilon^{(\alpha)}(x) \leqslant C_\alpha q_\varepsilon(x)(\varepsilon r |\varphi_2|)^{-|\alpha|}, \qquad |\alpha| = 0, 1, \ldots.$$

Let $t_\varepsilon = t(1 + \varepsilon)$, $a_{t,\varepsilon_1}(x, \xi) = |\xi|^2 + q_\varepsilon(x) - t_{\varepsilon_1}$. (25.8) yields

$$(25.10) \qquad \mathcal{N}_0(a_{t,-\varepsilon',w}, R^n) \leqslant N(t, A) \leqslant \mathcal{N}_0(a_{t,\varepsilon',w}, R^n),$$

where $\varepsilon' \to 0$ as $\varepsilon \to 0$.

In order to calculate the asymptotics of the functions on the left- and right-hand sides of inequalities (25.10), we must (see §24) find weight functions Ψ_t, ψ_t, $\overset{\circ}{q}_t$ and a subset $\Omega_t \subset R^n$ such that conditions (24.2)–(24.4) are satisfied.

Due to (25.9) [see also (24.7)–(24.9)], we must take $\Psi_t(x,\xi) = t^{1/2}$, $\psi_t(x,\xi) = \varepsilon r|\varphi_2|$, $\overset{\circ}{q}_t(x,\xi) = (|\xi|^2 + q(x) + t)^{1/2}$; therefore, (24.4) holds if we assume that

$$\Omega_t = v_n \cup \{x = x(r,\varphi) \mid r \geqslant 1, \varepsilon r|\varphi_2|t^{1/2} > \varepsilon^{-1}\}.$$

Let $M = \ln\ln t$, $\varepsilon = M^{-1}$, $\Omega_t^- = R^n \backslash \overline{\Omega}_t$. We divide Ω_t^-, with conical surfaces orthogonal to $\overset{\circ}{K}$, into the subsets $\Omega_{t,i}$, $i = 1, \ldots, N_t$ such that the sets $\Pi_{t,i} = int(K \cap \overline{\Omega}_{t,i}) \subset K$ satisfy the conditions

(25.11) $\operatorname{diam}\Pi_{t,i} < M^{-1}$; $\operatorname{mes}\Pi_{t,i} \geqslant cM^{1-n_1}$;
(25.12) $\exists C \mid \forall 0 < \nu < 1,$ $\operatorname{mes}(\partial\Pi_{t,i})_{\nu/M} \leqslant C\nu M^{1-n_1}$,

where $(\partial\Pi_{t,i})_\nu = \{x \in K \mid \operatorname{dist}(x, \partial\Pi_{t,i}) < \nu\}$;
(25.13)
     multiplicity of the covering $\{(\Pi_{t,i})_{\nu/M}\}$ is limited by the number C.

Here the constants c, C in (25.11)–(25.13) do not depend on t, ν, i.
Let $\Omega_{t,0} = \Omega_t$. Since $\Omega_{t,i} \cap \Omega_{t,j} = \emptyset$ for $j \neq i$, we have

(25.14) $$\mathcal{N}_0(a_{t,-\varepsilon',w}; R^n) \geqslant \sum_{0 \leqslant i \leqslant N_t} \mathcal{N}_0(a_{t,-\varepsilon',w}; \Omega_{t,i}).$$

25.3. In order to estimate the zero term in (25.14), we apply Theorem 7.5. On Ω_t,

$$q_\varepsilon(x) \geqslant cM^{2(a-a_1)}t^{-(a-a_1)/2}\langle r\rangle^{a_1} \overset{\text{def.}}{=} m(t,r);$$

obviously, we can change q_ε outside Ω_t so that $cm(t,r) \leqslant q_\varepsilon(x) \leqslant Cm(t,r)$ outside Ω_t, and the conditions of Theorem 7.5 hold with

$$g_{t,x,\xi}(y,\eta) = t^{-1}|\eta|^2 + (\max\{\langle r\rangle|\varphi_2|M^{-1}, Mt^{-1/2}\})^{-2}|y|^2,$$
$$\overset{\circ}{q}_t(x,\xi)^2 = |\xi|^2 + q_\varepsilon(x) + t, \qquad \overset{\circ}{a}_t(x,\xi) = |\xi|^2 + q_\varepsilon(x) - t_{-\varepsilon'},$$
$$a_t = \overset{\circ}{q}_t^{-1}\overset{\circ}{a}_t\overset{\circ}{q}_t^{-1}, \qquad d_t = h_{g_t} = h_t.$$

Let $\hat{a}(x,\xi) = |\xi|^2 + q(x)$. We denote by ε, ε_1, ε', etc., functions (depending on t) that vanish as $t \to \infty$. Due to (25.8), Theorem 7.5 yields
(25.15)
$$\mathcal{N}_0(a_{t,-\varepsilon',w}; \Omega_t) = \widehat{V}(\hat{a} - t, \Omega_t) + O(\widehat{V}(\hat{a} - t_{\varepsilon_1}, \Omega_t^1) - \widehat{V}(\hat{a} - t_{-\varepsilon_1}, \Omega_t^1))$$
$$+ O(\widehat{V}(\hat{a} - Ct, \Omega_t^c\backslash\Omega_t^C)) + O(\widehat{V}(\hat{a} - Ct, 2v_n)),$$

where $\Omega_t^c = \{x = x(r,\varphi) \mid |\varphi_2| > cM^2t^{-1/2}r^{-1}, r \geqslant 1\}$ and

$$\widehat{V}(\hat{a} - t, \Omega) = (2\pi)^{-n} \iint\limits_{\Omega \times R^n} N(t, \hat{a}(x,\xi))dx\,d\xi.$$

Let f_j be the eigenvalues of the matrix f and $\Omega^c_{t,r} = \{\varphi \mid (r,\varphi) \in \Omega^c_t\}$. Changing to the spherical coordinate system in R^n_x, integrating with respect to ξ, and then making the substitutions $r = t^{1/a} f_j(\varphi)^{-1/a} r_1$, we obtain

(25.16)

$$\widehat{V}(\hat{a} - t, \Omega^1_t)$$

$$= (2\pi)^{-n}|v_n| \int\limits_0^\infty dr \cdot r^{n-1} \int\limits_{\Omega^1_{t,r}} \sum_{1 \leqslant j \leqslant m} (t - r^a f_j(\varphi))^{n/2}_+ dS(\varphi) + O(t^{n/2})$$

$$= (2\pi)^{-n}|v_n| \int\limits_0^1 \int\limits_{\Omega^1_{t,r}} dr_1 \, dS(\varphi)(1 - r^a_1)^{n/2} r^{n-1}_1 \operatorname{Tr} f(\varphi)^{-n/a} t^{S_{n,a}} + O(t^{n/2}).$$

If $n_1/a_1 < n/a$, then

(25.17)

$$\int\limits_{S_{n-1}\backslash\Omega^1_{t,r}} \operatorname{Tr} f(\varphi)^{-n/a} dS(\varphi) \leqslant C \int\limits_{|\varphi_2| \leqslant CM^2 t^{-1/2}} |\varphi_2|^{-n(a-a_1)/a} d\varphi_2$$

$$\leqslant C_1 (M^2 t^{-1/2})^\omega,$$

where $\omega = n - n_1 - (a - a_1)n/a > 0$; thus, in (25.16) we can replace $\Omega^1_{t,r}$ with S_{n-1} by adding the term $o(t^{s_{n,a}})$ and can integrate with respect to r. We obtain

(25.18)

$$\widehat{V}(\hat{a} - t, \Omega_t) \sim c_f t^{s_{n,a}},$$

where

$$c_f = (2\pi)^{-n}|v_n| \cdot \tfrac{1}{a} B(\tfrac{n}{2} + 1, \tfrac{n}{a}) \int\limits_{S_{n-1}} \operatorname{Tr} f(\varphi)^{-n/a} dS(\varphi)$$

$$= \tfrac{1}{a}(2\pi)^{-n} \cdot \pi^{n/2}\Gamma(\tfrac{n}{2} + 1)^{-1} \cdot \Gamma(\tfrac{n}{2} + 1)\Gamma(\tfrac{n}{a})\Gamma(\tfrac{n}{2} + \tfrac{n}{a} + 1)^{-1}$$

$$\times \int\limits_{S_{n-1}} \operatorname{Tr} f(\varphi)^{-n/a} dS(\varphi)$$

$$= \gamma_{n,a} \int\limits_{S_{n-1}} \operatorname{Tr} q(x)^{-n/a} dS(x)$$

is the coefficient in formula (25.2′).

Due to (25.18), the second term in (25.15) is $o(t^{s_{n,a}})$; and, as in (25.16) and (25.17), we can show that this is also true for the third term. Since v_n is a bounded domain, then also

(25.19)

$$\widehat{V}(\hat{a} - Ct, 2v_n) = O(t^{n/2}) = o(t^{\overline{\overline{S}}_{n,a}}).$$

Therefore, under the condition $n_1/a_1 < n/a$,

$$(25.20) \qquad \mathcal{N}_0(a_{t,-\varepsilon',w}, \Omega_{t,0}) \sim c_f t^{s_{n,a}};$$

and formulas (25.10), (25.14), and (25.20) yield the estimate from below in (25.2′).

If $n_1/a_1 > n/a$, then $\omega < 0$ and (25.1) yields

$$\Omega_{t,r}^c \subset \{\varphi \mid \varphi_2 \in \tilde{\Omega}_{t,r_1}^{c_1,C}\},$$

where

$$\tilde{\Omega}_{t,r_1}^{c_1,C} = \{\varphi_2 \mid c_1 M^2 t^{-1/2-1/a} |\varphi_2|^{(a-a_1)/a} r_1^{-1} < |\varphi_2| < C\};$$

thus,

$$(25.21)$$

$$\int_{\Omega_{t,r}^c} \mathrm{Tr}\, f(\varphi)^{-n/a} d\varphi \leqslant C_1 \int_{\tilde{\Omega}_{t,r_1}^{c_1,C}} |\varphi_2|^{-n(a-a_1)/a} d\varphi_2$$

$$\leqslant C_2 \int_{c_2 z_{t,r_1}}^{\infty} dz \cdot z^{\omega-1} \leqslant C_3 M^k r_1^{-\omega a/a_1} t^{-\omega(a+2)/2a_1},$$

where $k = 2\omega a/a_1$, $z_{t,r_1} = M^{2a/a_1} t^{-(a+2)/2a_1} r_1^{-a/a_1}$.

Then, by integrating in (25.16) with respect to r_1, we obtain

$$\hat{V}(\hat{a} - t, \Omega_t^1) \leqslant CM^k t^\kappa,$$

where

$$\kappa = s_{n,a} - \omega(a+2)/2a_1$$
$$= n/2 + n/a - (n - n_1 - (a - a_1)na^{-1})(a+2)(2a_1)^{-1}$$
$$= n_1(2 = a)(2a_1)^{-1} = s_{n_1,\beta}.$$

The same estimates are valid also for the remaining terms in (25.15), and since $k < 0$ and $M \to \infty$ as $t \to \infty$, then

$$(25.22) \qquad \mathcal{N}_0(a_{t,-\varepsilon',w}, \Omega_{t,0}) = o(t^{s_{n_1,\beta}})$$

is contained in the o-term of formula (25.5).

Finally, if $n_1/a_1 = n/a$, then $\omega = 0$; thus, using (25.1) as in (25.21), we obtain

(25.23)

$$\int_{\Omega_{i,r}^c} \text{Tr } f(\varphi)^{-n/a} dS(\varphi) = O\left(\int_{M^{-1} < |\varphi_2| < C} d\varphi_2 |\varphi_2|^{-n(a-a_1)/a} \right)$$

$$+ (1 + \varepsilon_1) \int_K dK(\varphi_1) \int_{S_{n-n_1-1}} dS(\varphi_2) \text{Tr } \hat{f}(\varphi_1, \varphi_2)^{-n/a} \int_{cz_{t,r_1} < z < M^{-1}} z^{-1} dz$$

$$= I(\hat{f}) \frac{a+2}{2a_1} [\ln t + O(\ln M) + O(\ln \tfrac{1}{r_1})],$$

where $I(\hat{f})$ is the integral in the formula which defines the coefficient $\hat{c}_f$.

Thus,

(25.24) $$\widehat{V}(\hat{a} - t, \Omega_t) \sim \hat{c}_f t^{s_{n,a}} \ln t.$$

Now (25.24) implies that the second term in (25.15) is equal to $o(t^{s_{n,a}} \ln t)$. Due to (25.19), the same is true for the last term, and the third term is estimated as in (25.16) and (25.23):

$$\widehat{V}(\hat{a} - Ct, \Omega_i^c \backslash \Omega_i^c) = O(\ln M t^{s_{n,a}}) = o(t^{s_{n,a}} \ln t).$$

Therefore, under the condition $n_1/a_1 = n/a$,

(25.25) $$\mathcal{N}_0(a_{t,-\varepsilon',w}, \Omega_t) \sim \hat{c}_f t^{s_{n,a}} \ln t,$$

and formulas (25.10), (25.14), and (25.25) give the estimate from below in (25.7).

25.4. In order to estimate $\mathcal{N}_0(a_{t,\pm\varepsilon',w}, \Omega_{t,i})$ for $i > 0$, we choose the point $(\varphi_1^i, 0) \in K_{t,i} = K \cap \overline{\Omega}_{t,i}$ and denote

$$A_{t,\varepsilon}^i = -\Delta I_m + r^a \hat{f}(\varphi_1^i, \varphi_2) - t\varepsilon.$$

Then (25.1) and (25.11)$_1$ imply that

$$\mathcal{N}_0(A_{t,-\varepsilon_1}^i, \Omega_{t,i}) \leqslant \mathcal{N}_0(a_{t,\pm\varepsilon',w}, \Omega_{t,i}) \leqslant \mathcal{N}_0(A_{t,\varepsilon_1}^i, \Omega_{t,i});$$

therefore, by changing to a spherical coordinate system, we obtain

(25.26) $$\mathcal{N}_0(\widetilde{A}_{t,-\varepsilon_1}^i, \Omega_{t,i}') \leqslant \mathcal{N}_0(a_{t,\pm\varepsilon',w}, \Omega_{t,i}) \leqslant \mathcal{N}_0(\widetilde{A}_{t,\varepsilon_1}^i, \Omega_{t,i}'),$$

where

$$\Omega_{t,i}' = \{(r, \varphi) \mid \varphi_1 \in \Pi_{t,i}, |\varphi_2| < M^2 t^{-1/2} r^{-1}, r > 1\},$$
$$\widetilde{A}_{t,\varepsilon}^i = d_w - \Delta_\varphi - L(\varphi, \partial_\varphi) + r^{a+2} \hat{f}(\varphi_1^i, \varphi_2) - r^2 t_\varepsilon,$$

$L(\varphi, \partial_\varphi)$ is a DO of first order, $d(r, \tau) = r^2\tau^2 + c_1 r\tau + c_2$, and τ is a variable dual to the r. We make the substitution $z = r_1 t^{1/2}\varphi_2$, $r_1 = r$. Then $\Delta_{\varphi_2} = r_1^2 t\Delta_z$, $r\partial_r = r_1\partial_{r_1} + \langle z, \partial z\rangle$. But

$$|\langle L(\varphi, \partial_\varphi)u, u\rangle| + \|\langle z, \partial z\rangle u\|^2 \leqslant \varepsilon\langle(-\Delta_z + r^2 t)u, u\rangle;$$

therefore,

$$
(25.27) \quad
\begin{aligned}
\mathcal{N}_0(\tilde{\tilde{A}}^i_{t,-\varepsilon_2}, U_{t,i} \times B(M^2)) &\leqslant \mathcal{N}_0(\tilde{A}^i_{t,\pm\varepsilon_1}, \Omega'_{t,i}) \\
&\leqslant \mathcal{N}_0(\tilde{\tilde{A}}^i_{t,\varepsilon_2}, U_{t,i} \times B(M^2)),
\end{aligned}
$$

where $U_{t,i} = \Pi_{t,i} \times (1, \infty)$, $B(c) = \{z \mid |z| < c\}$,

$$\tilde{\tilde{A}}^i_{t,\varepsilon} = d_w - \Delta_{\varphi_1} - r^2 t\Delta_z - r^{2+a_1} t^{(a_1-a)/2}\hat{f}(\varphi^i_1, z) - r^2 t_\varepsilon.$$

Let $H_{t,0} = \overset{\circ}{H}{}^1(B(M^2))^m$. We construct $\tilde{r} \in C^\infty(R^1)$ to equal $|r|$ for $|r| \geqslant 1$ and $\frac{1}{2}$ for $|r| \leqslant \frac{1}{2}$, and realize $\tilde{\tilde{A}}^i_{t,\varepsilon}$ as a PDO in $R^{n_1}_{r,\varphi_1}$ with the Weyl operator-valued symbol

$$
\begin{aligned}
a^i_{t,\varepsilon}(\tilde{x}, \tilde{\xi}) = d(\tilde{r}, \tau) + |\zeta|^2 &- \tilde{r}^2 t\Delta_z + \tilde{r}^{a_1+2} t^{(a_1-a)/2}\hat{f}(\varphi^i_1, z) \\
&- \tilde{r}^2 t_\varepsilon : H_{t,0} \to H^*_{t,0},
\end{aligned}
$$

where $\tilde{x} = (r, \varphi_1)$ and $\tilde{\xi} = (\tau, \zeta)$ are variables dual to the $\tilde{x}$.
  Since

$$C_0^\infty(U_{t,i} \times B(M^2)) \subset C_0^\infty(U_{t,i}, H_{t,0}) \subset \overset{\circ}{H}{}^1(U_{t,i} \times B(M^2))^m$$

densely, then

$$\mathcal{N}_0(\tilde{\tilde{A}}^i_{t,\varepsilon}; U_{t,i} \times B(M^2)) = \mathcal{N}(a^i_{t,\varepsilon,w}; C_0^\infty(U_{t,i}; H_{t,0}));$$

and we can apply Theorem 22.1, assuming that

$$g_{t,\tilde{x},\tilde{\xi}}(\tilde{y}, \tilde{\eta}) = (t + \tilde{r}^{-2}|\tilde{\xi}|^2)^{-1}|\tilde{\eta}|^2 + \tilde{r}^{-2s}|\tilde{y}|^2,$$

$$H_t = L_2(B(M^2))^m, \qquad \Omega_t = U_{t,i}, \qquad \overset{\circ}{q}_t(\tilde{X}) = \hat{B}_t^i(X)^{1/2},$$

where $\tilde{X} = (\tilde{x}, \tilde{\xi})$, $\hat{b}_t^i(\tilde{X})$ is a Fridrichs extension of the operator

$$
\begin{aligned}
\tilde{b}_t^i(\tilde{X}) = \tilde{r}^2\tau^2 + |\zeta|^2 &- \tilde{r}^2 t\Delta_z + \tilde{r}^{a_1+2} t^{(a_1-a)/2}\hat{f}(\varphi^i_1, z) \\
&+ \tilde{r}^2 t : C_0^\infty(B(M^2))^m \to H_t,
\end{aligned}
$$

and $s \in (0,1)$ will be chosen later. Then for some $c > 0$,

$$(25.28) \qquad h_t(\tilde{X}) \leqslant \min\{t^{-1/2}\tilde{r}^{-s}c^{-1}(t + \tilde{r} + |\tilde{\xi}|)^{-c}\};$$

thus, condition (22.1) holds, and Lemmas 1.2, 1.3, 3.2, and 3.3 imply that g_t is a σ-temperate metric and $\hat{b}_t{}^i(\cdot)\hat{b}_t{}^i(0)^{-1} \in O_r(H_t, g_t)$ uniformly with respect to $t \geqslant 1$. Thus, condition A of Theorem 22.1 holds.

In order to verify the remaining conditions, we note that

$$(25.29) \qquad (1 - \varepsilon_1)\tilde{b}_t{}^i \leqslant a_{t,\varepsilon}^i + 2\tilde{r}^2 t \leqslant (1 + \varepsilon_1)\tilde{b}_t{}^i;$$

thus, we can assume that with the notations in §22,

$$\hat{\mathcal{A}}_{t,\tilde{X}}^{\varepsilon}[u] = \langle (\tilde{b}_t{}^i(\tilde{X}) - 2\tilde{r}^2 t_\varepsilon)u, u \rangle.$$

The operator $\tilde{b}_t{}^i(\tilde{X}) : C_0^\infty(r^{n-n_1})^m \to L_2(R^{n-n_1})^m$ is unitarily equivalent to the operator

$$b_t^i(\tilde{X}) = \tilde{r}^2 \tau^2 + |\zeta|^2 + \tilde{r}^{2+\beta}\Lambda(\varphi_1^i) + \tilde{r}^2 t;$$

hence, if C is large and either $\tilde{r} > Ct^{1/\beta}$ or $|\zeta| > C\tilde{r}t^{1/2}$, then (25.3) yields

$$(25.30) \qquad \mathcal{N}_0(\hat{\mathcal{A}}_{t,\tilde{X}}^{\varepsilon}, H_{t,0}) = 0.$$

If $s < 1/28$ and $\tilde{r} < Ct^{1/\beta}$, then (25.28) yields $h_t(0)^{1/7}\tilde{r}^s < t^{-\kappa}$, where $\kappa > 0$; therefore,

$$(25.31) \qquad \tilde{\tilde{\Omega}}_{t,i} \overset{\text{def.}}{=} \{\tilde{x} = (r, \varphi_1) \mid \text{dist}(\varphi_1, \partial\Pi_{t,i}) < cr^s h_t(0)^{1/7}, \ 1 < r < Ct^{1/\beta}\}$$
$$\subset (\partial\Pi_{t,i})_{t^{-\kappa}} \times (1, \infty),$$

and due to (25.30), the set $\mathcal{M}_t^\varepsilon$ introduced in §22 is contained in

$$\{(r, \varphi_1, \tilde{\xi}) \mid |\varphi_1| < C, \ \tilde{r} + |\tilde{\xi}| < Ct^C\}.$$

Therefore, if we denote

$$g_{t,\tilde{x},\tilde{\xi}}^1(\tilde{y}, \tilde{\eta}) = (t + |\tilde{\xi}|^{C_1} + \tilde{r}^{-2}|\tilde{\xi}|^2)^{-1}|\tilde{\eta}|^2 + \langle \tilde{x} \rangle^{-2s}|\tilde{y}|^2,$$

then conditions (22.2)–(22.5) hold. Condition (22.6) follows from (25.4).

Thus, all the conditions of Theorem 22.1 hold; and (22.7), (25.29), and (25.28) yield the estimate
(25.32)

$$\mathcal{N}(a_{t,\varepsilon,w}^i; C_0^\infty(U_{t,i}; H_{t,0}))$$
$$= \hat{V}(b_t^i - 2\tilde{r}^2 t, U_{t,i}) + O(\hat{V}(b_t^i - 2\tilde{r}^2 t_{\varepsilon_1}, U_{t,i}) - \hat{V}(b_t^i - 2\tilde{r}^2 t_{-\varepsilon_1}, U_{t,i}))$$
$$+ O(\hat{V}(b_t^i - C\tilde{r}^2 t, \Omega_{t,i}')) + O(\hat{V}(b_t^i - C\tilde{r}^2 t, \tilde{\tilde{\Omega}}_{t,i})),$$

where

$$\Omega'_{t,i} = \{\tilde{x} = (r, \varphi_1) \mid |\varphi_1 - \varphi_1^i| < CM^{-1}, |r - 1| < Ct^{-1/14}\}.$$

Obviously,

$$\mathcal{N}(b_i^i(\tilde{X}) - 2\tilde{r}^2 t_\varepsilon, H_{t,0}) \leqslant \mathcal{N}(-\Delta - C_1, H_{t,0}) \leqslant C_2 M^{2(n-n_1)};$$

therefore, taking (25.30) and (25.31) into account, we see that all the terms in (25.32) are estimated by
(25.33)

$$C \operatorname{mes} \Pi_{t,i} M^{2(n-n_1)} \int\limits_{-\infty}^{\infty} d\tau \int\limits_{1/2}^{ct^{1/\beta}} dr [r^2 (t - \tau^2)]^{(n_1-1)/2} \leqslant C_1 M^{2(n-n_1)} \operatorname{mes} \Pi_{t,i} t^{s_{n_1,\beta}}.$$

Due to condition (25.11), the number of terms in (25.14) is $O(M^{n_1-1})$; and, since $s_{n_1,\beta} < s_{n,a}$ for $n_1/a_1 < n/a$, $s_{n_1,\beta} = s_{n,a}$ for $n_1/a_1 = n/a$, and $M^N = o(\ln t)$ for any N, then (25.32), (25.33), (25.27), and (25.26) imply the following:
  if $n_1/a_1 < n/a$, then

$$(25.34) \qquad \sum_{1 \leqslant i \leqslant N_t} \mathcal{N}_0(a_{t,\pm\varepsilon',w}; \Omega_{t,i}) = o(t^{s_{n,a}});$$

  if $n_1/a_1 = n/a$, then

$$(25.35) \qquad \sum_{1 \leqslant i \leqslant N_t} \mathcal{N}_0(a_{t,\pm\varepsilon',w}; \Omega_{t,i}) = o(t^{s_{n,a}} \ln t).$$

Thus, in these cases, all the terms in (25.14), except for the zero term, are contained in the o-term of formulas (25.2′) and (25.7), respectively.

25.5. Now let $n_1/a_1 > n/a$. (25.30) and (25.31) imply that the last term in (25.32) is estimated by

$$CM^{2(n-n_1)} \operatorname{mes} \Pi_{t,i} t^{s_{n_1,\beta}-\kappa}, \qquad \kappa > 0;$$

obviously, the same is true for the third term. Hence, the sum with respect to i of the last two terms in (25.32) is $o(t^{s_{n_1,\beta}})$.

  Then (25.1), (25.11), and continuity of f yield

$$(1 - \varepsilon)\hat{f}(\varphi_1, z) \leqslant \hat{f}(\varphi_1^i, z) \leqslant (1 + \varepsilon)\hat{f}(\varphi_1, z) \quad \forall \varphi_1 \in \Pi_{t,i};$$

therefore, in (25.32) we can replace b_t^i with b_t, where b_t is defined as b_t^i, but with $\hat{f}(\varphi_1, z)$ in place of $\hat{f}(\varphi_1^i, z)$. Summing up with respect to $i \geqslant 1$, we obtain

$$(25.36) \qquad \sum_{1 \leqslant i \leqslant N_t} \mathcal{N}_0(a_{t,\pm\varepsilon',w}, \Omega_{t,i}) = F_1(t) + O(F_1(t_\varepsilon) - F_1(t_{-\varepsilon})) + o(t^{s_{n_1,\beta}}),$$

where

$$F_c(t) = (2\pi)^{-n_1} \int\limits_K dK(\varphi_1) \int\limits_c^\infty dr \int\limits_{-\infty}^\infty d\tau \int d\zeta\, \mathcal{N}_0(r^2\tau^2 + |\zeta|^2$$
$$- r^2 t \Delta_z + r^{a_1+2} t^{(a_1-a)/2} \hat{f}(\varphi_1, z) - tr^2; B(M^2)).$$

But

$$F_0(t) - F_1(t) \leqslant C \int\limits_0^1 dr \iint\limits_{R^{n_1}} d\tau\, d\zeta \mathcal{N}_0(r^2\tau^2 + |\zeta|^2 - r^2 t \Delta_z - r^2 t; B(M^2))$$
$$\leqslant C_1 M^{2(n-n_1)} t^{n_1/2} = o(t^{s_{n_1,\beta}});$$

therefore, in (25.36) we can replace $F_1(t)$ with $F_0(t)$. In the integral $F_0(t)$, we make the substitutions $\tau = t^{1/2}\tau_1$, $\zeta = rt^{1/2}\zeta_1$, $r = t^{1/\beta}r_1$, $z = r_1^{-\kappa}z_1$, where $\kappa = a_1(2 + a - a_1)^{-1}$; we have

(25.37)
$$F_0(t) = (2\pi)^{-n_1} t^{s_{n_1,\beta}} \int\limits_K dK(\varphi_1) \int\limits_0^\infty dr_1 \cdot r_1^{n_1-1}$$
$$\times \iint\limits_{R^{n_1}} d\tau_1\, d\zeta_1 \mathcal{N}_0(\tau_1^2 + |\zeta_1|^2 + r_1^\beta \Lambda(\varphi_1) - 1; B(M^2 r_1^{-\kappa})).$$

Now (25.3) implies that $\mathcal{F}(\tau_1, \zeta_1, \varphi_1, r_1, M)$—the integrand in (25.37)—is equal to zero, if $r_1 > c$; therefore, we can assume that in (25.37) the domain $B(M^2 r_1^{-\kappa})$ expands as $t \to \infty$. But then

$$\mathcal{F}(\tau_1, \zeta_1, \varphi_1, r_1, M) \to \mathcal{F}(\tau_1, \zeta_1, \varphi_1, r_1, \infty)$$

almost everywhere; therefore, passing over to the limit under the Lebesgue integral, in (25.37) we can replace $B(M^2 r_1^{-\kappa})$ with R^{n-n_1}. But the asymptotic function of the new integrals was already calculated in §25.1:

$$F_0(t) \sim ct^{s_{n_1,\beta}},$$

where c is the coefficient in (25.5). Hence,

$$F_0(t_\varepsilon) - F_0(t_{-\varepsilon}) = o(t^{s_{n_1,\beta}}),$$

and

(25.38)
$$\sum_{1 \leqslant i \leqslant N_t} \mathcal{N}_0(a_{t,\pm\varepsilon',w}; \Omega_{t,i}) = ct^{s_{n_1,\beta}} + o(t^{s_{n_1,\beta}}).$$

This proves the estimate from below in formula (25.5).

25.6. In order to obtain the estimates from above in (25.2′), (25.5), and (25.7), we construct the functions $\psi_{t,i}$, $0 \leqslant i \leqslant N_t$ such that

$$0 \leqslant \psi_{t,i} \leqslant 1, \qquad \sum_i \psi_{t,i}^2 = 1, \qquad \psi_{t,i} \in S(1, M^6| \cdot |_{R^n}^2),$$

$$\operatorname{supp} \psi_{t,i} \subset \Omega_{t,i}^+ = \{x \mid \operatorname{dist}(x, \Omega_{t,i}) < M^{-3}\}$$

[this is possible due to (25.13)]. Then

$$\langle (A - t)u, u \rangle \geqslant \sum_{0 \leqslant i \leqslant N_t} (1 - \varepsilon_1)\langle (A - t_\varepsilon)\psi_{t,i}u, \psi_{t,i}u \rangle;$$

therefore,

(25.39)
$$N(t, A) \leqslant \sum_{0 \leqslant i \leqslant N_t} N_0(a_{t,\varepsilon',w}, \Omega_{t,i}^+).$$

For any spherical layer Q which does not intersect $\bar{v}_n$, we have

$$\operatorname{mes}(Q \cap \Omega_{t,i}^+) = (1 + \varepsilon)\operatorname{mes}(Q \cap \Omega_{t,i});$$

therefore, the same arguments as in §§25.2–25.5 [see formulas (25.20), (25.22), (25.25), (25.34), (25.35), and (25.38)] show that all the terms in (25.39) have the same asymptotics (or admit the same estimates) as the terms in (25.14). Thus, the estimate from above for $N(t, A)$ has the same form as the estimate from below.

§26 Model Problems for Degenerate Differential Operators in a Bounded Domain

26.1. First, we shall consider the Dirichlet problem for an operator whose degeneracy is defined by power functions, and at the end of this section, we shall point out the changes necessary for studying a Neumann problem, and consider the simplest model example of a nonpower degeneracy.

Let $\Omega' \subset R^{n-1}$ be a bounded Lipschitz domain, $\Omega = \Omega' \times (0, 1)$ and $\ell, m \geqslant 1$ be integers. In $C_0^\infty(\Omega)^\ell$, we define the form

(26.1)
$$\mathcal{A}[u] = \sum_{\alpha, \beta \in I} \langle a_{\alpha\beta}(x)D^\alpha u, D^\beta u \rangle_{L_2(\Omega)^\ell},$$

where $I = \{\alpha \mid |\alpha| \leqslant m\}$, and $a_{\alpha\beta} = a_{\alpha\beta}^*$ are continuous in $\overline{\Omega}' \times (0, 1]$. In addition, uniformly with respect to $x' \in \overline{\Omega}'$ let there exist the limits

(26.2)
$$\lim_{x_n \to +0} a_{\alpha\beta}(x)x_n^{-\hat{k}_\alpha - \hat{k}_\beta} = \hat{a}_{\alpha\beta}(x').$$

For the sake of simplicity, we also assume that

$$(26.3) \qquad \|a_{\alpha\beta}^{(\gamma)}(x)\| \leqslant C_{\alpha\beta\gamma} x_n^{\hat{k}_\alpha + \hat{k}_\beta - \gamma_n}, \quad |\gamma| = 0, 1, \ldots,$$

and that the limit functions $\hat{a}_{\alpha\beta}$ are not only continuous, but also belong to $C^\infty(\overline{\Omega}'; \operatorname{End} \mathbb{C}^\ell)$. Transition to the general case will be realized at the end of the analysis.

Let $k_\alpha = \hat{k}_\alpha - \alpha_n$, $s_0 = \min_{\alpha \in I} k_\alpha$, $\alpha_{ij} = (j\delta_{is})_{s=1,\ldots,n}$, $k_{ij} = k_{\alpha_{ij}}$. We assume that the set $J = \{i \mid \exists j > 0 : k_{ij} = s_0\}$ is not empty and that the following conditions hold:

A. The points $(\alpha, \hat{k}_\alpha)$, $\alpha \in I$ lie on a lower surface of a convex polyhedron with vertices at some points $(\alpha, \hat{k}_\alpha)$ on coordinate planes $\alpha_i = 0, \forall_{i \neq j}$.

B. The function $(\tilde{I} = \{x \in (R_+)^n \mid \sum x_i \leqslant m\}) \ni x \to \hat{k}_x$ does not decrease on each coordinate axis and increases on each axis Ox_i, $i \notin J$.

C. If $J \neq \{n\}$, then $k_0 = s_0$, and the function $x \to k_x$ satisfies condition B.

We exclude the case of $k_0 < k_\alpha \; \forall (0 \neq) \alpha \in I$, since, by using an analogue of Theorem 7.1 for PDO with double symbols (see the author's work [2]), we can immediately obtain the classical Weyl formula for this case.

26.2. The form $\mathcal{A}$ will be described by the usual principal symbol

$$a^{-\infty}(x, \xi) = \sum_{|\alpha| = |\beta| = m} a_{\alpha\beta}(x) \xi^{\alpha+\beta}$$

and by a complete operator-valued symbol defined as follows.

If $n_2 = \operatorname{card} J < n$, then we assume that $\Omega = \Omega^1 \times \Omega^2$, where $\Omega^i \subset R_{x^i}^{n_i}$, $n_1 = n - n_2$, $x^2 = (x_j)_{j \in J}$ and let

$$\mathcal{A}_{x^1, \xi^1}[u] = \sum_{\alpha, \beta \in I} \xi^{1^{\alpha^1 + \beta^1}} \langle \hat{a}_{\alpha\beta}(x') x_n^{\hat{k}_\alpha + \hat{k}_\beta} D^{\alpha^2} u, D^{\beta^2} u \rangle_{L_2},$$

$$Q_{x^1, \xi^1}[u] = \sum_{\alpha \in I} \xi^{1^{2\alpha^1}} \|x_2^{\hat{k}_\alpha} D_{x^2}^{\alpha^2} u\|_{L_2(R^{n_2})^\ell}^2.$$

On the other hand, if $n_2 = n$, then we assume that

$$Q[u] = \sum_{\alpha \in I} \|x_n^{\hat{k}_\alpha} D^\alpha u\|_{L_2(R^n)^\ell}^2,$$

and define the form $\tilde{\mathcal{A}}$ by equality (26.1), replacing $a_{\alpha\beta}(x)$ with $\hat{a}_{\alpha\beta}(x') x_n^{\hat{k}_\alpha + \hat{k}_\beta}$; these forms will also be used in the case $n_2 < n$. Of course $\tilde{\mathcal{A}}$ can hardly be called a symbol. Below, we shall see that for the case where $n_2 = n$, only a weak degeneracy is possible; therefore, the symbol $\tilde{\mathcal{A}}$ is not needed to calculate the asymptotics of the spectrum, but without it we cannot formulate conditions that would guarantee the discreteness of the spectrum.

26.3. For $c \in (0, \infty]$, let $\Omega_c = \Omega' \times (0, c)$, $\Omega_c^i = \Omega^{i1} \times (0, c)$, if $\Omega^i = \Omega^{i1} \times (0, 1)$, and $\Omega_c^i = \Omega^i$ otherwise.

Theorem 26.1.   *Let there exist $c > 0$, such that, for all $(x, \xi) \in \Omega \times R^n$,*

(26.4)
$$a^{-\infty}(x, \xi) \geqslant cq^{-\infty}(x, \xi) \stackrel{\text{def.}}{=} c \sum_{|\alpha|=m} x_n^{2\hat{k}_\alpha} \xi^{2\alpha},$$

and

a) *if $n_2 < n$, then for all $(x^1, \xi^1) \in \Omega_c^1 \times R^{n_1}$,*

(26.5)
$$A_{x^1, \xi^1}[u] \geqslant cQ_{x^1, \xi^1}[u] \quad \forall u \in C_0^\infty(\Omega_c^2)^\ell;$$

b) *if $n_2 = n$, then*

(26.6)
$$\tilde{A}[u] \geqslant cQ[u] \quad \forall u \in C_0^\infty(\Omega_c)^\ell.$$

Then there exist $c_1, c_2 > 0$ such that

(26.7)
$$Q[u] \leqslant c_1 A[u] + c_2 \|u\|_{L_2}^2 \quad \forall u \in C_0^\infty(\Omega)^\ell.$$

Proof.   We fix $\theta \in C_0^\infty(R^1)$, $\theta|_{|t|<1/2} = 1$, $\theta|_{|t|\geqslant 1} = 0$, $0 \leqslant \theta \leqslant 1$, and let $\psi_\omega(x) = \theta(\omega^{-1}x)$, $\psi_{-\omega} = 1 - \psi_\omega$, $\varphi_{\pm\omega} = \psi_{\pm\omega}(\sum_{j=\pm\omega}\psi_j^2)^{-1/2}$, $u_j = \varphi_j u$. Then

$$A[u] = \sum_{j=\pm\omega} A[u_j] + \langle R_\omega u, u \rangle,$$

where

$$R_\omega = \sum D^{\hat{\beta}} r_{\alpha\beta\tilde{\alpha}_n\tilde{\beta}_n\omega}(x) D^{\hat{\alpha}},$$

and the summation is extended to all $\alpha = \hat{\alpha} + \tilde{\alpha}$, $\beta = \hat{\beta} + \tilde{\beta} \in I$, where $\tilde{\alpha} = (0, \tilde{\alpha}_n)$, $\tilde{\beta} = (0, \tilde{\beta}_n)$, $\tilde{\alpha}_n + \tilde{\beta}_n > 0$, and

$$\|r_{\alpha\beta\tilde{\alpha}_n\tilde{\beta}_n\omega}^{(\gamma)}(x)\| \leqslant c_\gamma x_n^{\hat{k}_\alpha+\hat{k}_\beta-\tilde{\alpha}_n-\tilde{\beta}_n-\gamma_n}, \quad |\gamma| = 0, 1, \dots.$$

Here $\omega/2 \leqslant x_n \leqslant \omega$ on the support of the coefficients of the operator R_ω. We construct $\rho = \rho_\omega \in C^\infty(R^1)$, such that $|\rho^{(k)}| \leqslant C_k \rho^{1-k}$, $\forall k$,

$$c\rho(x_n) \leqslant \min\{\max\{\omega/2, x_n\}, 1\} \leqslant C\rho(x_n),$$

fix a large t, denote

$$B_t = \sum_{\alpha \in I} D^\alpha \rho^{2\hat{k}_\alpha} D^\alpha + t, \qquad q_t(x, \xi)^2 = \sum_{\alpha \in I} \rho^{2\hat{k}_\alpha}\xi^{2\alpha} + t,$$

and find out how fast t should grow in order for the estimate $\pm R_\omega \leqslant \varepsilon_1 B_t$ to hold (from here on, ε, ε_1, etc., denote various functions satisfying the condition $\varepsilon \to 0$ as $t \to \infty$).

We shall use Lemma 8.4'; hence, we must select a metric g_t such that

$$(26.8) \qquad B_t \in \mathcal{L}I^+(g_t, q_t^2), \qquad R_\omega \in \mathcal{L}(g_t, \varepsilon q_t^2)$$

and

$$(26.9) \qquad h_t(x,\xi) = h_{g_t}(x,\xi) \to 0 \quad \text{as} \quad t \to \infty$$

uniformly with respect to (x,ξ).

We shall look for g_t in the form

$$g_{t,x,\xi}(y,\eta) = \sum_{1 \leqslant i \leqslant n} (\psi_{t,i}(x,\xi)^{-2}|y_i|^2 + \Psi_{t,i}(x,\xi)^{-2}|\eta_i|^2).$$

Obviously, we can assume that $\psi_{t,j}(x,\xi) = 1$, $1 \leqslant j \leqslant n-1$, $\psi_{t,n}(x,\xi) = \rho(x_n)$, and find the functions $\Psi_{t,j}$ from the condition

$$|\rho^{\hat{k}_\alpha}(x_n)\xi^{\alpha-\beta}| \leqslant c_{\alpha\beta}q_t(x,\xi)\Psi_t(x,\xi)^{-\beta},$$

for all $|\alpha| \leqslant m$, $\alpha_i \geqslant \beta_i \ (> 0$ for some $i)$.

For this we note that, due to condition A, there exist integers $0 \leqslant j_i^- \leqslant j_i^+ \leqslant m$ and $\tau_i^\pm \in [0,1]$, such that $\tau_i^- j_i^- + \tau_i^+ j_i^+ = \alpha_i$, $\sum_i(\tau_i^- + \tau_i^+) = 1$, $\hat{k}_\alpha = \sum_i(\hat{k}_{ij_i^-}\tau_i^- + \hat{k}_{ij_i^+}\tau_i^+)$ where $\hat{k}_{ij} = \hat{k}_{\alpha_{ij}}$. Therefore, there exist $\beta_i^\pm \geqslant 0$, $\beta_i^+ + \beta_i^- = \beta_i$ such that

$$|\rho(x_n)^{\hat{k}_\alpha}\xi^{\alpha-\beta}| \leqslant C \prod_i |\rho(x_n)^{\hat{k}_{ij_i^+}}\xi_i^{j_i^+}|^{\tau_i^+ - \beta_i^+/j_i^+}$$

$$\times \prod_i |\rho(x_n)^{\hat{k}_{ij_i^-}}\xi_i^{j_i^-}|^{\tau_i^- - \beta_i^-/j_i^-}\rho(x_n)^{p_1(\beta)}$$

$$\leqslant c_1(\sum_i(|\rho(x_n)^{\hat{k}_{ij_i^-}}\xi_i^{j_i^-}| + |\rho(x_n)^{\hat{k}_{ij_i^+}}\xi_i^{j_i^+}| + t^{1/2})\rho(x_n)^{p_1(\beta)}t^{-p_2(\beta)}$$

$$\leqslant c_2 q_t(\times,\xi)\rho(x_n)^{p_1(\beta)}t^{-p_2(\beta)},$$

where

$$p_2(\beta) = (\sum_i(\beta_i^-/j_i^- + \beta_i^+/j_i^+))/2, \qquad p_1(\beta) = \sum_i(\hat{k}_{ij_i^+}\beta_i^+/j_i^+ + \hat{k}_{ij_i^-}\beta_i^-/j_i^-).$$

Thus, we can assume that

$$\Psi_{t,i}(x,\xi) = \min_{0<j\leqslant m} t^{1/(2j)}\rho(x_n)^{-\hat{k}_{ij}/j},$$

and then

$$h_t(x,\xi) = \max_{1\leqslant i\leqslant n} \max_{0<j\leqslant m} t^{-1/2j}\rho(x_n)^{k_{ij}/j}.$$

If all k_{ij} were nonnegative, then condition (26.9) would hold for any $\omega \in (0,1)$; but if $s_0 < 0$, then we also need the condition

$$(26.10) \qquad \omega \geqslant Mt^{\kappa_0}, \quad \kappa_0 = 1/(2s_0),$$

where $M \to \infty$ as $t \to \infty$.

Obviously, for any $c < \frac{1}{2}$, there exists C, such that, if $|x_n - y_n| \leqslant c\rho(x_n)$, then $\rho(x_n)/\rho(y_n) + \rho(y_n)/\rho(x_n) \leqslant C$; thus, the metric g_t varies slowly, and the function q_t is g_t-continuous, due to Lemma 1.2. Since

$$g^{\sigma}_{t,x,\xi}(y,\eta) = \sum_i (\psi_{t,i}(x,\xi)^2 |\eta_i|^2 + \Psi_{t,i}(x,\xi)^2 |y_i|^2),$$

the σ-temperateness of the metric g_t and the σ,g_t-temperateness of the function q_t will be proved together with the inequalities

$$\rho(x_n)/\rho(y_n) \leqslant C(1 + \Psi_{t,n}(y,\xi)|x_n - y_n|)^N,$$
$$\rho(y_n)/\rho(x_n) \leqslant C(1 + \Psi_{t,n}(y,\xi)|x_n - y_n|)^N,$$
$$q_t(x,\xi)q_t(x,\eta)^{-1} \leqslant C(1 + \sum_i \psi_{t,i}(x,\eta)|\xi_i - \eta_i|)^N.$$

The first of them follows from the Lagrange formula and the conditions

$$|\partial\rho(x_n)| \leqslant C, \quad \rho(y_n)^{-1} \leqslant \Psi_{t,n}(y,\xi):$$
$$\rho(x_n)/\rho(y_n) \leqslant 1 + |\rho(x_n) - \rho(y_n)|\rho(y_n)^{-1}$$
$$\leqslant 1 + C\rho(y_n)^{-1}|x_n - y_n| \leqslant 1 + C\Psi_{t,n}(y,\xi)|x - y|,$$

and it suffices to prove the second inequality for the case $\rho(y_n) = M_1 t^{\kappa_0}$, $\rho(x_n) = M_2 t^{\kappa_0}$, $M \leqslant M_2 \leqslant M_1/2$, when it is equivalent to the inequality

$$M_2^{-1} t^{-\kappa_0} \leqslant C(1 + M_1^p)^N M_1^{-1} t^{-\kappa_0}, \quad p > 0;$$

but this inequality holds for $N > 1/p$, $C = 1$.

In order to prove the third inequality, we note that, due to condition A,

$$cq_t(x,\xi) \leqslant \sum_{1 \leqslant i \leqslant n} \sum_{0 \leqslant j \leqslant m} \rho(x_n)^{\hat{k}_{ij}} |\xi_i|^j + t^{1/2} \leqslant Cq_t(x,\xi);$$

thus,

$$q_t(x,\xi)q_t(x,\eta)^{-1} \leqslant C\sum_{i,j}(|\xi_i^j| + t^{1/2}\rho(x_n)^{-\hat{k}_{ij}})(|\eta_i|^j + t^{1/2}\rho(x_n)^{-\hat{k}_{ij}})^{-1}$$
$$\leqslant C_1(1 + \sum_{i,j}(\Psi_{t,i}(x_n,\eta)^{-1}|\xi_i - \eta_i|)^j),$$

but $\Psi_{t,i}^{-1} \leqslant \psi_{t,i}$, which proves the third inequality.

Above, we showed that $B_t \in \mathcal{L}I^+(g_t, q_t^2)$; the same arguments show that $R_\omega \in \mathcal{L}(g_t, h_t q_t^2)$. Hence, conditions (26.8) hold, and there exists ε_1 such that $\varepsilon_1 B_t \pm R_\omega \in \mathcal{L}I^+(g_t, q_t^2)$. By Lemma 8.4′, $\pm R_\omega \leqslant \varepsilon_1 B_t$.

On the support of the coefficients of the operator R_ω, we have $c\omega \leqslant x_n \leqslant C\omega$; thus, the last estimate holds if, in the definition of the operator B_t, we replace $\rho(x_n)$ with x_n. But then this estimate implies that

$$(26.11) \qquad \mathcal{A}[u] = \sum_{j=\pm\omega} \mathcal{A}[u_j] + o(Q[u] + t\|u\|_{L_2}^2).$$

Similarly,

$$(26.12) \qquad Q[u] = \sum_{j=\pm\omega} Q[u_j] + o(Q[u] + t\|u\|_{L_2}^2);$$

thus, (26.7) will be proved upon proof of the estimates

$$(26.13) \qquad \mathcal{A}[u_j] + t\|u_j\|_{L_2}^2 \geqslant c(Q[u_j] + t\|u_j\|_{L_2}^2).$$

In order to prove (26.13) for $j = -\omega$, we let $\omega = (\ln t)^{-1}$ and continue the coefficients $a_{\alpha\beta}$ from $\Omega \backslash \Omega_{\omega/2}$ to the functions $a_{\alpha\beta\omega} \in C^\infty(R^n)$, which satisfy the estimates

$$\|a_{\alpha\beta\omega}^{(\gamma)}\| \leqslant C_{\alpha\beta\gamma} \rho^{\hat{k}_\alpha + \hat{k}_\beta - |\gamma|}, \quad |\gamma| = 0, 1, \ldots,$$

$$\sum_{|\alpha|=|\beta|=m} a_{\alpha\beta\omega}(x)\xi^{\alpha+\beta} \geqslant cq_\omega^{-\infty}(x,\xi) \stackrel{\text{def.}}{=} c \sum_{|\alpha|=m} \rho(x_n)^{2\hat{k}_\alpha} \xi^{2\alpha}$$

[the last estimate is made possible by (26.4)]. Furthermore, let

$$A_\omega = \sum_{\alpha,\beta\in I} D^\beta a_{\alpha\beta\omega} D^\alpha, \qquad Q_\omega = \sum_{\alpha\in I} D^\alpha \rho(x_n)^{2\hat{k}_\alpha} D^\alpha,$$

$$q_t(x,\xi)^2 = q_\omega^{-\infty}(x,\xi) + t, \qquad g_{t,x,\xi}(y,\eta) = t^{-1/2m}|\eta|^2 + t^{1/4m}|y|^2.$$

Then

$$\mathcal{A}[u_{-\omega}] = \langle A_\omega u_{-\omega}, u_{-\omega}\rangle_{L_2}, \qquad Q[u_{-\omega}] \asymp \langle Q_\omega u_{-\omega}, u_{-\omega}\rangle,$$

and by the choice of $\omega = 1/\ln t$,

$$A_\omega + t \in \mathcal{L}I^+(g_t, q_t^2), \qquad Q_\omega + t \in \mathcal{L}I(g_t, q_t^2).$$

Therefore, if $c > 0$ is sufficiently small, then

$$A_\omega + t - c(Q_\omega + t) \in \mathcal{L}I^+(g_t, q_t^2),$$

and Lemma 8.4′ gives us estimate (26.13) for $j = -\omega$.

In order to prove (26.13) for $j = \omega$, we note that due to (26.2),

$$|\mathcal{A}[u_\omega] - \tilde{\mathcal{A}}[u_\omega]| \leqslant \varepsilon Q[u_\omega];$$

therefore, we can replace $\mathcal{A}$ with $\tilde{\mathcal{A}}$, and if $n_2 = n$, then estimate (26.13) coincides with (26.6).

26.4. If $n_2 < n$ and $n \in J$, then we construct $\tilde{\rho} = \tilde{\rho}_\omega \in C^\infty((0, \infty))$ such that

$$|\tilde{\rho}^{(k)}| \leqslant C_k \tilde{\rho}^{1-k}, \quad k = 0, 1, \ldots, \quad \tilde{p}(x_n) = \begin{cases} x_n, & x_n \leqslant 2\omega \\ 4\omega, & x_n \geqslant 4\omega, \end{cases}$$

and let

$$A_t = \sum_{\alpha, \beta \in I} D^\beta \hat{a}_{\alpha\beta}(x') \tilde{\rho}(x_n)^{\hat{k}_\alpha + \hat{k}_\beta} + t,$$

$$Q_t = \sum_{\alpha \in I} D^\alpha \tilde{\rho}(x_n)^{2\hat{k}_\alpha} D^\alpha + t.$$

Then the estimate in question can be written as $Q_t \leqslant C A_t$.

Let $H_t = L_2(\Omega_\omega^2)^\ell$, and let $H_{t,0}$ be the closure of $C_0^\infty(\Omega_\omega^2)^\ell$ with respect to the norm

$$\|u\|_{H_{t,0}} = \Big(\sum_{\alpha = (0, \alpha^2) \in I} \|\tilde{\rho}(x_n)^{\hat{k}_\alpha} D^{\alpha^2} u\|_{H_t}^2 \Big)^{1/2}.$$

We continue $\hat{a}_{\alpha,\beta}$ to the functions of the class $C^\infty(R^{n-1}, \operatorname{End} C^\ell)$, bounded, as well as all of their derivatives, and realize A_t, Q_t as PDO on R^{n_1} with symbols taking values in $\operatorname{Hom}(H_{t,0}, H_{t,0}^*)$.

We fix an isomorphism: $J : H_t \to H_{t,0}$ and let

$$\hat{A}_t = J^* A_t J, \qquad \hat{Q}_t = J^* Q_t J, \qquad q_t = \sigma^w(\hat{Q}_t)^{1/2}.$$

Conditions A and B yield $\hat{k}_\alpha > \hat{k}_{(\alpha^1 - \beta^1, \alpha^2)}$, if $\alpha_i^1 \geqslant \beta_i^1$, $(> 0$ for some $i)$; therefore, there exist $c, C > 0$ such that

$$\|\tilde{\rho}(x_n)^{\hat{k}_\alpha} \xi^{1^{\alpha^1 - \beta^1}} D^{\alpha^2} u\| \leqslant C \omega^{c|\beta^1|} \|\tilde{\rho}(x_n)^{\hat{k}_{(\alpha^1 - \beta^1, \alpha^2)}} D^{\alpha^2} u\|.$$

Hence, $\hat{Q}_t, \hat{A}_t \in \mathcal{L}(g_t, q_t^*, q_t)$, where

$$g_{t, x^1, \xi^1}(y^1, \eta^1) = |y^1|^2 + \omega^{2c} |\eta^1|^2.$$

Obviously, the metric g_t is σ-temperate, and q_t is g_t-continuous due to Lemma 1.2. Furthermore, for the same α', β' as before,

$$\tilde{\rho}(x_n)^{2\hat{k}_\alpha} |\xi^{1^{2\alpha^1}} - \eta^{1^{2\alpha^1}}| \leqslant C \sum_{\beta^1} |\xi^1 - \eta^1|^{2|\beta_i^1|} |\eta^1|^{2(\alpha^1 - \beta^1)} \tilde{\rho}(x_n)^{2\hat{k}_\alpha}$$

$$\leqslant C(1 + |\xi^1 - \eta^1|)^{2m} \sum_\alpha \eta^{1^{2\alpha^1}} \tilde{\rho}(x_n)^{2\hat{k}_\alpha};$$

thus,

$$\sigma^w(\widehat{Q}_t)(x^1,\xi^1) \leqslant C(1 + |\xi^1 - \eta^1|)^{2m}\sigma^w(\widehat{Q}_t)(y^1,\eta^1),$$

and, therefore, $q_t \in O_r(H_t, g_t)$.

Then (26.5) yields $\hat{A}_t \in \mathcal{L}I^+(g_t, q_t^*, q_t)$; hence, for some $c_1 > 0$, $\hat{A} - c_1\widehat{Q}_t \in \mathcal{L}I^+(g_t, q_t^*, q_t)$. But $h_t = \omega^c \to 0$ as $\omega \to 0$; thus, Lemma 8.4' yields $\hat{A}_t \geqslant c_1\widehat{Q}_t$ for $\omega \leqslant \omega_0$.

Hence,

$$(26.14) \qquad \langle A_t u, u\rangle \geqslant c_1\langle Q_t u, u\rangle \quad \forall u \in C_0^\infty(\Omega_\omega)^\ell,$$

and (26.7) is proved.

26.5. If $n \notin J$, then (26.14) can be proved in almost exactly the same way as in §26.4, but using PDO calculus with double symbols. In order to manage with the usual ones, we make the substitution $y' = x'$, $y_n = x_n^{-1}$, $u(x) = y_n v(y', y_n)$. Then $\forall u \in C_0^\infty(\Omega_\omega)^\ell$,

$$(26.15) \qquad \langle A_t u, u\rangle_{L_2(\Omega_\omega)^\ell} = \langle \widetilde{A}_t v, v\rangle_{L_2(\Omega_\omega^-)^\ell},$$

where $\Omega_\omega^- = \Omega' \times (\omega^{-1}, \infty)$, $\widetilde{A}_t = A_t' + A_t''$,

$$(26.16) \quad A_t' = \sum_{\alpha,\beta \in I} (-1)^{\alpha_n + \beta_n} D^\beta \hat{a}_{\alpha\beta}(y')\hat{\rho}(y_n)^{-k_\alpha - k_\beta + \alpha_n + \beta_n} D^\alpha + t,$$

$$|\hat{\rho}^{(k)}| \leqslant c_k \hat{\rho}^{1-k}, \quad k = 0, 1, \ldots, \qquad \hat{\rho}(y_n) = \begin{cases} (4\omega)^{-1}, & |y_n| < (4\omega)^{-1}, \\ |y_n|, & |y_n| > (2\omega)^{-1}, \end{cases}$$

$$(26.17) \qquad\qquad A_t'' = \sum D^{\hat{\beta}} a''_{\alpha\beta\tilde{\alpha}_n\tilde{\beta}_n}(y) D^{\hat{\alpha}}.$$

The summation in (26.17) extends to all $\alpha = \hat{\alpha} + \tilde{\alpha}$, $\beta = \hat{\beta} + \tilde{\beta} \in I$ such that $\tilde{\beta} = (0, \tilde{\beta}_n)$, $\tilde{\alpha} = (0, \tilde{\alpha}_n)$ and $\tilde{\beta}_n + \tilde{\alpha}_n > 0$, and the coefficients satisfy the estimates

$$(26.18) \qquad \|a''_{\alpha\beta\tilde{\alpha}_n\tilde{\beta}_n}(y)\| \leqslant C\hat{\rho}(y_n)^{-k_\alpha - k_\beta + \hat{\alpha}_n + \hat{\beta}_n}.$$

Defining $\widetilde{Q}_t$ similarly, we also obtain (26.15) and the equality $\widetilde{Q}_t = Q_t' + Q_t''$, where Q_t' is defined by equality (26.16) with $\hat{a}_{\alpha\beta}(y') = 0$, $\forall \alpha \neq \beta$, $\hat{a}_{\alpha\alpha}(y') = I$, and Q_t'' has the form (26.17) with coefficients satisfying condition (26.18).

Conditions A–C yield $-k_\alpha - k_\beta < -k_{\hat{\alpha}} - k_{\hat{\beta}}$, if $\tilde{\alpha}_n + \tilde{\beta}_n > 0$; therefore, there exists $\delta > 0$ such that

$$(26.19) \qquad\qquad \langle \widetilde{A}_t v, v\rangle = \langle A_t' v, v\rangle + O(\omega^\delta \langle Q_t' v, v\rangle),$$

$$(26.20) \qquad\qquad \langle \widetilde{Q}_t v, v\rangle = \langle Q_t' v, v\rangle + O(\omega^\delta \langle Q_t' v, v\rangle).$$

Therefore, it suffices to prove that for small ω,

$$(26.21) \qquad\qquad A'_t \geqslant cQ'_t.$$

Let $H_t = L_2(\Omega^2)^\ell$, $H_{t,0} = \overset{\circ}{H}{}^m(\Omega^2)^\ell$. We realize A'_t, Q'_t as PDO on R^{n_1} with symbols taking values in $\mathrm{Hom}(H_{t,0}; H^*_{t,0})$. Let $J : H_t \to H_{t,0}$ be a fixed isomorphism and let

$$\hat{A}'_t = J^* A'_t J, \qquad \widehat{Q}'_t = J^* Q'_t J, \qquad q_t = \sigma^w(\widehat{Q}'_t)^{1/2}.$$

In order to find out what metric can describe the operators $\hat{A}'_t$, $\widehat{Q}'_t$, we note that conditions A–C imply that for all $0 \leqslant \beta^1_i \leqslant \alpha^1_i$ such that $\beta^1_i < \alpha^1_i$ for some $i \notin J$, we have $k_{(\beta^1,\alpha^2)} - k_\alpha = -c_{\alpha\beta^1} < 0$; therefore, there exists $c > 0$ such that for all such α^1, β^1,

$$|\hat{\rho}(y_n)^{-k_\alpha + \alpha_n} \xi^{1^{\beta^1}}| \leqslant C \hat{\rho}(y_n)^{-c|\alpha^1 - \beta^1| + \alpha_n - \beta_n} \hat{\rho}(y_n)^{-k_{(\beta^1,\alpha^2)} + \beta_n} |\xi^{1^{\beta^1}}|,$$

and $\hat{A}'_t, \widehat{Q}'_t \in \mathcal{L}(g_t, q^*_t, q_t)$, where

$$g_{t,y^1,\eta^1}(z^1, \zeta^1) = |z^{11}|^2 + \hat{\rho}(y_n)^{-2}|z_n|^2 + \hat{\rho}(y_n)^{-2c}|\zeta^{11}|^2 + \hat{\rho}(y_n)^{2-2c}|\zeta_n|^2$$

[here $y^1 = (y^{11}, y_n)$]. Since $h_t(y, \zeta) = \hat{\rho}(y_n)^{-c}$, then due to Lemmas 1.2, 3.2, and 3.3, the metric g_t is σ-temperate and q_t is a right σ, g_t-temperate function. (26.5) yields $\hat{A}'_t \in \mathcal{L}I^+(g_t, q^*_t, q_t)$; therefore, for some $c_1 > 0$, also $\hat{A}'_t - c_1 \widehat{Q}'_t \in \mathcal{L}I^+(g_t, q^*_t, q_t)$. Applying Lemma 8.4', we see that for $\omega \leqslant \omega_0$ estimate (26.21) holds and, hence, estimates (26.14) and (26.7) hold. This proves Theorem 26.1. ∎

26.6. We assume that the conditions of Theorem 26.1 and, therefore estimate (26.7), hold. Then, in particular, the form $\mathcal{A}$ is semibounded from below. Let A be an operator associated with the variational triple $\mathcal{A}$, $C^\infty_0(\Omega)^\ell$, $L_2(\Omega)^\ell$.

Theorem 26.2. *The spectrum of the operator A is discrete if and only if $s_0 < 0$.*

Proof. For $s_0 < 0$, discreteness of the spectrum will be proved together with the theorem for its asymptotics. On the other hand, if $s_0 \geqslant 0$, we fix a nonzero $\varphi \in C^\infty_0(\Omega')^\ell$ and denote

$$u_j(x) = 2^{j/2}\theta((x_n - 2^{-j})2^{j+2})\varphi(x').$$

Then $\langle u_i, u_j \rangle = 0 \ \forall i \neq j$ and there exists $c > 0$ such that for all i, $\|u_i\|_{L_2} \geqslant c$, $\mathcal{A}[u_i] \leqslant c^{-1}$. Hence, the energy space of the operator A is not compactly imbedded into $L_2(\Omega)^\ell$, and the spectrum is nondiscrete. ∎

26.7. At this point, we shall construct a set of symbols which will help us to calculate asymptotic coefficients.

Let π_i be a plane in (α, k)-space defined by the conditions $\alpha_j = 0 \ \forall j \neq i$, and P_i be a broken line passing through the points $(\alpha, k_\alpha) \in \pi_i$ with vertices at some of them. This is a downward convex broken line in the plane π_i. For $s \leqslant s_0$, we let ℓ_{is} denote a straight line passing through the point $(0, s)$ and also through at least one point of the broken line P_i and not lying above P_i (i.e., a straight line, supporting P_i), and we let γ_{is} denote a slope of this straight line, and construct a hyperplane Q_s passing through the straight lines ℓ_{is}.

Furthermore, let

$$I_s = \{|\alpha| \leqslant m \mid (\alpha, k_\alpha) \in Q_s\}, \qquad \tilde{I}_s = \{x \in (R^1_+)^n \mid (x, k_x) \in Q_s\},$$

$$a^s(x, \xi) = \sum_{\alpha, \beta \in I_s} \hat{a}_{\alpha\beta}(x') x_n^{\hat{k}_\alpha + \hat{k}_\beta} \xi^{\alpha+\beta}, \qquad q^s(x, \xi) = \sum_{\alpha \in I_s} x_n^{2\hat{k}_\alpha} \xi^{2\alpha}.$$

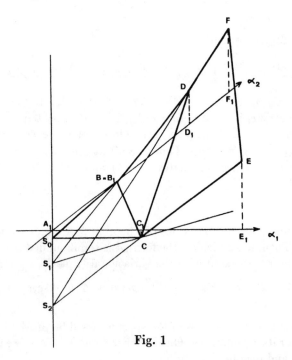

Fig. 1

For each $s \leqslant s_0$, we have two possibilities: 1) there are no less then two points (α_{ij}, k_{ij}) with $\alpha_{ij} \neq 0$ on one of the edges ℓ_{is} (the number of such s is, obviously, finite: $s_0 > s_1 > \cdots > s_{\hat{n}}, \ \hat{n} \geqslant 0$); 2) there is exactly one such point on each edge ℓ_{is}.

Let $s_{\hat{n}+1} = -\infty$. We fix $s_p^+ \in (s_p, s_{p+1})$ and note that 1) $\forall s \in (s_p, s_{p+1})$, $\tilde{I}_s = \tilde{I}_{s_p^+} \ (p \leqslant \hat{n})$; so, the arbitrariness in the choice of s_p^+ is not essential;

2)

(26.22) $$\tilde{I}_{s_{p-1}^+} \cup \tilde{I}_{s_p^+} \subset \tilde{I}_{s_p}, \quad p = 1, \ldots, \hat{n}.$$

In Fig. 1, $n = 2$, $J = \{1\}$, $\tilde{I}_{s_0}$ and $\tilde{I}_{s_1}$ are the triangles $A_1 B_1 C_1$, and $B_1 C_1 D_1$, respectively; $\tilde{I}_{s_2}$ is the quadrangle $C_1 D_1 F_1 E_1$, and $\tilde{I}_{s_0^+}$, $\tilde{I}_{s_1^+}$, $\tilde{I}_{s_2^+}$ are the intervals $B_1 C_1$, and $C_1 D_1$, $E_1 F_1$, respectively.

In addition to the functions a^s, q^s, we shall need the forms $\mathcal{A}^0$, Q^0 which are defined similarly to $\tilde{\mathcal{A}}$, Q, but with summation over $\alpha, \beta \in I_{s_0}$ instead of over $\alpha, \beta \in I$; and in the case of $n_2 < n$, we need the forms

$$\mathcal{A}^0_{x^1, \xi^1}[u] = \sum_{\alpha, \beta \in I_{s_0}} \xi^{1^{\alpha^1 + \beta^1}} \langle \hat{a}_{\alpha\beta}(x') x_n^{\hat{k}_\alpha + \hat{k}_\beta} D^{\alpha^2} u, D^{\beta^2} u \rangle_{L_2(R^{n_2})^\ell},$$

$$Q^0_{x^1, \xi^1}[u] = \sum_{\alpha \in I_{s_0}} \xi^{1^{2\alpha^1}} \| x_n^{\hat{k}_\alpha} D^{\alpha^2} u \|^2_{L_2(R^{n_2})^\ell}.$$

26.8.

Theorem 26.3. *If condition (26.7) holds, then there exists $c > 0$, such that*
a)

(26.23) $$a^{-\infty}(x, \xi) \geqslant c q^{-\infty}(x, \xi) \quad \forall (x, \xi) \in \Omega \times (R^n \backslash O);$$

b) *for all $s \in (-\infty, s_0)$*

(26.24) $$a^s(x, \xi) \geqslant c q^s(x, \xi) \quad \forall (x, \xi) \in \Omega_\infty \times R^n;$$

c) *if $n_2 < n$, then $\forall (x^1, \xi^1) \in \Omega^1_\infty \times R^{n_1}$,*

(26.25) $$\forall u \in C_0^\infty (\Omega^2_\infty)^\ell, \quad \mathcal{A}^0_{x^1, \xi^1}[u] \geqslant c Q^0_{x^1, \xi^1}[u];$$

d) *if $n_2 = n$, then*

(26.26) $$\forall u \in C_0^\infty (\Omega_\infty)^\ell, \quad \mathcal{A}^0[u] \geqslant c Q^0[u].$$

Proof. a) (26.7) and Lemma 13.1 imply that (26.23) holds on $\Omega' \times [\varepsilon, 1] \times R^n$ [with $c = c(\varepsilon)$] for any ε, and (26.24), with $s = s_{\hat{n}}^+$, yields (26.23) on $\Omega'_\varepsilon \times R^n$ if $\varepsilon > 0$ is sufficiently small.

b) Since the functions in (26.24) are quasi-homogeneous with respect to (x_n, ξ), it is sufficient to prove this inequality for a fixed $(\overset{\circ}{x}_n, \overset{\circ}{\xi}) \neq 0$, provided that we can choose $c > 0$ to be the same for all $\overset{\circ}{x}' \in \Omega'$.

We fix $v \in \mathbf{C}^{\ell}$, $|v| = 1$, and let $\varepsilon = (\ln t)^{-1}$,

$$\gamma^s = (\gamma_{1s}, \ldots, \gamma_{ns}), \qquad \hat{\gamma}^s = \gamma^s + (0, 0, \ldots, 0, 1), \qquad \alpha^{(s)} = \langle \gamma^s, \alpha \rangle,$$
$$u_t(x) = \theta(\varepsilon^{-1}|x' - \overset{\circ}{x}'|)\theta(\varepsilon^{-1}|tx_n - \overset{\circ}{x}_n|) \exp(i \sum_j t^{\hat{\gamma}_{js0}} \xi_j x_j)v,$$

where $\theta \in C_0^{\infty}(R^1)$ is the same as in §26.3 Then

(26.27)
$$A[u_t] = \sum_{\alpha, \beta \in I} t^{(\alpha+\beta)^{(s)} + \alpha_n + \beta_n} \overset{\circ}{\xi}^{\alpha+\beta} \langle a_{\alpha\beta}(x)u_t, u_t \rangle_{L_2}$$
$$+ \sum t^{(\hat{\alpha}+\hat{\beta})^{(s)} + \alpha_n + \beta_n} \varepsilon^{-|\tilde{\alpha}| - |\tilde{\beta}|} B_{\alpha\beta\tilde{\alpha}\tilde{\beta}}[u_t].$$

The second sum is taken over $\alpha = \hat{\alpha} + \tilde{\alpha}, \beta = \hat{\beta} + \tilde{\beta} \in I$ with $\tilde{\alpha} + \tilde{\beta} > 0$, and

$$|B_{\alpha\beta\tilde{\alpha}\tilde{\beta}}[u_t]| \leqslant C_{|\xi^0|} \operatorname{mes}\{x' \mid |x' - \overset{\circ}{x}'| < \varepsilon\}$$
$$\times \operatorname{mes}\{x_n \mid t^{-1}(\overset{\circ}{x}_n - \varepsilon) < x_n < t^{-1}(\overset{\circ}{x}_n + \varepsilon)\}t^{-\hat{k}_\alpha - \hat{k}_\beta}$$
$$\leqslant c'_{|\xi^0|}\varepsilon^n t^{-\hat{k}_\alpha - \hat{k}_\beta - 1}.$$

For $s < s_0$, $\gamma_{is} > 0$ for any i, hence, the terms in the second sum are estimated by $Ct^{r_{\alpha\beta s} - 1 - \delta}$, where $\delta > 0$, and $r_{\alpha\beta s} = (\alpha + \beta)^{(s)} - k_\alpha - k_\beta$; similarly, the terms of the first sum are $O(t^{r_{\alpha\beta s} - 1}\varepsilon^n)$. Furthermore, condition (26.2) implies that as $t \to \infty$,

$$\langle a_{\alpha\beta}(x)u_t, u_t \rangle_{L_2} = c_\theta \varepsilon^n t^{-\hat{k}_\alpha - \hat{k}_\beta - 1}((\hat{a}_{\alpha\beta}(\overset{\circ}{x}')v, v)\overset{\circ}{x}_n^{\hat{k}_\alpha + \hat{k}_\beta} + o(1));$$

thus, estimate (26.27) can be rewritten in the form

(26.27′)
$$A[u_t] = \sum_{\alpha, \beta \in I_s} c_\theta \varepsilon^n t^{r_{\alpha\beta s} - 1} \langle \hat{a}_{\alpha\beta}(\overset{\circ}{x}')\overset{\circ}{\xi}^{\alpha+\beta}v, v \rangle \overset{\circ}{x}_n^{\hat{k}_\alpha + \hat{k}_\beta}$$
$$+ o\left(\sum_{\alpha, \beta \in I_s} \varepsilon^n t^{r_{\alpha\beta s} - 1} \right) + O\left(\sum_{\alpha, \beta \in I \backslash I_s} t^{r_{\alpha\beta s} - 1} \right).$$

In order to estimate the last two terms, we shall show that, for any $x \in \tilde{I}_s$, $y \in \tilde{I} \backslash \tilde{I}_s$

(26.28)
$$-s = x^{(s)} - k_x > y^{(s)} - k_y.$$

Since the equality is valid for $x = 0$, $x \in I_s \cap \pi_i$, $i = 1, \ldots, n$, it also holds for the rest of the x, but then it suffices to prove the inequality for $y = \tau x$, $\tau \in R_+^1$. Noting that the graph of the function $f_x : R_+^1 \ni \tau \to k_{\tau x} \in R^1$ is a downward convex broken line and that the straight line $k = s + \langle \gamma^s, \tau x \rangle$ supports it, we see that the inequality holds indeed.

(26.27) and (26.28) imply that as $t \to \infty$,

$$A[u_t] + C\|u_t\|_{L_2}^2 = \varepsilon^n t^{-2s-1}(c_\theta \langle a^s(\overset{\circ}{x}, \overset{\circ}{\xi})v, v \rangle + o(1)),$$

where $c_\theta > 0$ and the constant in the o-term do not depend on $\overset{\circ}{x}'$.

Similarly,

$$Q[u_t] = \varepsilon^n t^{-2s-1} c_\theta (q^s(\overset{\circ}{x}, \overset{\circ}{\xi}) + o(1)).$$

By substituting these two equalities into (26.7), by dividing by $\varepsilon^n t^{-2s-1}$ and by passing to the limit as $t \to \infty$, we obtain (26.24).

c) In order to prove (26.25) for the case $n \in J$, we note that the substitution $x_n = \omega y_n$, $\xi_i = \omega^{-\gamma_{1s0}} \eta_i$, $i \notin J$ induces the same factor ω^{2s_0} on the right- and left-hand sides of estimate (26.25); therefore, it suffices to prove it for a fixed $(x^{10}, \xi^{10}) \in \Omega^1 \times (R^{n_1} \backslash O)$.

Let $\varphi \in C_0^\infty(\Omega_\infty^2)^\ell$, $\varphi_t(x^2) = \varphi(x^{21}, tx_n)$,

$$u_t(x) = \theta(\varepsilon^{-1}|x^1 - x^{10}|) \exp(i \sum_{j \notin J} t^{\gamma_{js0}} \xi_j^0 x_j) \varphi_t(x^2).$$

(26.2) implies that as $t \to \infty$,

$$(26.29) \qquad A[u_t] = \tilde{A}[u_t] + o(Q[u_t]);$$

therefore, by making the substitution $x_n = t^{-1}y_n$ in the forms $A[u_t]$, $Q[u_t]$ and by taking (26.28) into account, we obtain

$$A[u_t] + c\|u_t\|_{L_2}^2 = c_\theta \varepsilon^{n_1} t^{-2s_0-1}(A_{x^{10}, \xi^{10}}^0[\varphi] + o(1)),$$
$$Q[u_t] = C_\theta \varepsilon^{n_1} t^{-2s_0-1}(Q_{x^{10}, \xi^{10}}^0[\varphi] + o(1)).$$

By substituting these two equalities in (26.7), by dividing by $\varepsilon^{n_1} t^{-2s_0-1}$, and by passing to the limit as $t \to \infty$, we obtain (26.25).

If $n \notin J$, then we obtain the same equalities by choosing arbitrary $\varphi \in \Omega^2$, $(x^{10}, \xi^{10}) \in \Omega_\infty^1 \times (R^{n_1} \backslash O)$ and assuming that

$$x^1 = (x^{11}, x_n),$$
$$u_t(x) = \theta(\varepsilon^{-1}|x^{110} - x^{11}|)\theta(\varepsilon^{-1}|tx_n - \overset{\circ}{x}_n|) \exp(i \sum_{j \notin J} t^{\hat{\gamma}_{js0}} \xi_j^0 x_j) \times \varphi(x^2).$$

d) Without loss of generality, we can assume that $\Omega' \ni \{0\}$. We fix $v \in C_0^\infty(\Omega_\infty)^\ell$ and let $y_i = x_i t^{\hat{\gamma}_{is0}}$, $u_t(x) = v(y)$. Then, using (26.29) and (26.28), we obtain

$$A[u_t] + C\|u_t\|_{L_2}^2 = t^p(A^0[v] + o(1)),$$
$$Q[u_t] = t^p(Q^0[v] + o(1)), \quad \text{where } p = -2s_0 - \sum \hat{\gamma}_{iso}.$$

By substituting these equalities into (26.7), dividing by t^p, and by passing to the limit, we obtain (26.26).

This proves Theorem 26.3.   ∎

26.9. If $J \neq \{n\}$, then (26.7) and Poincaré inequality imply that, for some $c, C > 0$,

$$\mathcal{A}[u] + C\|u\|_{L_2}^2 \geqslant c\langle x_n^{2s_0} u, u \rangle_{L_2} \quad \forall u \in C_0^\infty(\Omega)^\ell,$$

and, if we assume that the Hardy inequality holds for all exponents, then this estimate is also valid in the case of $J = \{n\}$.

In accordance with the formal scheme of §24.3, the asymptotics of $N(t, A)$ is defined by the function

$$a_c(x, \xi) = a(x, \xi) + c x_n^{2s_0}, \quad \text{where } a(x, \xi) = \sum_{\alpha, \beta \in I} a_{\alpha\beta}(x) \xi^{\alpha+\beta}$$

is a full symbol of the form $\mathcal{A}$. At this point, we shall study the properties of the function a_C.

Lemma 26.1. *Let* $M = \ln \ln t$, $\kappa_p = 1/(2s_p)$, $p \leqslant \hat{n}$, $\kappa_{\hat{n}+1} = 0$, $\tau_p = t^{\kappa_p}$. *Then, if* $c, c_1, C_1 > 0$ *do not depend on* t *and*

$$c_1 t I \leqslant a_c(x, \xi) \leqslant C_1 t I,$$

then as $t \to \infty$,
   a) *if* $x_n \in (M^{-1}, 1)$, *then*

$$(26.30) \qquad\qquad a_c(x, \xi) = a^{-\infty}(x, \xi)(1 + o(1));$$

   b) *if* $x_n \in (M^{-1}\tau_p, M\tau_p)$, $0 < p \leqslant \hat{n}$, *then*

$$(26.31) \qquad\qquad a_c(x, \xi) = a^{s_p}(x, \xi)(1 + o(1));$$

   c) *if* $x_n \in (M\tau_p, M^{-1}\tau_{p+1})$, $p \leqslant \hat{n}$, *then*

$$(26.32) \qquad\qquad a_c(x, \xi) = a^{s_p^+}(x, \xi)(1 + o(1));$$

   d) *if* $x_n \in (0, M\tau_0)$ *and* c *is sufficiently large, then*

$$(26.33) \qquad\qquad a_c(x, \xi) = (a^{s_p^+}(x, \xi) + c x_n^{2s_0})(1 + o(1)).$$

Proof. Estimate (26.24), with $s = s_0^+$, implies that, if $C > 0$ is sufficiently large and $c > 0$ is sufficiently small, then

$$a^{s_0^+}(x, \xi) + C x_n^{2s_0} \geqslant c(q^{s_0}(x, \xi) + C x_n^{2s_0}).$$

In addition, due to (26.23) and (26.24), there exist $c, C > 0$, such that, for all $s \in [-\infty, s_0)$,

$$ca^s(x,\xi) \leqslant q^s(x,\xi) \leqslant Ca^s(x,\xi).$$

These estimates imply that, for some $c, C', C > 0$,

$$cq_C \leqslant a_C \leqslant C'q_C,$$

if the lemma is valid with q instead of a, but then it is also valid for a.

Thus, we are proving parts (a)–(d), replacing a with q.

a) Here $|\xi| < M^N t^{1/2m}$; therefore $|q_C(x,\xi) - q^{-\infty}(x,\xi)| = o(t)$.

b), d) The substitution $x_n = t^{\kappa_p} y_n$, $\xi_i = t^{-\kappa_p \hat{\gamma}_{i,p}} \eta_i$ and (26.28) yield

$$t^{-1} q_C(x,\xi) = q^{s_p}(y,\eta) + O(t^{-\omega} |\eta|^{2m}),$$

where $\omega > 0$, and both terms are positive. Therefore, we need $|\eta| \leqslant M^N$, and the second term is infinitely small.

c) If (26.32) is false, then there exist $\alpha \notin I_{s_p^+}$, $c_1 > 0$, C_1 and a sequence $t_j \to \infty$ such that

$$(26.34) \qquad c_1 t_j \leqslant x_n^{2\hat{k}_\alpha} \xi^{2\alpha}, \qquad q_{s_p^+}(x,\xi) \leqslant c_1 t_j,$$

where (x,ξ) depend on $j = 1, 2, \ldots$. We shall show that (26.34) contradicts the condition of part (c).

We note that the straight line $\ell_{i s_p^+}$ contains one and only one point $(j(i,p), k(i,p)) \in P_i$ with $j(i,p) > 0$, and $I_{s_p^+}$ has the form $I_{s_p^+} = \{\alpha \mid \sum_i \alpha_i/j(i,p) = 1\}$. Therefore, $\alpha \notin I_{s_p^+}$ if either $\sum_i \alpha_i/j(i,p) < 1$, or $\sum_i \alpha_i/j(i,p) > 1$.

In the last case, we assume that $j_i = j(i,p)$, $k_i = \hat{k}(i,p)$ and make the substitution $\xi_i = t^{1/2j_i} x_n^{-k_i/j_i} \eta_i$ $(i = 1, \ldots, n)$ to obtain

$$(26.35) \qquad q^{s_p^+}(x,\xi) = tq^{s_p^+}(1,\eta), \qquad x_n^{2\hat{k}_\alpha} \xi^{2\alpha} = t\eta^{2\alpha} x_n^{2\omega'_\alpha} t^{\omega_\alpha},$$

where

$$\omega_\alpha = \sum_i \alpha_i/j_i - 1 < 0, \qquad \omega'_\alpha = \hat{k}_\alpha - \sum_i k_i \alpha_i/j_i.$$

Let $\gamma_i = (k_i - s_p)/j_i$; then $k = s_p + \sum_i \gamma_i \alpha_i$ is an equation of the hyperplane Q_{s_p}; and since $\alpha \notin I_{s_p}$, the point $(\alpha, \hat{k}_\alpha)$ lies above this hyperplane, i.e.,

$$\hat{k}_\alpha > s_p + \sum_i \gamma_i \alpha_i = s_p + \sum_i k_i \alpha_i/j_i - s_p \sum_i \alpha_i/j_1,$$

$$\omega'_\alpha = \hat{k}_\alpha - \sum_i k_i \alpha_i/j_i > -s_p (\sum_i \alpha_i/j_i - 1) = -s_p \omega_\alpha.$$

Since $x_n > Mt^{1/2s_p}$, then $x_n^{2s_p} < M^{2s_p}t$ and $x_n^{2\omega_\alpha'}t^{\omega_\alpha} < M^{-2s_p\omega_\alpha} \to 0$ as $t \to \infty$. But due to the conditions of the lemma, $|\eta| \leqslant C$ in (26.35); therefore, (26.35) contradicts (26.34).

Now, let $\sum_i \alpha_i/j_i > 1$. Here we also obtain (26.35), but with $\omega_\alpha > 0$. The same arguments as above, but with hyperplane $Q_{s_{p+1}}$ instead of Q_{s_p}, yield $\omega_\alpha' \geqslant -s_{p+1}\omega_\alpha$. Since $x_n \leqslant M^{-1}t^{1/2s_{p+1}}$, then $t \leqslant (x_n M)^{2s_{p+1}}$ and $x_n^{2\omega_\alpha'}t^{\omega_\alpha} < M^{2s_{p+1}\omega_\alpha} \to 0$ as $t \to \infty$. Hence, (26.35) contradicts (26.34) also in the case of $\sum_i \alpha_i/j_i > 1$.

Thus, (26.34) is false and (26.32) is proved. ∎

Lemma 26.1 implies the next lemma:

Lemma 26.2. *There exist $c, C > 0$ such that $a_C \geqslant cq_C$.*

26.10. At this point, we shall predict the asymptotic function of $N(t, A)$ for the cases of a weak and an intermediate degeneracy and an order of asymptotic functions for the case of a strong degeneracy. For this, we have to analyze asymptotics of the function $\widehat{V}(a_C - t, \Omega)$.

For $p = 0, \ldots, \hat{n}$, let

$$\sigma_p' = \sum_i 1/(2j(i,p)), \quad \sigma_p = -\kappa_p \sum_i \gamma_i s_p, \quad d_p = \sum_i k(i,p)/j(i,p).$$

Since $s_0 < 0$ and the points $(0, s_0), (j(i,0), k(i,0)), \ldots, (j(i,\hat{n}), k(i,\hat{n}))$ lie on the downward convex broken line P_i, then $d_0 < \cdots < d_{\hat{n}}$.

Theorem 26.4. *Let $C > 0$ be sufficiently large. Then as $t \to \infty$,*

a) *if $d_{\hat{n}} < 0$, then*

(26.36) $\widehat{V}(a_C - t, \Omega) \sim t^{n/2m}\widehat{V}(a^{-\infty} - 1, \Omega)$;

b) *if $d_{p-1} < 0 < d_p$, $1 \leqslant p \leqslant \hat{n}$, then*

(26.37) $\widehat{V}(a_C - t, \Omega) \sim t^{\sigma_p}\widehat{V}(a^{s_p} - 1, \Omega' \times (0, \infty))$;

c) *if $d_p = 0$, then*

(26.38) $\widehat{V}(a_C - t, \Omega) \sim t^{\sigma_p'} \ln t (\kappa_{p+1} - \kappa_p)\widetilde{V}(a^{s_p^+} - 1, \Omega')$,

where

$$\widetilde{V}(a, \Omega') = (2\pi)^{-n} \iint\limits_{\Omega' \times R^n} N(0, a(x', 1, \xi))dx'\, d\xi;$$

d) *if $d_0 > 0$, then formula (26.37) holds with $p = 0$ and $a_C^{s_0}$ instead of a^{s_0}.*

Proof. In parts (a)–(d) we use formulas (26.30)–(26.33), respectively, and in each of the points, Lemma 26.2.

a) The condition $d_{\hat{n}} < 0$ and homogeneity of $a^{-\infty}$ with respect to ξ yield

$$F(t) \overset{\text{def.}}{=} \widehat{V}(a^{-\infty} - t, \Omega) = t^{n/2m} \widehat{V}(a^{-\infty} - 1, \Omega),$$
$$F_C^1(t) \overset{\text{def.}}{=} \widehat{V}(q^{-\infty} - Ct, \Omega' \times (0, M^{-1})) = o(t^{n/2m});$$

therefore,

$$\widehat{V}(a_C - t, \Omega) = F(t) + O(F(t_\varepsilon) - F(t_{-\varepsilon})) + O(F_{C_1}^1(t))$$

satisfies (26.36).

b) Let $F_M(t) = \widehat{V}(a^{s_p} - t, \Omega' \times (\tau_p/M, M\tau_p))$. Then

$$\widehat{V}(a_C - t, \Omega) = F_M(t) + O(F_M(t_\varepsilon) - F_M(t_{-\varepsilon})) + O(\widehat{V}(q^{s_{p-1}^+} - C_1 t, (0, \tau_p/M))$$
$$+ O(\widehat{V}(q^{s_p^+} - C_1 t; \Omega' \times (M\tau_p, \infty))).$$

By making the substitution $x_n = \tau_p y_n$, $\xi_i = \tau_p^{-\hat{\gamma}_{i \cdot p}} \eta_i$, we obtain $F_M(t) = t^{\sigma_p} M_{\cdot 1}(1)$ and due to (26.22), obtain estimates for the last two terms through

$$c_2 t^{\sigma_p} \int_0^{M^{-1}} dx_n \, \mathrm{mes}\{\xi \mid \sum_i \xi_i^{2j(i,p-1)} x_n^{2\hat{k}(i,p-1)} < c_2\}$$

$$= c_3 t^{\sigma_p} \int_0^{M^{-1}} x_n^{-d_p-1-1} dx_n = c_4 t^{\sigma_p} M^{d_p-1},$$

$$c' t^{\sigma_p} \int_M^\infty x_n^{-d_p-1} dx_n = c_1' t^{\sigma_p} M^{-d_p},$$

respectively. Similarly, Lemma 26.2 also yields

$$|F_M(1) - \widehat{V}(a^{s_p} - 1, \Omega' \times (0, \infty)| \leqslant c(M^{-d_p} + M^{d_{p-1}}),$$

and (26.37) is proved.

Under the conditions of part (c), we have

$$\widehat{V}(a_C - t, \Omega) = F(t) + O(F(t_\varepsilon) - F(t_{-\varepsilon})) + O(\widehat{V}(q^{s_p} - C_1 t, \Omega' \times (c\tau_0, M\tau_p)))$$
$$+ O(\widehat{V}(q^{s_{p+1}} - c_1 t, \Omega' \times (M^{-1}\tau_{p+1}, 1)))$$
$$= F(t) + O(F(t_\varepsilon) - F(t_{-\varepsilon})) + O(F_1(t)) + O(F_2(t)),$$

where

(26.39)
$$F(t) = t^{\sigma_p'} \widehat{V}(a^{s_p^+} - 1, \Omega' \times (M\tau_p, M^{-1}\tau_{p+1}))$$
$$= t^{\sigma_p'} \widetilde{V}(a^{s_p^+} - 1, \Omega') \int_{(M\tau_p, M^{-1}\tau_{p+1})} x_n^{-1} dx_n$$
$$= t^{\sigma_p'} \ln t(\kappa_{p+1} - \kappa_p) \widetilde{V}(a^{s_p^+} - 1, \Omega')(1 + o(1)).$$

Furthermore, if $p = \hat{n}$, then $\kappa_{p+1} = 0$, $\sigma_p' = j/2m$, and

$$F_2(t) \leqslant Ct^{\sigma_p'} \int_{M^{-1}}^{1} x_n^{-1} dx_n = o(\ln t \, t^{\sigma_p'});$$

and if $p < \hat{n}$, then similarly to part (b),

(26.40) $$F_2(t) = CM^C t^{b_{p+1}}(1 + o(1)).$$

On the other hand, $I_{S_p^+} \subset I_{S_{p+1}}$; hence,

(26.41) $$F_2(t) \leqslant \hat{V}(q^{S_p^+} - Ct, \Omega' \times (M^{-1}\tau_{p+1}, 1)) \leqslant c \ln t \, t^{\sigma_p'}.$$

Since $M^C = o(\ln t)$, then (26.40) and (26.41) imply that $\sigma_{p+1} \leqslant \sigma_p'$, and, finally,

(26.42) $$F_2(t) = o(\ln t \, t^{\sigma_p'}).$$

Furthermore, $\tau_0 < M^{-1}\tau_p$; therefore, similarly to part (b),

$$F_1(t) = cM^C t^{\sigma_p}(1 + o(1)),$$

but $I_{s_p^+} \subset I_{s_p}$; hence, similarly to (26.39), we obtain estimate (26.41) and, therefore, estimate (26.42) for $F_1(t)$.

Estimate (26.38) follows from (26.39) and (26.42). d) is proved similarly to (b). ∎

Corollary 26.1. a) *If* $d_0 \leqslant 0$, *then for all* $N, C_1 > 0$,

$$M^N t^{\sigma_0} = o(\hat{V}(a_C - C_1 t, \Omega)).$$

b) *If* $d_0 > 0$, *then for all* C_1,

$$\hat{V}(a_C - C_1 t, \Omega' \times (M\tau_0, 1)) = o(t^{\sigma_0}).$$

Proof. Quasi-homogeneity of the symbol $a_C^{S_0}$ implies that

$$ct^{\sigma_0} \leqslant \hat{V}(a_C^{S_0} - C_1 t, \Omega' \times (0, M\tau_0)) \leqslant CM^N t^{\sigma_0},$$

due to Lemma 26.2,

$$\hat{V}(a_C^{S_0} - C_1 t, \Omega' \times (0, M\tau_0)) \leqslant \hat{V}(a_C - C_2 t, \Omega' \times (0, M\tau_0));$$

and in Theorem 26.4 we actually showed that for $d_0 \leqslant 0$, the last function is $o(\hat{V}(a_C - C_2 t, \Omega))$ [it was included in the o-term of the formulas (26.36)–(26.38)]. But $\hat{V}(a_C - C_1 t, \Omega)$ has a power and a logarithmic-power asymptotic function; thus, the exponent of the latter is not smaller than σ_0 (and smaller than σ_0 for an asymptotic power function). This proves the statement of part (a); part (b) is proved similarly. ∎

26.11.

Theorem 26.5. *If $d_0 \leqslant 0$, then asymptotic function of $N(t, A)$ is defined by the formulas (26.30)–(26.32); and if $d_0 > 0$, then as $t \to \infty$,*

$$(26.43) \quad N(t, A) \sim (2\pi)^{-n_1} t^{\sigma_0} \iint\limits_{\Omega_\infty^1 \times R^{n_1}} \mathcal{N}(\mathcal{A}^0_{x^1, \xi^1} - \| \cdot \|^2_{L_2}; C_0^\infty(\Omega_\infty^2)^\ell) dx^1 d\xi^1.$$

Remark. a) If $d_0 > 0$, then $n_2 < n$ and the forms $\mathcal{A}^0_{x^1, \xi^1}$ are defined. b) $\mathcal{A}^0_{x^1, \xi^1}$ is a principal operator-valued symbol which was defined in §24 [this will be proved during the proof of formula (26.43)]. c) Later we shall show that the coefficient in formula (26.43) is finite if and only if $d_0 > 0$.

Proof of Theorem 26.5. Due to Glazmann's lemma,

$$N(t, A) = \mathcal{N}_0(\overset{\circ}{A}_t, \Omega), \quad \text{where } \overset{\circ}{A}_t = \sum_{\alpha, \beta \in I} D^\beta a_{\alpha\beta}(x) D^\alpha - t.$$

In §26.3 we showed that if $(x, \xi) \in (\Omega \backslash \overline{\Omega}_\omega) \times R^n$, where $\omega < 1$ satisfies condition (26.10), then $\forall \alpha, \beta$,

$$(26.44) \quad \|(q_t^{-1} \sigma^w(\overset{\circ}{A}_t)^{(\alpha)}_{(\beta)} q_t^{-1})(x, \xi)\| \leqslant C_{\alpha\beta}(\Psi_t^{-\alpha} \psi_t^{-\beta})(x, \xi),$$

where $\Psi_{t,i}$, $\psi_{t,i}$, q_t are the functions constructed in §26.3. Since (26.9) is valid, then in accordance with the general scheme of analysis of degenerate operators in §24, we can assume that $\Omega_t = \Omega \backslash \overline{\Omega}_\omega$, where $\omega = M t^{\kappa_0}$.

Obviously,

$$(26.45) \quad \mathcal{N}_0(\overset{\circ}{A}_t, \Omega) \geqslant \mathcal{N}_0(\overset{\circ}{A}_t, \Omega_\omega) + \mathcal{N}_0(\overset{\circ}{A}_t, \Omega_t),$$

and (26.11)–(26.13) yield

$$(26.46) \quad \mathcal{N}_0(\overset{\circ}{A}_t, \Omega) \leqslant \mathcal{N}_0(\overset{\circ}{A}_{t_\epsilon}, \Omega_{2\omega}) + \mathcal{N}_0(\overset{\circ}{A}_{t_\epsilon}, \Omega_t).$$

26.12. In order to calculate the asymptotics of the second terms in (26.45) and (26.46), we note that $x_n^{2S_0} = o(t)$ on $\Omega_t \times R^n$, and, therefore, that due to Lemma 26.2, in the same domain,

$$(26.47) \quad \sigma^w(\overset{\circ}{A}_t) = a_C - t + o(q_t^2), \quad \sigma^w(\overset{\circ}{A}_t) + 2t \geqslant c q_t^2.$$

In the proof of (26.47), we used the composition theorem; it can be applied to $\overset{\circ}{A}_t \in \mathcal{L}(g_t, q_t^2, \Omega_t \times R^n)$, since $\overset{\circ}{A}_t$ is a differential operator for which the asymptotic sum in Theorem 3.1 is finite.

We change the coefficients of the differential operator $\overset{\circ}{A}_t$ outside of Ω_t so that $\overset{\circ}{A}_t = \overset{\circ}{A}_t^* \in \mathcal{L}(g_t, q_t^2)$, and $(26.47)_2$ would hold everywhere. As was shown in §26.3, the metric g_t is σ-temperate, $q_t \in O(C, g_t)$ uniformly with respect to t, but these conditions, together with (26.9) and (26.47), are the conditions of Theorem 7.5. Due to (26.47), this theorem yields
(26.48)

$$\mathcal{N}_0(\overset{\circ}{A}_t, \Omega_t)$$

$$= \widehat{V}(a_C - t, \Omega_t)(1 + o(1)) + O(\widehat{V}(a_C - t_{\varepsilon_1}, \Omega_t)$$

$$- \widehat{V}(a_C - t_{-\varepsilon_1}, \Omega_t)) + O(\widehat{V}(a_C - t_\varepsilon; (\partial\Omega')_{\varepsilon_2} \times (\omega, 1)))$$

$$+ O(\widehat{V}(a_C - t_{\varepsilon_1}; \Omega' \times (\omega/2, 2\omega))) + O(\widehat{V}(a_C - t_\varepsilon, \Omega'_{\varepsilon_2} \times (1 - \varepsilon_3, 1 + \varepsilon_3))$$

$$= F(t) + \sum_{i=1,2,3} O(F_i(t)),$$

where $\Omega'_\varepsilon = \{x' \mid \operatorname{dist}(x', \Omega') < \varepsilon\}$, and $\varepsilon_i(t) \to 0$ as $t \to \infty$.

Lemma 26.2 and (26.48) yield

$$(26.49) \qquad\qquad \mathcal{N}_0(\overset{\circ}{A}_t, \Omega_t) = o(t^{\sigma_0}), \quad \text{if } d_0 > 0.$$

Furthermore, the same arguments as in part (a) of Lemma 26.2 show that, in all cases, $F_3(t) = o(t^{n/2m}) = o(F(t))$; and if $d_0 \leqslant 0$, then Lemmas 26.1 and 26.2 yield

$$\widehat{V}(a_C - c_1 t, \Omega'_{\varepsilon_2} \times (0, c\omega)) = O(M^N t^{\sigma_0}) = o(F(t)),$$

and in (26.48) we can replace Ω_t with Ω. In addition, the last estimate yields $F_2(t) = o(F(t))$.

Since $\operatorname{mes}(\partial\Omega')_{\varepsilon_2} \to 0$ as $t \to \infty$, then we have

$$\widehat{V}(a_C - c_1 t, (\partial\Omega')_{\varepsilon_2} \times (0,1)) \leqslant \widehat{V}(q_C - c_2 t, (\partial\Omega')_{\varepsilon_2} \times (0,1))$$

$$= \varepsilon_3 \widehat{V}(q_C - c_2 t, \Omega)$$

$$\leqslant \varepsilon_3 \widehat{V}(a_C - c_3 t, \Omega) = \varepsilon_4 F(c_4 t),$$

but, as we showed in Theorem 26.4, the asymptotics of $F(t)$ is a power or a logarithmic-power asymptotics; therefore, $F(C_2 t) \leqslant C_1 F(t)$ and $F_1(t) + F_2(t) = o(F(t))$. Thus,

$$(26.50) \qquad \mathcal{N}_0(\overset{\circ}{A}_t, \Omega) = \widehat{V}(a_C - t, \Omega)(1 + o(1)), \quad \text{if } d_0 \leqslant 0.$$

26.13. We shall calculate asymptotics of $\mathcal{N}_0(\overset{\circ}{A}_t, \Omega_{c\omega})$.

Let A_t, Q_t be the operators constructed in §26.4. As it was shown in §26.4 and §26.5, on the space $C_0^\infty(\Omega_{c\omega})^\ell$,

$$(26.51) \qquad A_t - 2t - \varepsilon Q_t \leqslant \overset{\circ}{A}_t \leqslant A_t - 2t + \varepsilon Q_t, \qquad cQ_t \leqslant A_t \leqslant CQ_t;$$

therefore, it suffices to analyze asymptotics of the function $\mathcal{N}_0(A_t - 2t, \Omega_{c\omega})$. For this, we shall use Theorem 22.1.

26.14. If $n_2 < n$, $n \in J$, then we realize $A_t - 2t$ as a PDO on R^{n_1} with an operator-valued symbol. Let H_t, $H_{t,0}$, g_t, q_t, $J : H_t \to H_{t,0}$ be the same as in §26.4. In the same section we showed that $J^*(A_t - 2t)J \in \mathcal{L}(g_t, q_t^*, q_t)$ and $h_t \leqslant \omega^c \to 0$ as $t \to \infty$; thus, condition A in §22 holds.

By applying the Hardy inequality and by recalling the definition of the set J, we see that there exists $k' < \hat{k}_{ij} \quad \forall \alpha_{ij} \in I_{s_0}$, $i \notin J$ such that

$$Q_{x^1, \xi^1}[u] \geqslant c(\|x_n^{k'} u\|^2 + \sum_{\alpha_{ij} \in I_{s_0}, i \notin J} \|x_n^{\hat{k}_{ij}} \xi_i^j u\|_{L_2}^2 + \sum_{\alpha_{ij} \in I_{s_0}, i \in J} \|x_n^{\hat{k}_{ij}} D_i^j u\|_{L_2}^2).$$

Let $\widehat{\mathcal{A}}_{t,X^1}^\varepsilon$ be the form introduced in §22. The last estimate and estimate (26.5) imply that for all $u \in C_0^\infty(\Omega_{c\omega}^2)^\ell$,

$$\widehat{\mathcal{A}}_{t,X^1}^\varepsilon[u] = \langle (\sigma^w(A_t) - 2t - \varepsilon \sigma^w(Q_t))(x^1, \xi^1)u, u \rangle$$
$$\geqslant c\langle (|\xi^1|^r - Ct)u, u \rangle$$

for some $r > 0$, and

$$\mathcal{N}(\widehat{\mathcal{A}}_{t,X^1}^{\varepsilon_1}, H_{t,0}) \leqslant \mathcal{N}(|\xi^1|^r - Ct, H_{t,0}) = 0,$$

if $|\xi^1| < \omega^{-C_1}$, and C_1 is sufficiently small. Hence, the set $\mathcal{M}_i^\varepsilon$ introduced in §22 here is contained in

$$\{X^1 \mid |x^1| < C, |\xi^1|^{r_1} < \omega^{-c}\}, \quad r_1 = c/C_1 > 0,$$

the condition (22.5) is satisfied, and conditions (22.2)–(22.4) can be satisfied, if we let

$$g_{t,x^1,\xi^1}^1(y^1, \eta^1) = \langle x^1 \rangle^{-2}|y^1|^2 + (|\xi^1|^{r_1} + \omega^{-c})^{-2}|\eta^1|^2.$$

Estimate (26.5) yields the estimate

$$\mathcal{N}(\widehat{\mathcal{A}}_{t,X^1}^{\varepsilon_1}, H_{t,0}) \leqslant \mathcal{N}_0(B_\varepsilon - C_\varepsilon t, \Omega_1^2),$$

where

$$B_\varepsilon = x_n^{2s_0 + \varepsilon} + \sum_{\alpha_{ij} \in I_{s_0}, i \in J} D_i^j x_n^{2\hat{k}_{ij} + 2\varepsilon} D_i^j,$$

for any $\varepsilon > 0$. By taking the inverse and then applying Theorem 7.1, it is easy to show that the asymptotics of the last function is a power asymptotics, so the last condition (22.6) of Theorem 22.1 is satisfied. The details are left to the reader as an easy exercise. Besides, the same will be done below for operators with operator-valued symbols (when we analyze the case of $n \notin J$).

By using (26.5) and the same arguments as those for estimating $F_2(t)$ in §26.12, we obtain

$$V(\widehat{\mathcal{A}}_t^{\pm\varepsilon}; H_{t,0}; (\partial\Omega')_{\varepsilon_1} \times R^{n_1}) = \varepsilon_3 V(\widehat{\mathcal{A}}_t^{\varepsilon_4}, H_{t,0}, \Omega' \times R^{n_4}).$$

Therefore, (22.7), (26.5), and (26.51) yield

$$(26.52) \qquad \mathcal{N}_0(\overset{\circ}{\mathcal{A}}_t, \Omega_\omega) = (2\pi)^{-n_1} F(t)(1 + o(1)) + O(F(t_\varepsilon) - F(t_{-\varepsilon})),$$

where

$$F(t) = \iint\limits_{\Omega^1 \times R^{n_1}} dX^1 \mathcal{N}(\mathcal{A}_{X^1} - t\|\cdot\|^2, C_0^\infty(\Omega_\omega^2)^\ell).$$

The substitution $x_n = t^{\kappa_0} y_n$, $\xi_i = t^{-\kappa_0 \gamma_{i \circ 0}} \eta_i$, $i \notin J$ yields

$$F(t) = t^{\sigma_0} \widehat{F}(t),$$

where

$$\widehat{F}(t) = \iint\limits_{\Omega^1 \times R^{n_1}} dX^1 \mathcal{N}(\mathcal{A}_{t,X^1} - \|\cdot\|^2, C_0^\infty(\Omega_M^2)^\ell),$$

$$\mathcal{A}_{t,X^1} = \mathcal{A}_{X^1}^0 + \sum_{(\alpha,\beta) \in I^2 \setminus I_{s_0}^2} t^{-\delta_{\alpha\beta}} B_{\alpha\beta,X^1},$$

and $\delta_{\alpha\beta} > 0$, due to (26.28).

Condition (26.5) implies that for some C_1,

$$(26.53) \qquad\qquad\qquad C_1 \mathcal{A}_{t,X^1} \geqslant \mathcal{A}_{X^1}^0;$$

therefore,

$$\mathcal{N}(\mathcal{A}_{t,X^1} - \lambda\|\cdot\|^2, C_0^\infty(\Omega_M^2)^\ell) \leqslant \mathcal{N}(\mathcal{A}_{X^1}^0 - C_1\lambda\|\cdot\|^2, C_0^\infty(\Omega_M^2)^\ell).$$

Due to (26.25), the last function is estimated by $\mathcal{N}_0(P_{X^1} - C_2\lambda, \Omega_M^2)$, where

$$P_{X^1} = \sum_{i \notin J} x_n^{2k(i,0)} \xi_i^{2j(i,0)} + \sum_{i \in J} D_i^{j(i,0)} x_n^{2\hat{k}(i,0)} D_i^{j(i,0)};$$

therefore, similarly to (26.46),
(26.54)
$$\mathcal{N}(\mathcal{A}_{t,X^1} - \lambda\|\cdot\|^2, C_0^\infty(\Omega_M^2)^\ell) \leqslant \mathcal{N}_0(P_{X^1} - C_2\lambda, \Omega_M^2)$$
$$\leqslant \mathcal{N}_0(P_{X^1} - 2C_2\lambda, \Omega_{2C}^2) + \mathcal{N}_0(P_{X^1} - 2C_2\lambda, \Omega_M^2\backslash\overline{\Omega}_C^2),$$

if C is sufficiently large.

The arguments, used in §26.14 when we estimated the forms $\hat{\mathcal{A}}_{t,X^1}^\varepsilon$, show that, uniformly with respect to X^1,

(26.55)
$$\mathcal{N}_0(P_{X^1} - 2C_2\lambda, \Omega_{2C}^2) \leqslant C_3(C_2\lambda, C),$$

and

(26.56)
$$\mathcal{N}_0(P_{X^1} - 2C_2\lambda, \Omega_{2C}^2) = 0, \quad \text{if } |\xi^1| > C_4(C_2\lambda, C).$$

For fixed $C_2\lambda$, the second term in (26.54) can be estimated by using Theorem 7.5, replacing x_n in the definition of the operator P_{X^1} with

$$\rho(x_n) = \begin{cases} x_n, & |x_n| > C/2 \\ C/4, & |x_n| < C/4 \end{cases}, \qquad |\rho^{(S)}| \leqslant C_S\rho^{1-S} \quad \forall S,$$

and letting

(26.57)
$$g_{C,x^2,\xi^2}(y^2,\eta^2) = \rho(x_n)^{-2c}|\eta^{21}|^2 + \rho(x_n)^{2-2c}\eta_n^2 + |y^{21}|^2 + \rho(x_n)^{-2}y_n^2,$$

(26.58)
$$\overset{\circ}{q}_{C,x^1,\xi^1}(x^2,\xi^2)^2 = \sum_i \rho(x_n)^{2k(i,0)}\xi_i^{2j(i,0)} + 1,$$

where $c = \min\{-k(i,0)/j(i,0) \mid i \in J\} > 0$. The metric (26.57) and the function (26.58) are σ-temperate and σ, g_C-temperate, respectively, uniformly with respect to $C \geqslant 1, X^1$ (the proof is the same as in §26.3), and as $C \to \infty$, $h_C \leqslant (C/4)^{-c} \to 0$; therefore, Theorem 7.5 can be applied, and for $C \geqslant C_0$,

(26.59)
$$\mathcal{N}_0(P_{X^1} - 2C_2\lambda, \Omega_M^2\backslash\overline{\Omega}_c^2) \leqslant C_3\hat{f}_{M,C}(X^1, \mu),$$

where $\mu = \mu(C_2, \lambda)$, and

$$\hat{f}_{M,C}(X^1, \mu) = \text{mes}\{(x_n, \xi^2) \mid x_n^{2\hat{k}(i,0)}\xi_i^{2j(i,0)} < \mu, \quad C < x_n < M\}.$$

Obviously,

$$\iint\limits_{\Omega^1 \times R^{n_1}} dX^1 \hat{f}_{M,C}(X^1, \mu) \leqslant C(\mu)M^N,$$

hence, (26.54)–(26.56) and (26.59) yield the estimate $\widehat{F}(t) \leqslant CM^N$, and (26.52) yields

$$(26.60) \qquad \mathcal{N}_0(\overset{\circ}{A}_t, \Omega_\omega) \leqslant C_1 M^N t^{\sigma_0}.$$

If, on the other hand, $d_0 > 0$, then, uniformly with respect to M,

$$\underset{\Omega^1 \times R^{n_1}}{\iint} \hat{f}_{M,C}(X^1, \mu) dX^1 \leqslant C_1 \int\limits_C^\infty dx_n \, \mathrm{mes}\{\xi \mid \sum_i x_n^{2k(i,0)} \xi_i^{2j(i,0)} < \mu\}$$

$$(26.61)$$

$$\leqslant C_2 \int\limits_C^\infty x_n^{-d_0-1} dx_n = C_2',$$

hence $\widehat{F}(t) \leqslant C''$, and (26.60) holds for $N = 0$.

Moreover, if we show that for almost all X^1, as $t \to \infty$,

$$(26.62) \qquad \mathcal{N}(\mathcal{A}_{t,X^1} - \| \cdot \|^2, C_0^\infty(\Omega_M^2)^\ell) \to \mathcal{N}(\mathcal{A}_{X^1}^0 - \| \cdot \|^2, C_0^\infty(\Omega_\infty^2)^\ell),$$

then, due to (26.61), we can pass to the limit in the integral $\widehat{F}(t)$ and obtain

$$\widehat{F}(t) = \widehat{F}(\infty)(1 + o(1)).$$

This yields

$$(26.63) \qquad \mathcal{N}_0(\overset{\circ}{A}_t, \Omega_\omega) = (2\pi)^{-n_1} \widehat{F}(\infty) t^{\sigma_0}(1 + o(1)).$$

Since $\mathcal{A}_{X^1}^0$ is a polynomial with respect to ξ^1, then any λ, for almost all X^1, is a point of continuity of the function $\mathcal{F}(\lambda, \mathcal{A}^0) = \mathcal{N}_0(\mathcal{A} - \lambda \| \cdot \|^2, \Omega_\infty^2)$; therefore, it suffices to prove (26.62) only for such X^1.

We shall prove (26.62), omitting the index X^1. Let λ be a point of continuity. If $\varepsilon > 0$ is sufficiently small, then $\mathcal{F}(\lambda - 2\varepsilon, \mathcal{A}^0) = \mathcal{F}(\lambda + 2\varepsilon, \mathcal{A}^0) = N_0$, and there exists $L \subset C_0^\infty(\Omega_\infty^2)^\ell$, such that $\dim L = N_0$ and $\mathcal{A}^0[u] < (\lambda - \varepsilon)\|u\|^2$ $\forall (0 \neq)u \in L$. But, for $t \geqslant t(L)$, $L \subset C_0^\infty(\Omega_M^2)^\ell$, $|\mathcal{A}_t[u] - \mathcal{A}^0[u]| \leqslant \varepsilon\|u\|_{L_2}^2$, $\forall u \in L$, and hence, $N_0 \leqslant \mathcal{N}(\mathcal{A}_t - \lambda \| \cdot \|^2, C_0^\infty(\Omega_M^2)^\ell)$.

Furthermore, if C is sufficiently large, then similar to (26.46),

$$\mathcal{N}(\mathcal{A}_t - \lambda \| \cdot \|^2, C_0^\infty(\Omega_M^2)^\ell) \leqslant \mathcal{N}(\mathcal{A}_t - (\lambda + \nu)\| \cdot \|^2, C_0^\infty(\Omega_C^2)^\ell)$$

$$+ \mathcal{N}(\mathcal{A}_t - (\lambda + \nu)\| \cdot \|^2, C_0^\infty(\Omega_M^2 \backslash \overline{\Omega}_{C/2}^2)^\ell),$$

where $\nu \to 0$ as $C \to \infty$. By using (26.5), it is easy to show that if C is sufficiently large, then

$$\mathcal{A}_{t,X^1}[u] \geqslant 2\lambda\|u\|^2 \quad \forall u \in C_0^\infty(\Omega_M^2 \backslash \overline{\Omega}_{C/2}^2);$$

thus, for large C the second term is equal to zero.

Hence, in order to complete the proof of (26.62), it suffices to show that, if C is large, then for $t \geqslant t_0$,

$$\mathcal{N}(\mathcal{A}_t - (\lambda + \varepsilon)\| \cdot \|^2, C_0^\infty(\Omega_C^2)^\ell) \leqslant \mathcal{N}_0(\mathcal{A}^0 - (\lambda + \tfrac{3}{2}\varepsilon)\| \cdot \|^2, C_0^\infty(\Omega_C^2)^\ell).$$

By integrating by parts in the form $\mathcal{A}_t - \mathcal{A}^0$, we obtain: $\forall u \in C_0^\infty(\Omega_C^2)^\ell$, $\forall C_1$ $(\mathcal{A}_t - \mathcal{A}^0)[u, u_t] \to 0$ as $t \to \infty$, uniformly with respect to $u_t \in C_0^\infty(\Omega_C^2)^\ell$ such that $\mathcal{A}_t[u_t] \leqslant C_1$.

By replacing, if necessary, $\mathcal{A}_t$ and $\mathcal{A}^0$ by $\mathcal{A}_t[\cdot] + C_0\| \cdot \|^2$ and $\mathcal{A}^0[\cdot] + C_0\| \cdot \|^2$, respectively, we can assume that the forms $\mathcal{A}_t$, $\mathcal{A}^0$ are positive, and therefore, that the operators A_t and A^0 associated with the variational triples $\mathcal{A}_t$, $C_0^\infty(\Omega_C^2)^\ell$, $L_2(\Omega_C^2)^\ell$ and $\mathcal{A}^0$, $C_0^\infty(\Omega_C^2)^\ell$, $L_2(\Omega_C^2)^\ell$ have bounded inverses A_t^{-1} and A^{0-1}, respectively. The arguments given above show that $A_t^{-1} \to A^{0-1}$ strongly and that if $\varepsilon > 0$ is sufficiently small and C is sufficiently large,

$$N_0 = \mathcal{N}_0(\mathcal{A}^0 - \lambda\| \cdot \|^2, C_0^\infty(\Omega_C^2)^\ell) = N(\lambda, A^0)$$
$$= N(\lambda - \varepsilon, A^0) = N(\lambda + \varepsilon, A^0).$$

Therefore, $-\lambda^{-1}$ is a point of continuity of the function $N(\mu, -A^{0-1})$, and if $\nu > 0$ is sufficiently small and $\mu \in (-\lambda^{-1} - \nu, -\lambda^{-1} + \nu)$, then as $t \to \infty$, $N(\mu, -A_t^{-1}) \to N(\mu, -A^{0-1})$. Hence, for a sufficiently small $\varepsilon > 0$,

$$\mathcal{N}(\mathcal{A}_t - (\lambda + \varepsilon)\| \cdot \|^2, C_0^\infty(\Omega_C^2)^\ell) = N(\lambda + \varepsilon, A_t) \to N(\lambda + \varepsilon, A^0)$$

as $t \to \infty$, and (26.62) is proved.

Thus, (26.63) is proved. We note that (26.61), and therefore (26.63), hold only provided $d_0 > 0$.

Formulas (26.60) and (26.63) hold also for $c\omega$ instead of ω, and the asymptotic coefficients in formulas (26.63) and (26.43) are equal; therefore, (26.45), (26.46), (26.50), (26.60), and (26.63) imply that

if $d_0 > 0$, then

(26.64) $$\mathcal{N}_0(\overset{\circ}{A}_t, \Omega) = (2\pi)^{-n_1} \widehat{F}(\infty) t^{\sigma_0} + o(t^{\sigma_0});$$

if $d_0 \leqslant 0$, then

(26.65) $$\mathcal{N}_0(\overset{\circ}{A}_t, \Omega) = \widehat{V}(a_C - t, \Omega)(1 + o(1)) + O(M^N t^{\sigma_0}).$$

Now (26.64) implies (26.43); and (26.65) and Corollary 26.1 imply the statement of Theorem 26.5, provided that $d_0 \leqslant 0$.

26.15. If $n_2 < n$ and $n \notin J$, then we make the same transformations as in §26.5, and by using (26.19) and (26.20), obtain

(26.66) $$\mathcal{N}_0(A_t' - 2t_{-\varepsilon}, \Omega_\omega^-) \leqslant \mathcal{N}_0(\overset{\circ}{A}_t, \Omega_\omega) \leqslant \mathcal{N}_0(A_t' - 2t_\varepsilon, \Omega_\omega),$$

where $\Omega_\omega^- = \Omega^{11} \times (\omega^{-1}, \infty)$, and A_t' is PDO, constructed in §26.5, in R^{n_1} with an operator-valued symbol. Let H_t, $H_{t,0}$, $J : H_t \to H_{t,0}$, g_t, q_t be spaces, operator metric, and function, constructed in §26.5. Then $J^*(A_t' - 2t)J \in \mathcal{L}(g_t, q_t^*, q_t)$ and $h_t \leqslant (4\omega)^\delta \to 0$ as $t \to \infty$; therefore, condition A of Theorem 22.1 is satisfied.

Due to condition (26.5), the form introduced in §22,

$$\hat{\mathcal{A}}_{t,Y^1}^{\varepsilon_1}[u] = \langle \sigma^w(A_t') - 2t_\varepsilon - \varepsilon_1 \sigma^w(Q_t'))(Y^1)u, u \rangle_{L_2}$$
$$\geqslant c\langle ((\sum_{i \notin J} y_n^{-2\hat{k}(i,0)} |\eta_i|^{2j(i,0)} + y_n^{-2s_0} B - Ct)u, u \rangle_{L_2},$$

where

(26.67) $$B = I + \sum_{i \in J} D_i^{2j(i,0)} > I.$$

Hence,

$$\mathcal{N}_0(\hat{\mathcal{A}}_{t,Y_1}^{\varepsilon_1}, \Omega^2) \leqslant \mathcal{N}_0(B - Ct^C, \Omega^2) \leqslant C_1 t^{C_1}$$

(the last estimate can be obtained with the help of Theorem 7.2, as in §13), and if $|y_n| + |\eta^1| > \omega^{-C}$, where C is sufficiently large, then $\hat{\mathcal{A}}_{t,Y^1}^{\varepsilon_1} \geqslant 0$, so that conditions (22.5), (22.6) hold, and conditions (22.2)–(22.4) can be satisfied, provided

$$g_{t,y^1,\eta^1}^1(z^1, \zeta^1) = |z^{11}|^2 + \hat{\rho}(y_n)^{-2}|z_n|^2 + (\hat{\rho}(y_n) + \langle y^1 \rangle$$
$$+ \langle \eta^1 \rangle^r)^{-2\delta}|\zeta^{11}|^2 + (\hat{\rho}(y_n) + \langle y^1 \rangle + \langle \eta^1 \rangle^r)^{2-2\delta}|\zeta_n|^2,$$

where $r = 1/C$; σ-temperateness of the metric g_t^1 follows from Lemmas 1.3 and 3.3.

Using (26.5) and arguments similar to those of §26.12 when we estimated $F_2(t)$, we can easily show that

$$V(\hat{A}_t^{\pm\varepsilon}, H_{t,0}, (\partial\Omega^{11})_{\varepsilon_1} \times (c\omega^{-1}, \infty) \times R^{n_1})$$
$$\leqslant \varepsilon_2 V(\hat{A}_t^{\varepsilon_3}, H_{t,0}, \Omega^{11} \times (c\omega^{-1}, \infty) \times R^{n_1});$$

therefore, (22.7) and (26.66) yield

(26.68) $$\mathcal{N}_0(\overset{\circ}{A}_t, \Omega_\omega) = (2\pi)^{-n_1} F(t)(1 + o(1)) + O(F(t_\varepsilon) - F(t_{-\varepsilon}))$$
$$+ O(V(\hat{A}_t^{\varepsilon'}; H_{t,0}; \Omega^{11} \times (\omega^{-1}/2, 2\omega^{-1}) \times R^{n_1})),$$

where

$$F(t) = \iiint\limits_{\Omega^{11} \times (\omega^{-1}, \infty) \times R^{n_1}} dY^1 \mathcal{N}_0(\check{A}_{Y^1} - t, \Omega^2),$$

$$\check{A}_{Y^1} = \sum_{\alpha, \beta \in I} (-1)^{\alpha_n + \beta_n} D^{\alpha^2} y_n^{-k_\alpha - k_\beta + \alpha_n + \beta_n} \hat{a}_{\alpha\beta}(y') D^{\beta^2} \eta^{1^{\alpha^1 + \beta^1}}.$$

We make the substitution

$$y_n = t^{-\kappa_0} x_n^{-1}, \qquad \eta_n = -t^{-\kappa_0(\gamma_{n s_0}-1)} x_n^2 \xi_n, \qquad \eta_i = t^{-\kappa_0 \gamma_{i s_0}} \xi_i,$$

$J \not\ni i < n$ and obtain $F(t) = t^{\sigma_0} \widehat{F}(t)$, where $\widehat{F}(t) = \widetilde{F}(t, 1)$,

$$\widetilde{F}(t, \lambda) = \iint\limits_{\Omega^1_\infty \times R^{n_1}} f_M(t, X^1, \lambda) dX^1,$$

$$f_M(t, X^1, \lambda) = \chi_M(x_n) \mathcal{N}_0(A_{t, X^1} - \lambda \| \cdot \|^2; \Omega^2),$$

$$A_{t, X^1} = A^0_{X^1} + \sum_{(\alpha, \beta) \in I^2 \backslash I^2_{s_0}} t^{-\delta_{\alpha\beta}} B_{\alpha\beta},$$

where $\delta_{\alpha\beta} > 0$ due to (26.28), and χ_M is a characteristic function of the interval $(0, M)$.

Condition (26.5) implies (26.53) and, therefore, (26.54) holds with $\Omega^2_M = \Omega^2$. If C is sufficiently large and $x_n^{-1} > C$, or if $C^{-1} < x_n < M$ and $|\xi^1| > M^C$, then due to the conditions $s_0 < 0$, $k(i, 0) > s_0 \ \forall i \notin J$, we have $f_M^+(X^1, \mu) = 0$; therefore $\widetilde{F}(t, \lambda) < M^N$ and the first terms in (26.68) are estimated by $M^N t^{\sigma_0}$. Similarly, we can obtain the same estimate for the last term; therefore, estimate (26.60) holds.

If $d_0 > 0$, then it can be sharpened in exactly the same way as in §26.14. It suffices to show that, for any μ,

(26.69)
$$G(\mu) = \iint\limits_{\Omega^1_\infty \times R^{n_1}} f_M^+(X^1, \mu) dX^1 < \infty.$$

(26.26) yields $G(\mu) \leqslant G_1(C\mu)$, where

$$G_1(\mu) = \iint\limits_{\Omega^1_\infty \times R^{n_1}} \mathcal{N}_0(\sum_{i \notin J} x_n^{2\widetilde{k}(i,0)} \xi_i^{2j(i,0)} + x_n^{2k_0} B - \mu, \Omega^2),$$

and B is defined by equality (26.67). Using any of Theorems 7.1, 7.2, and 7.5, we can easily show that as $\mu \to \infty$,

$$\mathcal{N}_0(B - \mu, \Omega^2) \sim C\mu^p, \quad \text{where } p = \sum_{t \in J} 1/(2j(i, 0));$$

therefore, by taking into account the positive-definiteness of the operator B and by making the substitutions $\xi_i = x_n^{-\gamma_{i s_0}} \eta_i$, we obtain

$$G_1(\mu) \leqslant \int\limits_{R^{n_1}} d\xi^1 \int\limits_c^\infty dx_n x_n^{-1-\sum \gamma_{i s_0}} [(\mu x_n^{-2s_0} - \sum_{i \notin J} \xi_i^{2j(i,0)})_+^p + 1]$$

$$\leqslant c_\mu \int\limits_c^\infty dx_n x_n^{-1-q},$$

where

$$q = \sum \gamma_{is_0} + s_0 \sum_{i \notin J} \frac{1}{j(i,0)} + 2s_0 p$$

$$= \sum_{i \notin J} [(k(i,0) - s_0)j(i,0)^{-1} + s_0/j(i,0)] + s_0 \sum_{i \in J} j(i,0)^{-1} = d_0 > 0.$$

Thus, integral (26.69) converges; therefore, the same arguments as in §26.14 show that in the integral $\widetilde{F}(t, \lambda)$ we can pass to the limit and prove that the sum of the first two terms in (26.68) is

$$(2\pi)^{-n_1} \widehat{F}(\infty) t^{\sigma_0} + o(t^{\sigma_0});$$

similarly, the last term is $o(t^{\sigma_0})$, and we obtain (26.63).

Now the proof of Theorem 26.5 in the case of $n \notin J$ is completed in exactly the same way as in §26.14.

We note that the condition $d_0 > 0$ is sufficient and necessary for the passage to the limit in the integral $\widetilde{F}(t, \lambda)$.

26.16. We shall consider the last case: $n_2 = n$. (26.6), (26.26) and the Poincaré inequality yield, for any $\varepsilon > 0$,

$$\mathcal{N}_0(\overset{\circ}{A}_t, \Omega_\omega) \leqslant \mathcal{N}_0(B^\varepsilon - Ct\| \cdot \|^2, \Omega_\omega),$$

where

$$B^\varepsilon[u] = \sum_{\alpha \in I_{s_0}^+} \|x_n^{\hat{k}_\alpha + \varepsilon \alpha_n} D^\alpha u\|_{L_2}^2 + \|x_n^{s_0} u\|_{L_2}^2.$$

If we define a set J^ε for the form B^ε, then we see that $J^\varepsilon = \{1, \ldots, n-1\}$; therefore, the arguments in §26.15 yield

$$\mathcal{N}_0(\overset{\circ}{A}_t, \Omega_\omega) \leqslant \mathcal{N}_0(B^\varepsilon - Ct\| \cdot \|^2, \Omega_\varepsilon) \leqslant C_\varepsilon t^{-\kappa_0 \varepsilon}.$$

Since $\varepsilon > 0$ is arbitrary, then $\mathcal{N}_0(\overset{\circ}{A}_t, \Omega_\omega)$ is contained in the o-terms of the asymptotic formulas of Theorem 26.5, and the proof is completed as in §26.14. ∎

26.17. Now let condition (26.3) not hold. Using condition (26.2) and continuity of the functions $a_{\alpha\beta}$, we can, for any $\varepsilon > 0$, construct functions $a_{\alpha\beta\varepsilon}$, such that

$$(26.70) \qquad \|a_{\alpha\beta\varepsilon}^{(\gamma)}(x)\| \leqslant C_{\alpha\beta\gamma} \varepsilon^{-|\gamma|} x_n^{\hat{k}_\alpha + \hat{k}_\beta - \gamma_n}, \quad |\gamma| = 0, 1, \ldots,$$

(26.71) $$\|a_{\alpha\beta\varepsilon}(x) - a_{\alpha\beta}(x)\| \leqslant \varepsilon' x_n^{\hat{k}_\alpha + \hat{k}_\beta},$$

(26.72) $\quad \hat{a}_{\alpha\beta\varepsilon} \in C^\infty(\overline{\Omega}'; \operatorname{End} C^\ell), \qquad \|\hat{a}_{\alpha\beta\varepsilon}(x') - \hat{a}_{\alpha\beta}(x')\| \leqslant \varepsilon'',$

where $\varepsilon', \varepsilon'' \to 0$ as $\varepsilon \to 0$.

Let us define all the forms and symbols ini terms of the functions $a_{\alpha\beta\varepsilon}$ instead of the functions $a_{\alpha\beta}$, supplying the latter with the index ε. Then (26.71) yields

$$|\mathcal{A}_\varepsilon[u] \quad - \mathcal{A}[u]\| \leqslant \varepsilon_1 Q[u] \quad \forall u \in C_0^\infty(\Omega)^\ell,$$

where $\varepsilon_1 = \varepsilon_1(\varepsilon) \to 0$ as $\varepsilon \to 0$. Hence,

(26.73) $$\mathcal{A}_{(-\varepsilon)} \overset{\text{def.}}{=} \mathcal{A}_\varepsilon - \varepsilon_1 Q \leqslant \mathcal{A} \leqslant \mathcal{A}_{(\varepsilon)} \overset{\text{def.}}{=} \mathcal{A}_\varepsilon + \varepsilon_1 Q,$$

and similar inequalities hold for the forms $\mathcal{A}_{x^1,\xi^1(\pm\varepsilon)}$, $\mathcal{A}_{x^1,\xi^1}$; $\mathcal{A}^0_{x^1,\xi^1(\pm\varepsilon)}$, $\mathcal{A}^0_{x^1,\xi^1}$; $\mathcal{A}^0_{(\pm\varepsilon)}$, $\mathcal{A}^0$ and the symbols $a^s_{(\pm\varepsilon)}$, a^s.

Thus, for small ε, the forms $\mathcal{A}_{(\pm\varepsilon)}$ satisfy the same conditions as the form $\mathcal{A}$, but conditions (26.70) and (26.72)$_1$ mean that the forms $\mathcal{A}_{(\pm\varepsilon)}$ satisfy all the conditions in §26.1, so that all the arguments in §§26.2–26.16 apply to them. In particular, the forms $\mathcal{A}_{(\pm\varepsilon)}$ are semi-bounded from below; (26.73) implies that the form $\mathcal{A}$ is also semi bounded from below. Let $A_{\pm\varepsilon}$ be operators associated with the variational triples $\mathcal{A}_{(\pm\varepsilon)}$, $C_0^\infty(\Omega)^\ell$, $L_2(\Omega)^\ell$. Glazmann's lemma and (26.73) yield

(26.74) $$N(t, A_\varepsilon) \leqslant N(t, A) \leqslant N(t, A_{-\varepsilon});$$

and due to Theorem 26.5,

(26.75) $$N(t, A_{\pm\varepsilon}) = c_{\pm\varepsilon} f(t) + o(f(t)), \quad t \to \infty,$$

where the coefficients $c_{\pm\varepsilon}$ and the function f are defined by the numbers $\hat{k}_\alpha$ and the coefficients $a_{\alpha\beta\varepsilon}$ as in Theorems 26.4 and 26.5. In addition, the constant in the o-term depends, in general, on ε.

The coefficients $c_{\pm\varepsilon}$ in (26.75) and the coefficient c in the similar formal asymptotic function for $N(t, A)$ can be written as

$$c_{\pm\varepsilon} = c' \int_{\mathcal{M}} N(1, b_{\pm\varepsilon}(z)) dz, \qquad c = c' \int_{\mathcal{M}} N(1, b(z)) dz,$$

where

(26.76) $$-\varepsilon' b(z) \leqslant b_{\pm\varepsilon}(z) - b(z) \leqslant \varepsilon' b(z),$$

$\varepsilon' \to 0$ as $\varepsilon \to 0$, and $b_{\pm\varepsilon}$, b polynomially depend on some z_i. But then (26.76) and part (d) of Remark 23.1 imply that $c_{\pm\varepsilon} \to c$ as $\varepsilon \to 0$, and (26.74) and (26.75) yield Theorem 26.5 for $N(t, A)$. ∎

26.18. If all $\hat{k}_\alpha > -\frac{1}{2}$ and $J = \{n\}$, the form $\mathcal{A}$ can be defined on $C_0^\infty([0,1);$ $C_0^\infty(\Omega'))^\ell \subset C^\infty(\overline{\Omega})^\ell$; the formulations and proofs change in an obvious way. In particular, in Theorem 26.5 only the asymptotic coefficient in formula (26.43) is changed: instead of $C_0^\infty(\Omega_\infty^2)^\ell$, we have to consider $C_0^\infty([0,\infty))^\ell$. ∎

26.19. Let us consider the simplest model example of essentially nonpower degeneracy. Let Ω be the same as above, $a_{\alpha\beta} = a_{\alpha\beta}^* \in C^\infty(\overline{\Omega}; \operatorname{End} C^\ell)$, and let, for all $\xi_\alpha \in C^\ell$, $x \in \overline{\Omega}$,

$$(26.77) \qquad \sum_{|\alpha|=|\beta|=m} \langle a_{\alpha\beta}(x)\xi_\alpha, \xi_\beta \rangle_{C^\ell} \geqslant c \sum_{|\alpha|=m} |\xi_\alpha|_{C^\ell}^2,$$

and $p \in C^\infty(R_+^1)$ satisfy the condition

$$(26.78) \qquad |p^{(s)}(\tau)| \leqslant C_s p(\tau)\tau^{-s}, \quad s = 0, 1, 2, \dots.$$

Obviously, the form

$$\mathcal{A}[u] = \sum_{|\alpha|=|\beta|=m} \langle a_{\alpha\beta}(x)p(x_n)^2 D^\alpha u, D^\beta u \rangle_{L_2(\Omega)^\ell}$$

is coercive:

$$(26.79) \qquad \mathcal{A}[u] \geqslant c \sum_{|\alpha|=m} \|p(x_n)D^\alpha u\|_{L_2(\Omega)^\ell}^2 \quad \forall u \in C_0^\infty(\Omega)^\ell.$$

Let A be an operator, associated with the variational triple $\mathcal{A}$, $C_0^\infty(\Omega)^\ell$, $L_2(\Omega)^\ell$ and let

$$a(x,\xi) = \sum_{|\alpha|=|\beta|=m} a_{\alpha\beta}(x)p(x_n)^2\xi^{\alpha+\beta}, \qquad a_C(x,\xi) = a(x,\xi) + Cp^2(x_n)x_n^{-2m}.$$

Theorem 26.6. *Let conditions* (26.77) *and* (26.78) *hold and, for some* $c_1, c > 0$, *let*

$$(26.80) \qquad c_1 t^{m/n-c} \leqslant p(t), \quad t \leqslant 1.$$

Then as $t \to \infty$,

$$(26.81) \qquad N(t, A) \sim t^{n/2m}\widehat{V}(a-1, \Omega).$$

Proof. Let $\omega \to +0$ as $t \to \infty$ such that for all $x_n \in (\omega, 1)$,

$$(26.82) \qquad t^{1/2m}p(x_n)^{-1/m}x_n \geqslant M_1 \to \infty.$$

By using (26.78) and by repeating the arguments of Lemma 1.2, we obtain

(26.83) $c \leqslant p(\tau)/p(\tau_1) \leqslant C, \quad \text{if } |\tau - \tau_1| < \tau/2;$

therefore, we can construct the functions $\rho_\omega, p_\omega \in C^\infty(R^1)$ such that

(26.84) $|\rho_\omega^{(s)}| \leqslant C_s \rho_\omega^{1-s}, \qquad |p_\omega^{(s)}| \leqslant C_s p_\omega \rho_\omega^{-s}, \qquad s = 0, 1, \ldots,$

$$\rho_\omega(\tau) = \begin{cases} 2, & |\tau| > 2, \\ |\tau|, & \omega < |\tau| < 1, \\ \omega/2, & |\tau| < \omega/2, \end{cases} \qquad p_\omega(\tau) = \begin{cases} p(2), & |\tau| > 2, \\ p(|\tau|), & \omega < |\tau| < 1, \\ p(\omega/2), & |\tau| < \omega/2, \end{cases}$$

and to obtain, similar to (26.45) and (26.46), the estimates

(26.85) $N(t, A) \geqslant \mathcal{N}_0(A_t, \Omega \backslash \overline{\Omega}_\omega) + \mathcal{N}_0(\mathcal{A} - t\|\cdot\|^2, \Omega_\omega),$

(26.86) $N(t, A) \leqslant \mathcal{N}_0(A_{t_\varepsilon}, \Omega \backslash \overline{\Omega}_\omega) + \mathcal{N}_0(\mathcal{A} - t_\varepsilon\|\cdot\|^2, \Omega_{2\omega}),$

where $t_\varepsilon = t(1 + \varepsilon)$, $\varepsilon \to 0$ as $t \to \infty$, and

$$A_t = \sum_{|\alpha|=|\beta|=m} D^\beta a_{\alpha\beta}(x) p_\omega(x_n)^2 D^\alpha - t.$$

Let $q_t(x, \xi)^2 = p_\omega(x_n)^2 |\xi|^{2m} + t$, $g_{t,x,\xi}(y, \eta) = t^{-1/m} p_\omega(x_n)^{+2/m} |\eta|^2 + p_\omega(x_n)^{-2} |y|^2$. (26.84) and Lemmas 1.2 and 1.3 imply that the metric g_t varies slowly and that the function q_t is g_t-continuous; and the same arguments as those in §26.3 show that g_t is σ-temperate and $q_t \in O(\mathbb{C}^1, g_t)$. Extending $a_{\alpha\beta}$ on R^n to functions that are bounded, along with their derivatives, and that satisfy condition (26.77) uniformly with respect to $x \in R^n$, we have that uniformly with respect to $t \geqslant 1$, $A_t \in \mathcal{L}(g_t, q_t^2)$ and $A_t \in \mathcal{L}I^+(g_t, q_t^2)$ for any t.

Thus, all the conditions of Theorem 7.5 are satisfied, since due to (26.82) and (26.83),

(26.87) $h_t(X) \to 0$ as $t \to \infty$, uniformly with respect to $X \in R^n$.

By using (26.87) and the composition theorem, we obtain

(26.88) $a + t \asymp q_t^2, \qquad \sigma^w(A_t) + t - a = o(q_t^2);$

therefore, Theorem 7.5 yields

(26.89) $\mathcal{N}_0(A_t, \Omega \backslash \overline{\Omega}_\omega) = F_\omega(t) + O(F_\omega(t_\varepsilon) - F_\omega(t_{-\varepsilon})) + O\left(\sum_{i=1,2,3} G(t_\varepsilon, \Omega_t^i)\right),$

where $\varepsilon \to 0$ as $t \to \infty$, and $G(t, \Omega) = \widehat{V}(a - t, \Omega)$,

$$F_\omega(t) = G(t, \Omega \backslash \overline{\Omega}_\omega), \qquad \Omega_t^1 = (\partial \Omega')_\varepsilon \times (\omega, 1),$$
$$\Omega_t^2 = (\Omega')_\varepsilon \times (1 - \varepsilon, 1 + \varepsilon), \qquad \Omega_t^3 = (\Omega')_\varepsilon \times (\omega/2, 2\omega).$$

Now $(26.88)_1$ and (26.77) yield

(26.90) $G(t_\varepsilon, \Omega_t^2) = o(t^{n/2m}), \qquad F_\omega(t) = t^{n/2m} F_\omega(1),$

(26.91) $$F_\omega(1) \asymp \mathcal{F}(\omega) \overset{\text{def.}}{=} \int\limits_\omega^1 p(\tau)^{-n/m} d\tau,$$

and since $\text{mes}(\partial \Omega')_\varepsilon = \varepsilon'$, we have $G(t_\varepsilon, \Omega_t^1) = o(t^{n/2m}) \mathcal{F}(\omega)$. By using $(26.88)_1$ again, we obtain

$$G(t_\varepsilon, \Omega_t^3) = O(t^{n/2m})(\mathcal{F}(\omega/2) - \mathcal{F}(2\omega)),$$

and (26.87)–(26.91) give the estimate
(26.92)
$$\mathcal{N}_0(A_{t_\varepsilon}, \Omega \backslash \overline{\Omega}_\omega) = t^{n/2m} \widehat{V}(a - 1, \Omega \backslash \overline{\Omega}_\omega)(1 + o(1)) + O(t^{n/2m})(\mathcal{F}(\omega/2) - \mathcal{F}(2\omega)).$$

Due to (26.79) and (26.80), we can repeat the arguments in §26.14 to obtain

(26.93)
$$\mathcal{N}_0(\mathcal{A} - t_\varepsilon \| \cdot \|^2, \Omega_{2\omega}) \leqslant C \int d'\xi \, \mathcal{N}_0(p(x_n^2)|\xi'|^{2m}$$
$$+ D_n^m p(x_n)^2 D_n^m - Ct, (0, 2\omega)).$$

Moreover, in (26.93) we can replace $p(x_n)$ with $x_n^{-c+m/n}$ and, therefore, estimate the right-hand side in (26.93) by $\varepsilon_1 t^{n/2m}$.

Finally, the same condition (26.80) implies that as $\omega \to 0$,

(26.94) $\mathcal{F}(\omega) \to \mathcal{F}(0), \qquad \widehat{V}(a - 1, \Omega \backslash \overline{\Omega}_\omega) \to \widehat{V}(a - 1, \Omega),$

and (26.92)–(26.94), (26.85), and (26.86) yield the formula (26.81). ∎

26.20.

Theorem 26.7. *Let conditions* (26.77) *and* (26.78) *hold, and let there exist a constant* $s < m$ *and a function* f *such that as* $\tau \to +0$,

(26.95) $f(\tau) \to \infty, \qquad \mathcal{F}(\tau) - \mathcal{F}(\tau f(\tau)) = o(\mathcal{F}(\tau));$
(26.96) $p(\tau) \to 0$ *and* $p(\tau)\tau^{-s} \to \infty$ *monotonically.*

Then, a) if $\mathcal{F}(0) < \infty$, then formula (26.81) holds;
   b) if $\mathcal{F}(0) = \infty$, then as $t \to \infty$,

$$(26.97) \qquad N(t,A) \sim (2\pi)^{-n} t^{n/2m} \Phi(t) \int_{\Omega'} dx' \int_{R^n} d\xi\, N(1, \tilde{a}(x',\xi)),$$

where

$$\tilde{a}(x',\xi) = \sum_{|\alpha|=|\beta|=m} a_{\alpha\beta}(x',0)\xi^{\alpha+\beta}, \qquad \Phi(t) = \mathcal{F}(\omega_1(t)),$$

and $\omega_1(t)$ is a solution of the equation $p(\tau)^2 \tau^{-2m} = t$.

Remarks. a) Formulas (26.81) and (26.97) are realizations of the generalized Weyl formula

$$(26.98) \qquad N(t,A) \sim \widehat{V}(a_C - t, \Omega).$$

b) Conditions (26.95) and (26.96) hold if, for example, in some neighborhood of zero, $p(\tau) = \tau^{m/n} |\ln \tau|^\alpha (\ln |\ln \tau|)^\beta \ldots$.

Example. If $p(\tau) = (\tau(\ln \frac{2}{\tau})^\alpha)^{m/n}$, then
1) for $\alpha > 1$, formula (26.81) holds;
2) for $\alpha = 1$ (for $\alpha < 1$), formula (26.97) holds with $\Phi(t) = \ln \ln t$ (with $\Phi(t) = \frac{1}{1-\alpha}(\frac{n}{2m(n-1)} \ln t)^{1-\alpha}$).
   In particular, for $\alpha = 0$, we obtain the well-known formula.

Proof of Theorem 26.7. When proving estimates (26.92) and (26.93), we really used only conditions (26.77) and (26.78) and the condition $p(\tau) \geqslant c\tau^{m-c}$, but this last condition follows from (26.96)$_2$. Hence, estimates (26.92) and (26.93) hold here, too. They yield the estimate

$$(26.99) \qquad \mathcal{N}_0(A_t, \Omega\backslash\overline{\Omega}_\omega) \sim t^{n/2m} \widehat{V}(a-1, \Omega\backslash\overline{\Omega}_\omega).$$

(26.96)$_2$ implies that $\omega_1 = \omega_1(t) \to 0$ as $t \to \infty$ (and is well-defined for $t \geqslant t_0$), and if $M \to \infty$ as $t \to \infty$ and $\omega = M\omega_1$, then condition (26.82) holds; and therefore, formulas (26.99) and (26.93) hold for this ω.

Let us make the substitution $\xi_i = \omega_1^{-1}\eta_i$ $(1 \leqslant i \leqslant n-1)$, $x_n = y_n\omega_1$ in (26.93); we obtain

$$(26.100) \qquad \mathcal{N}_0(\mathcal{A} - t\| \cdot \|^2; \Omega_{2\omega}) \leqslant C_1 \omega_1^{1-n} \int d\xi\, \mathcal{N}_0(\tilde{p}_\omega(x_n)^2 |\xi'|^{2m}$$
$$+ D_n^m \tilde{p}_\omega(x_n)^2 D_n^m - C; (0, 2M)),$$

where

$$\tilde{p}_\omega(x_n) = p(\omega_1 x_n)/p(\omega_1) = \frac{\omega_1^s}{p(\omega_1)} \cdot \frac{p(\omega_1 x_n)}{(\omega_1 x_n)^s} x_n^s$$

$$\geqslant \frac{\omega_1^s}{p(\omega_1)} \cdot \frac{p(\omega_1 \cdot 2M)}{(\omega_1 \cdot 2M)^s} x_n^s = \frac{p(\omega_1 2M)}{p(\omega_1)} \cdot \frac{x_n^s}{(2M)^s} \geqslant \frac{x_n^s}{(2M)^s}$$

due to (26.96). But, $s < m$; therefore, the integral in (26.100) can be estimated as in §26.14 to obtain

$$(26.101) \qquad \mathcal{N}_0(\mathcal{A} - t\| \cdot \|^2; \Omega_{2\omega}) \leqslant C(M)\omega_1^{1-n}.$$

Let $p_1(\tau) = p(\tau)\tau^{-m/n}$. Then for $\tau = \omega_1$,

$$p_1(\tau)^2 \tau^{2m/n - 2m} = t, \qquad p_1(\tau)^{-2} = t^{-1} \cdot \tau^{2m(1-n)/n},$$

$$t^{-n/2m}\omega_1^{1-n} = p_1(\omega_1)^{-n/m}, \qquad \mathcal{F}(\omega) = \int_\omega^1 \tau^{-1} p_1(\tau)^{-n/m} d\tau,$$

and (26.83) yields

$$\mathcal{F}(\omega_1) - \mathcal{F}(2\omega_1) = \int_{\omega_1}^{2\omega_1} \tau^{-1} p_1(\tau)^{-n/m} d\tau \geqslant c p_1(\omega_1)^{-n/m}.$$

Taking (26.95) into account, we obtain

$$\omega_1^{1-n} = t^{n/2m} p_1(\omega_1)^{-n/m} = \varepsilon t^{n/2m} \mathcal{F}(\omega_1),$$

where $\varepsilon \to 0$ as $t \to \infty$ does not depend on M.

Choosing $M \to \infty$ such that $C(M)\varepsilon \to 0$ as $t \to \infty$, we derive from (26.99) and (26.101),

$$(26.102) \qquad N(t, A) \sim t^{n/2m} \widehat{V}(a - 1, \Omega \backslash \overline{\Omega}_\omega).$$

If $\mathcal{F}(0) < \infty$, then (26.102) yields (26.81), and, if $\mathcal{F}(0) = \infty$ and $M \to \infty$ sufficiently slowly, then

$$\widehat{V}(a - 1, \Omega \backslash \overline{\Omega}_{M^{-1}}) = o(\mathcal{F}(\omega))$$

and

$$\widehat{V}(a - 1, \Omega_{M^{-1}} \backslash \Omega_\omega) \sim \widehat{V}(\hat{a} - 1, \Omega \backslash \Omega_\omega),$$

where

$$\hat{a}(x, \xi) = \tilde{a}(x', \xi)p(x_n)^2;$$

thus, (26.102) implies (26.97).

In order to prove (26.98), we note that on $\Omega \backslash \overline{\Omega}_\omega$, $p(x_n)^2 x_n^{-2m} = o(t)$; therefore, we can replace a with a_c in (26.102). Finally, (26.96)$_2$ yields $\widehat{V}(a_c - 1, \Omega_{c_1\omega_1}) = 0$, if c_1 is sufficiently small; therefore,

$$\widehat{V}(a_c - 1, \Omega_\omega) \leqslant C(\mathcal{F}(c_1\omega_1) - \mathcal{F}(\omega_1)) = o(\mathcal{F}(\omega)),$$

and (26.102) yields (26.98).

§27 Degenerate Differential Operators in a Bounded Domain

27.1. For the sake of simplicity, we consider operators here that degenerate on the entire boundary $\Gamma = \partial\Omega \in C^m$ of the bounded domain $\Omega \subset R^n$. Degeneracy on the submanifold $\Gamma \subset \partial\Omega$, $\Gamma \in C^m$ is studied similarly; here pointedness of Ω in a neighborhood of Γ is admissible. For the reason mentioned above, we consider only the Dirichlet problem for the case of isotropic degeneracy (i.e., degeneracy of derivatives of one order is defined by one weight function); the reader can easily use the same method to obtain the asymptotics for anisotropic degeneracy and for the Neumann problem.

27.2. Let the points $(j, \hat{k}_j)$, $j = 0, \ldots, m$ lie on a downward convex broken line (or segment) $\mathcal{L}$, where $\hat{k}_j > \hat{k}_i$ $\forall j > i$.

Let us consider the following form in $C_0^\infty(\Omega)^\ell$,

$$\mathcal{A}[u] = \sum_{|\alpha|,|\beta| \leqslant m} \langle a_{\alpha\beta}(x)D^\alpha u, D^\beta u\rangle_{L_2(\Omega)^\ell},$$

where $a_{\alpha\beta} = a_{\beta\alpha}^* \in C(\Omega; \text{End}\,C^\ell)$. We can consider a small neighborhood of Γ in Ω as a neighborhood of $\Gamma \times \{0\}$ in $\Gamma \times R_+^1$. Let $(\tilde{x}, y)$ be coordinates in $\Gamma \times R_+^1$. Suppose that, uniformly with respect to $\tilde{x} \in \Gamma$, there exist the limits

$$\lim_{y \to +0} a_{\alpha\beta}(x(\tilde{x}, y))y^{-\hat{k}_{|\alpha|}-\hat{k}_{|\beta|}} = \hat{a}_{\alpha\beta}(\tilde{x}).$$

Let $s_0 = \min_j\{\hat{k}_j - j\}$, $I^0 = \{j \mid \hat{k}_j - j = s_0\}$, $j_1 = \max\{j \mid j \in I^0\}$, let us assume that $j_1 > 0$, and let $(j_1, k^1), \ldots, (j_p, k^p) = (j_m, \hat{k}_m)$ be vertices of the polygon $\mathcal{L}$ that do not lie on the left from $(j_1, \hat{k}_{j_1})$. Let $(j_0, k^0) = (0, s_0)$, and let γ_i be slopes of straight lines which connect (j_i, k^i) with (j_{i+1}, k^{i+1}). Let s_i denote ordinates of the points of intersection of these lines with the axis Ok, and let $I^i = \{j \mid j_i \leqslant j \leqslant j_{i+1}\}$, $i \leqslant p - 1$.

Furthermore, we denote

$$a^{-\infty}(x, \xi) = \sum_{|\alpha|=|\beta|=m} a_{\alpha\beta}(x)\xi^{\alpha+\beta}, \qquad a_i(\tilde{x}, \xi) = \sum_{|\alpha|=|\beta|=j_i} \hat{a}_{\alpha\beta}(\tilde{x})\xi^{\alpha+\beta},$$

$$a_{(i)}(\tilde{x}, y, \xi) = \sum_{\alpha,\beta \in I^i} \hat{a}_{\alpha\beta}(\tilde{x})y^{\hat{k}_\alpha+\hat{k}_\beta}\xi^{\alpha+\beta},$$

$$\mathcal{A}_{\tilde{x},\xi}[u] = \sum_{|\alpha|,|\beta| \leqslant m} \tilde{\xi}^{\tilde{\alpha}+\tilde{\beta}}\langle \hat{a}_{\alpha\beta}(\tilde{x})y^{\hat{k}_\alpha+\hat{k}_\beta}D^{\alpha_n}u, D^{\beta_n}u\rangle_{L_2},$$

$$\|u\|_{r,s}^2 = \sum_{|\alpha| \leqslant r} \|\rho(x)^{\hat{k}_\alpha+s}D^\alpha u\|_{L_2}^2, \qquad \rho(x) = \text{dist}(x, \Gamma),$$

where $\alpha = (\tilde{\alpha}, \alpha_n)$, and define $\mathcal{A}_{\tilde{x},\xi}^0$ similarly to $\mathcal{A}_{\tilde{x},\xi}$, replacing summation over $|\alpha|, |\beta| \leqslant m$ with summation over $|\alpha|, |\beta| \in I^0$.

Theorem 27.1.   1) *For any $x \in \Omega$, let there exists $c_x > 0$ such that*

$$(27.1) \qquad a^{-\infty}(x,\xi) \geqslant c_x |\xi|^{2m} \qquad \forall \xi \in R^n;$$

2) *let $\exists c, C, C_1 > 0 \mid \forall (\tilde{x}, \tilde{\xi}) \in T^*\Gamma, \forall u \in C_0^\infty ((0,c))^\ell$,*

$$(27.2) \qquad C\|u\|_{L_2}^2 + C_1 A_{\tilde{x}, \tilde{\xi}}[u] \geqslant \sum_{|\alpha| \leqslant m} \tilde{\xi}^{2\tilde{\alpha}} \|y^{\hat{k}_\alpha} D^{\alpha_n} u\|_{L_2}^2.$$

Then there exist $c, C > 0$ such that

$$(27.3) \qquad A[u] \geqslant c\|u\|_{m,0}^2 - C\|u\|_{L_2}^2 \quad \forall u \in C_0^\infty (\Omega)^\ell.$$

Proof. We consider Ω to be a union of two nonintersecting sets $\tilde{\Omega}_i$, with $i = 0, 1, \ldots, N_0$, such that $\overline{\tilde{\Omega}_0} \cap \Gamma = \emptyset$, and, in a local coordinate system that straightens Γ, we can write $\Omega_i = \operatorname{int} \tilde{\Omega}_i$ as $\Omega_i' \times (0, c_i)$, where $i \geqslant 1$ and $\Omega_i' \subset R^{n-1}$ is a Lipschitz domain. We note that straightening diffeomorphisms can be chosen to be of the class $\mathbf{C}^m$.

Let $(\Omega_i)_\varepsilon = \{x \mid \operatorname{dist}(x, \Omega_i) < \varepsilon\}$. We construct, for any $\varepsilon > 0$, the functions $\varphi_{\varepsilon,i}$, $i = 0, \ldots, N_0$ such that

$$\operatorname{supp} \varphi_{\varepsilon,i} \subset (\Omega_i)_\varepsilon, \qquad \sum_i \varphi_{\varepsilon,i}^2 = 1, \qquad |\varphi_{\varepsilon,i}^{(s)}| \leqslant c_s \varepsilon^{-s}, \quad \forall s,$$

and denote $u_{\varepsilon,i} = \varphi_{\varepsilon,i} u$. Obviously,

$$A[u] = \sum A[u_{\varepsilon,i}] + B_\varepsilon^1[u], \qquad \|u\|_{m,0}^2 = \sum \|u_{\varepsilon,i}\|_{m,0}^2 + B_\varepsilon^2[u],$$

where

$$(27.4) \qquad |B_\varepsilon^j[u]| \leqslant C\varepsilon^{-N} \|u\|_{m-1,s} \|u\|_{m,0},$$

$N = 2m$, and $s > 0$, due to the condition $\hat{k}_j > \hat{k}_i \quad \forall j > i$.

By applying the same procedure to the forms B_ε^j and by integrating a sufficient number of times, we obtain

$$A[u] = \sum_i (A[u_{\varepsilon,i}] + B_{i,\varepsilon}^1[u_{\varepsilon,i}]) + O(\varepsilon^{-N} \|u\|_{L_2}^2),$$

$$\|u\|_{m,0}^2 = \sum_i (\|u_{\varepsilon,i}\|_{m,0}^2 + B_{i,\varepsilon}^2[u_{\varepsilon,i}]) + O(\varepsilon^{-N} \|u\|_{L_2}^2),$$

where $B_{i,\varepsilon}^j$ satisfies estimate (27.4) and, therefore, the estimate

$$|B_{i,\varepsilon}^j[u]| \leqslant \varepsilon(\|u\|_{m,0}^2 + C\varepsilon^{-2N-2} \|u\|_{m-1,s}^2).$$

Thus,

(27.5)
$$\mathcal{A}[u] \geqslant \sum_i (\mathcal{A}[u_{\varepsilon,i}] - \varepsilon \|u_{\varepsilon,i}\|_{m,0}^2 - C\varepsilon^{-N_1} \|u_{\varepsilon,i}\|_{m-1,s}^2) - C\varepsilon^{-N_1} \|u\|_{L_2}^2,$$

$$\|u\|_{m,0}^2 \leqslant C \sum_i (\|u_{\varepsilon,i}\|_{m,0}^2 + C\varepsilon^{N_1} \|u_{\varepsilon,i}\|_{m-1,s}^2) + C\varepsilon^{-N_1} \|u\|_{L_2}^2;$$

and therefore, it suffices to prove (27.3) for the pair of forms

$$\mathcal{A}_\varepsilon = \mathcal{A} - \varepsilon \| \cdot \|_{m,0}^2 - C\varepsilon^{-N_1} \| \cdot \|_{m-1,s}^2, \| \cdot \|_{m,0}^2,$$

which are defined in $C_0^\infty((\Omega_i)_\varepsilon)^\ell$, $i = 0, \ldots, N_0$.

As in §26.17, we can proceed to forms with C^∞ coefficients and then, in the case of $i = 0$, prove (27.3) as the estimate (26.13) for $j = -\omega$, and in the case of $i \geqslant 1$, as the estimate (26.13) for $j = \omega$ (in the last case we must change to a local coordinate system that straightens Γ).   ∎

27.3. Let A be an operator associated with the variational triple $\mathcal{A}$, $C_0^\infty(\Omega)^\ell$, $L_2(\Omega)^\ell$.

Theorem 27.2. *The spectrum of the operator A is discrete if and only if $s_0 < 0$.*

Proof.   The discreteness of the spectrum in the case of $s_0 < 0$ will be proved together with the theorem on its asymptotics. For $s_0 \geqslant 0$, the discreteness of the spectrum can be proven in exactly the same way as it was done in Theorem 26.2 by choosing $u_i \in C_0^\infty(\Omega_1)^\ell$ and writing the form $\mathcal{A}$ in a local coordinate system that straightens Γ.   ∎

If $\varepsilon > 0$ is sufficiently small, then (27.5) and Glazmann's lemma yield the estimates

(27.6) $$N(t, A) \leqslant \sum_i \mathcal{N}_0(\mathcal{A}_\varepsilon - (t + C_\varepsilon)\| \cdot \|_{L_2}^2, (\Omega_i)_\varepsilon),$$

(27.7) $$N(t, A) \geqslant \sum_i \mathcal{N}_0(\mathcal{A} - t\| \cdot \|_{L_2}^2, \Omega_i).$$

As shown in §26.17, we can calculate the asymptotics of the terms in (27.6) and (27.7), assuming that the coefficients are infinitely smooth. But then Theorem 13.1 yields

(27.8)
$$\mathcal{N}_0(\mathcal{A}_\varepsilon - (t + C_\varepsilon)\| \cdot \|_{L_2}^2, (\Omega_0)_\varepsilon)$$
$$= t^{n/2m} \widehat{V}(a^{-\infty} - 1, \Omega_0) + O(\varepsilon t^{n/2m}) + \varepsilon_1 t^{n/2m},$$

since $\mathrm{mes}((\Omega_0)_\varepsilon \backslash \Omega_0) \leqslant C\varepsilon$;   and the same formula is also valid for $\mathcal{N}_0(\mathcal{A} - t\| \cdot \|_{L_2}^2, \Omega_0)$. In (27.8) $\varepsilon_1 = \varepsilon_1(\varepsilon, t) \to 0$ as $t \to \infty$ for any fixed $\varepsilon > 0$.

After changing to a local coordinate system that straightens Γ, the asymptotics of the remaining terms in (27.6) and (27.7) is calculated with the help of Theorem 26.5 (the case where $J = \{n\}$). Then summing over i and taking the limit as $\varepsilon \to 0$ as in §26.17, we obtain the asymptotic formula for $N(t, A)$.

In order to calculate the asymptotic coefficients and the exponents of the asymptotic formulas, we note that the constants and functions introduced in §26 are defined here by the equalities $\kappa_p = 0$, $\kappa_i = 1/(2s_i)$, $\sigma_i = (1 - n\gamma_i)\kappa_i$, $0 \leqslant i < p$, $a^{s_i} = a_{(i)}$, $a^{s_i^+} = a_{i+1}$, $\sigma'_i = n/2j_{i+1}$, $d_i = (nk^{i+1} - j_{i+1})/j_{i+1}$; and, therefore, (27.8) and Theorem 26.5 yield the following theorem:

Theorem 27.3. *Let conditions (27.1) and (27.2) hold, and let $s_0 < 0$. Then as $t \to \infty$,*

a) *if $n\hat{k}_m < m$, then*

$$N(t, A) \sim t^{n/2m} \hat{V}(a^{-\infty} - 1, \Omega);$$

b) *if $nk^i < j_i$, $nk^{i+1} > j_{i+1}$, $0 < i < p$, then*

$$N(t, A) \sim t^{\sigma_i}(2\pi)^{-n} \int_\Gamma d\tilde{x} \int_{R_+} dy \int_{R^n} d\xi \, N(1, a^i(\tilde{x}, y, \xi));$$

c) *if $n\hat{k}_{j_i} = j_i$, then*

$$N(t, A) \sim t^{n/(2j_i)} \ln t (\kappa_i - \kappa_{i-1})(2\pi)^{-n} \int_\Gamma d\tilde{X} \int_{R^n} d\xi \, N(1, a_i(\tilde{x}, \xi));$$

d) *if $nk_{j_1} > j_1$, then*

$$N(t, A) \sim t^{\sigma_0}(2\pi)^{1-n} \iint_{T \cdot \Gamma} \mathcal{N}(\mathcal{A}^0_{\tilde{x}, \xi} - \| \cdot \|_{L_2}^2, C_0^\infty(R_+)) d\tilde{x} \, d\tilde{\xi}.$$

§28 Degenerate Differential Operators in an Unbounded Domain

28.1. By analogy with degenerate differential operators in a bounded domain, an operator in an unbounded domain can be called degenerate if there exists a sequence $\{(x_i, \xi_i)\} \subset \Omega \times R^n$ such that $|x_i| + |\xi_i| \to \infty$ but $a(x_i, \xi) \not\to \infty$ (here a is a Weyl symbol of the operator). The spectrum of degenerate operators in a domain conic at infinity can be discrete only if some coefficients increase at infinity; for operators in a domain narrowing at infinity, this condition can be weakened. In general, discreteness of the spectrum (and its asymptotics) is defined by the relation between the speed of growth (possibly, of decrease) of the coefficients and "the speed of narrowing" at infinity of the set Ω.

Degenerate operators in an unbounded domain can be studied by using the general method in §24; here, for the sake of simplicity, we consider a problem that with a change of variables can be reduced to the model problem in §26. At the end of this section, we consider the simplest problem on the spectrum of the quotient of forms.

28.2. Let points $(j, \tilde{k}_j)$ lie on an upward convex broken line (or the segment) $\mathcal{L}$, where

(28.1) $$\tilde{k}_j \leqslant \tilde{k}_i + \gamma(j - i) \quad \forall j > i, \quad \text{where } \gamma \in (-\infty, 1),$$

and let $\Omega' \subset R^{n-1}$ be a bounded Lipschitz domain, and $\widetilde{\Omega} = \{x = (x', x_n) \mid x_n > 1, x_n^{-\gamma} x' \in \Omega'\}$.

Also, for $|\alpha|, |\beta| \leqslant m$, let $\tilde{a}_{\alpha\beta} = \tilde{a}_{\beta\alpha}^* \in C(\overline{\widetilde{\Omega}}; \operatorname{End} C^\ell)$, and uniformly with respect to $\varphi \in \Omega'$, let there exist the limits

(28.2) $$\lim_{r \to \infty} \tilde{a}_{\alpha\beta}(r^\gamma \varphi, r) r^{-\tilde{k}_{|\alpha|} - \tilde{k}_{|\beta|}} = \hat{a}_{\alpha\beta}(\varphi).$$

We consider the form

(28.3) $$\tilde{A}[u] = \sum_{|\alpha|, |\beta| \leqslant m} \langle \tilde{a}_{\alpha\beta}(x) D^\alpha u, D^\beta u \rangle_{L_2(\widetilde{\Omega})^\ell}.$$

Below, we shall formulate the conditions that ensure semi-boundedness from below of the form $\tilde{A}$ in $L_2(\widetilde{\Omega})^\ell$; let $\tilde{A}$ be an operator associated with the variational triple $\tilde{A}$, $C_0^\infty(\widetilde{\Omega})^\ell$, $L_2(\widetilde{\Omega})^\ell$.

In order to study the spectrum of the operator $\tilde{A}$, we let $\Omega = \Omega' \times (0, 1)$,

$$u(x', x_n) = x_n^{-\gamma(n-1)/2 - 1} v(x_n^{-\gamma} x', x_n^{-1}),$$

and we make the change of variables $x_n = r_1^{-1}$, $x' = r_1^{-\gamma} \varphi$. Then $\partial_{x'} = r_1^\gamma \partial_\varphi$, $\partial_{x_n} = -\gamma r_1 \langle \varphi, \partial_\varphi \rangle - r_1^2 \partial_{r_1}$, and due to (28.1),

(28.4) $$x_n^{\tilde{k}_{|\alpha|}} D_x^\alpha = (-1)^{\alpha_n} r_1^{\hat{k}_\alpha} D_{\varphi, r_1}^\alpha + \sum_{|\beta| \leqslant |\alpha|} r_1^{\hat{k}_\beta} b_{\alpha\beta}(\varphi, r_1) D_{\varphi, r_1}^\beta,$$

where

$$\hat{k}_\alpha = -\tilde{k}_{|\alpha|} + \gamma|\alpha'| + 2\alpha_n, \qquad |b_{\alpha\beta}| \leqslant c_{\alpha\beta} r_1^{\min\{1, 1-\gamma\}}.$$

Thus,

(28.5) $$\|u\|_{L_2(\widetilde{\Omega})^\ell} = \|v\|_{L_2(\Omega)^\ell}, \qquad \tilde{A}[u] = A[v],$$

where

$$\mathcal{A}[v] = \sum_{|\alpha|,|\beta| \leqslant m} \langle a_{\alpha\beta}(\varphi, r_1) D^\alpha v, D^\beta v \rangle_{L_2(\Omega)^\ell},$$

and due to (28.2) and (28.4), the coefficients of the form $\mathcal{A}$ satisfy the following condition: uniformly with respect to $\varphi \in \Omega'$, there exist the limits

$$(28.6) \qquad \lim_{r_1 \to +0} (-1)^{\alpha_n + \beta_n} a_{\alpha\beta}(\varphi, r_1) r_1^{-\hat{k}_\alpha - \hat{k}_\beta} = \hat{a}_{\alpha\beta}(\varphi).$$

In addition, if $\Phi : T^*\widetilde{\Omega} \to T^*\Omega$ is a symplectomorphism, induced by our change of variables, and if $\tilde{a}^{-\infty}$, $a^{-\infty}$ are principal symbols of the forms $\widetilde{\mathcal{A}}$, $\widetilde{\mathcal{A}}$, then due to invariance of the principal symbol,

$$(28.7) \qquad a^{-\infty}(\Phi(x,\xi)) = \tilde{a}^{-\infty}(x,\xi).$$

Now (28.6) implies that the form $\mathcal{A}$ satisfies the conditions in §26 with $J = \{1, \ldots, n-1\}$, if

$$s_0 = \min_{0 < j \leqslant m} (-\tilde{k}_j + \gamma j) \leqslant -\tilde{k}_0,$$

which will be assumed below (if $s_0 \geqslant -\tilde{k}_0$, then the asymptotics of the spectrum of the operator $\tilde{A}$ can be calculated by directly applying Theorem 7.5).

Due to condition (28.5), $\tilde{A}$ is unitarily equivalent to the operator A, which is associated with the variational triple $\mathcal{A}$, $C_0^\infty(\Omega)^\ell$, $L_2(\Omega)^\ell$; therefore, the conditions of the coerciveness discreteness of the spectrum coincide with their asymptotic formulas for the spectrum. Thus, we could formulate them in terms of the coefficients $a_{\alpha\beta}$ and the exponents $\hat{k}_\alpha$, but it is more natural to use $\tilde{a}_{\alpha\beta}$ and $\tilde{k}_{|\alpha|}$.

In order to obtain them, we note that the constants s_j, κ_j, s_j, d_j and the sets I_s, which describe the operator A, can be defined as follows. Let $I^0 = \{j \mid \tilde{k}_j - \gamma j = -s_0\}$, $j_1 = \max\{j \mid j \in I_0\}$. (28.1) and the condition $s_0 \leqslant -\tilde{k}_0$ yield $s_0 = -\tilde{k}_0$ and $j_1 > 0$. Let $(j_0, k^0) = (0, -s_0)$, $(j_1, k^1), \ldots, (j_p, k^p) = (j_m, \tilde{k}_m)$ be vertices of the broken line $\mathcal{L}$, let $(0, -s_i)$ be points of intersection with the axis $O\tilde{k}$, and let $-\gamma_i$ be the slopes of the straight lines passing through the points (j_i, k^i), (j_{i+1}, k^{i+1}), $i = 0, 1, \ldots, p-1$. Then $\gamma_{j s_i} = \gamma_i + \gamma j = 1, \ldots, n-1$, $\gamma_{n s_i} = \gamma_i + 1$, $\kappa_i = 1/(2s_i)$, $i < p$, $\kappa_p = 0$, $\sigma_i = -(n\gamma_i + s)\kappa_i$, $i < p$, where $s = (n-1)\gamma + 1$,

$$d_i = (-k^{i+1} + \gamma j_{i+1})(n-1)/j_{i+1} + (-k^{i+1} + j_{i+1})/j_{i+1}$$
$$= (-nk^{i+1} + sj_{i+1})/j_{i+1},$$
$$I_{s_i} = \{\alpha \mid |\alpha| \in I^i\}, \qquad I_{s_i^+} = \{\alpha \mid |\alpha| = j_{i+1}\}.$$

In (26.5) and (26.43), we make the substitutions $x_n = r^{-1}$, $\xi_n = -r^2\tau$, and in (26.37) and (26.38), the substitutions $x_n \to x_n^{-1}$, $\xi_n \to -x_n^2\xi_n$, $\xi' \to x_n^\gamma\xi'$,

$x' \to x_n^{-\gamma} x'$. Then the functions a^{s_i}, $a^{s_i^+}$ and the forms $\mathcal{A}_{x^1, \xi^1}$, $\mathcal{A}_{x^2, \xi^1}^0$ become the functions

(28.8)
$$\tilde{a}^1(x, \xi) = \sum_{|\alpha|, |\beta| \in I^i} \hat{a}_{\alpha\beta}(\varphi) r^{\tilde{k}_\alpha + \tilde{k}_\beta} \xi^{\alpha+\beta},$$

(28.9)
$$\tilde{a}_i(x, \xi) = \sum_{|\alpha|=|\beta|=j_{i+1}} \hat{a}_{\alpha\beta}(\varphi) r^{\tilde{k}_\alpha + \tilde{k}_\beta} \xi^{\alpha+\beta},$$

where $r = r(x) = x_n$, $\varphi = \varphi(x) = x_n^{-\gamma} x'$, and the forms

(28.10)
$$\tilde{\mathcal{A}}_{r,\tau}[u] = \sum_{|\alpha|, |\beta| \leqslant m} r^{\tilde{k}_{|\alpha|} + \tilde{k}_{|\beta|} - \gamma(|\alpha'| + |\beta'|)} \tau^{\alpha_n + \beta_n} \langle \hat{a}_{\alpha\beta}(\varphi) D_\varphi^{\alpha'} u, D_\varphi^{\beta'} u \rangle_{L_2(\Omega')^\ell},$$

and $\tilde{\mathcal{A}}_{r,\tau}^0$; the latter is defined by (28.10) with the summation over $|\alpha|, |\beta| \in I^0$.

Also by (28.7) taking into account, we see that Theorems 26.1, 26.2, and 26.5 yield the following theorems.

Theorem 28.1. *Let there exist $c > 0$ such that*
a) $\tilde{a}^{-\infty}(x, \xi) \geqslant c x_n^{2\tilde{k}_m} |\xi|^{2m} \quad \forall (x, \xi) \in \Omega \times R^n$;
b) $\forall r > c^{-1}, \forall \tau \in R, \forall u \in C_0^\infty(\Omega')^\ell$,

$$\tilde{\mathcal{A}}_{r,\tau}[u] \geqslant c \sum_{|\alpha| \leqslant m} r^{2(\tilde{k}_{|\alpha|} - \gamma|\alpha'|)} \tau^{2\alpha_n} \|D_\varphi^{\alpha'} u\|_{L_2(\Omega')^\ell}^2.$$

Then there exist $C, c > 0$ such that $\forall u \in C_0^\infty(\tilde{\Omega})^\ell$,

$$\tilde{\mathcal{A}}[u] + C\|u\|_{L_2(\tilde{\Omega})^\ell}^2 \geqslant c \sum_{|\alpha| \leqslant m} \|x_n^{\tilde{k}_{|\alpha|}} D^\alpha u\|_{L_2(\tilde{\Omega})^\ell}^2.$$

Theorem 28.2. *The spectrum of the operator $\tilde{A}$ is discrete if and only if $s_0 < 0$.*

Theorem 28.3. *As $t \to \infty$,*
a) *if $n\tilde{k}_m > sm$, then*

$$N(t, \tilde{A}) \sim t^{n/2m} \hat{V}(\tilde{a}^{-\infty} - 1, \tilde{\Omega});$$

b) *if $nk^{i+1} < sj_{i+1}$, $nk^i > sj_i$, $i = 0, \ldots, p-1$, then*

(28.11)
$$N(t, \tilde{A}) \sim t^{\sigma_i} \hat{V}(\tilde{a}^i - 1, \tilde{\tilde{\Omega}}),$$

where $\tilde{\tilde{\Omega}} = \{x = (x', x_n) \mid x_n > 0, x_n^{-\gamma} x' \in \Omega'\}$;

c) *if* $nk^i = sj_i$, *then*

(28.12) $N(t, \tilde{A}) \sim t^{n/2j_i} \ln t(\kappa_i - \kappa_{i-1}) \tilde{V}(\hat{a}_i - 1, \Omega' \times R^n)$,

where $\hat{a}_i(\varphi, \xi) = a_{i-1}(\varphi, 1, \xi)$;
  d) *if* $nk^1 < sj_1$, *then*

$$N(t, \tilde{A}) \sim t^{\sigma_0 (2\pi)^{-1}} \iint\limits_{R_+ \times R^1} \mathcal{N}_0(\tilde{\mathcal{A}}_{r,\tau}^0 - \| \cdot \|_{L_2}^2, C_0^\infty (\Omega')^\ell) dr \, d\tau.$$

28.3. Using the general method in §24, we can also study an operator in a domain conic at infinity, that is defined by a form of the type in (28.3) with coefficients satisfying the following condition: uniformly with respect to $\varphi \in S_{n-1}$, there exist the limits

$$\lim_{r \to \infty} \tilde{a}_{\alpha\beta}(x(r, \varphi)) r^{-\tilde{k}_{|\alpha|} - \tilde{k}_{|\beta|}} = \hat{a}_{\alpha\beta}(\varphi),$$

where (r, φ) are spherical coordinates, and the exponents $\tilde{k}_j$ satisfy condition (28.1) with $\gamma = 1$. Under natural conditions of coerciveness (similar to the conditions of Theorem 26.1 for the case where $n_2 = n$), we can obtain analogies of Theorems 28.1–28.3; everywhere, we must assume that $\gamma = 1$, so that $s = n$ and the strong degeneracy $nk^1 < sj_1$ is possible only for an operator with a nondiscrete spectrum.

In formulas (28.11) and (28.12), $\tilde{\tilde{\Omega}}$ is a cone coinciding with the domain $\tilde{\Omega}$ outside some ball $\Omega' = \tilde{\tilde{\Omega}} \cap S_{n-1}$, and $\tilde{a}^i$, $\tilde{a}_i$ are defined by equalities (28.8) and (28.9), where (r, φ) are spherical coordinates.

In the following section, part of these statements will be proved for the simplest problem on the spectrum of quotient of forms.

28.4. Let $a_{\alpha\beta} = a_{\beta\alpha}^*$ satisfy the conditions

(28.13) $\|a_{\alpha\beta}^{(\gamma)}(x)\| \leqslant c_\gamma \langle x \rangle^{2k - |\gamma|}, \quad |\gamma| = 0, 1, \ldots,$

(28.14) $\sum\limits_{|\alpha| = |\beta| = m} \langle a_{\alpha\beta}(x) \xi_\alpha, \xi_\beta \rangle_{\mathbf{C}^\ell} \geqslant c \langle x \rangle^{2k} \sum\limits_{|\alpha| = m} |\xi_\alpha|_{\mathbf{C}^\ell}^2 \quad \forall x, \forall \xi_\alpha \in \mathbf{C}^\ell.$

Also let $k - k' > m$ and let A be an operator associated with the variational triple

$$\mathcal{A}[u] = \sum\limits_{|\alpha| = |\beta| = m} \langle a_{\alpha\beta}(x) D^\alpha u, D^\beta u \rangle_{L_2}, \qquad C_0^\infty (R^n)^m, \qquad \|\langle x \rangle^{k'} u\|_{L_2}^2.$$

Theorem 28.4. *As $t \to \infty$,*

$$(28.15) \qquad N(t, A) \sim t^{n/2m} \hat{V}(a^{-\infty} - 1, R^n),$$

where

$$a^{-\infty}(x, \xi) = \sum_{|\alpha| = |\beta| = m} a_{\alpha\beta}(x) \langle x \rangle^{-2k'} \xi^{\alpha + \beta}.$$

Proof. By Glazmann's lemma, $N(t, A) = \mathcal{N}_0(A_t, R^n)$, where

$$A_t = \sum_{|\alpha| = |\beta| = m} D^\beta a_{\alpha\beta}(x) D^\alpha - t \langle x \rangle^{2k'}.$$

We fix $\kappa \in (0, 1/(2(k - k' - m)))$ and write

$$\Omega_{c,t} = \{ x \in R^n \mid |x| < ct^\kappa \}, \qquad \Omega_{-c,t} = R^n \backslash \Omega_{c,t},$$
$$\langle x \rangle_t = \langle x \rangle \theta(t^{-\kappa} |x|/2) + (1 - \theta(t^{-\kappa}|x|/2)) t^\kappa,$$
$$a_{t\alpha\beta}(x) = a_{\alpha\beta}(x) \theta(t^{-\kappa}|x|/2) + (1 - \theta(t^{-\kappa}|x|/2)) t^{2k\kappa},$$
$$q_t(x, \xi)^2 = \langle x \rangle_t^{2k} |\xi|^{2m} + t \langle x \rangle_t^{2k'},$$
$$g_{t,x,\xi}(y, \eta) = \langle x \rangle_t^{-2(k'-k)/m} t^{-1/m} |\eta|^2 + \langle x \rangle_t^{-2} |y|^2,$$
$$\tilde{A}_t = \sum_{|\alpha| = |\beta| = m} D^\beta a_{t\alpha\beta}(x) D^\alpha - t \langle x \rangle_t^{2k'},$$

where $\theta \in C_0^\infty(R^1)$, $0 \leqslant \theta \leqslant 1$, $\theta\big|_{|t| < 1/2} = 1$, $\theta\big|_{|t| > 1} = 0$.

Then $h_t(x, \xi) \leqslant c^{-1} t^{-c}$ for some $c > 0$, the metric g_t is σ-temperate, $q_t \in O(C^1, g_t)$, $\tilde{A}_t \in \mathcal{L}(g_t, q_t^2)$ uniformly with respect to $t \geqslant 1$, and

$$(28.16) \qquad N(t, A) \geqslant \mathcal{N}_0(\tilde{A}_t, \Omega_{1,t}) + \mathcal{N}_0(A_t, \Omega_{-1,t}).$$

By using the metric g_t, we can obtain, similar to (26.46),

$$(28.17) \qquad N(t, A) \leqslant \mathcal{N}_0(\tilde{A}_{t_\varepsilon}, \Omega_{1,t}) + \mathcal{N}_0(A_{t_\varepsilon}, \Omega_{-1/2,t}),$$

where $t_\varepsilon = t(1 + \varepsilon)$, $\varepsilon \to 0$ as $t \to \infty$.

According to the composition theorem, for all $(x, \xi) \in \Omega_{1,t} \times R^n$,

$$\sigma^w(\tilde{A}_t)(x, \xi) = (a^{-\infty}(x, \xi) - t) \langle x \rangle^{2k'} + o(q_t(x, \xi)^2),$$

and condition (28.14) yields, for the same (x, ξ),

$$c(a^{-\infty}(x, \xi) + t) \leqslant q_t^2(x, \xi) \langle x \rangle^{-2k'} \leqslant C(a^{-\infty}(x, \xi) + t);$$

therefore, Theorem 7.5 yields the estimate

$$\mathcal{N}_0(\tilde{A}_{t_\varepsilon}, \Omega_{1,t}) = F(t) + O(F(t_{\varepsilon_1}) - F(t_{-\varepsilon_1})) + O(F_1(t)),$$

where $\varepsilon_1 = \varepsilon_1(\varepsilon) \to 0$ as $t \to \infty$,

$$F(t) = \hat{V}(a^{-\infty} - t, R^n) = t^{n/2m} \hat{V}(a^{-\infty} - 1, R^n),$$
$$F_1(t) = \operatorname{mes}\{(x, \xi) \mid |x| > t^\kappa/2, |x|^{2k}|\xi|^{2m} < Ct|x|^{2k'}\}$$
$$\leqslant c_1 t^{n/2m} \int\limits_{t^\kappa/2}^{\infty} |x|^{n(k'-k)/m} dx = c_2 t^{n/2m-c},$$

where $c > 0$ due to the choice of κ. Hence,

$$(28.18) \qquad \mathcal{N}_0(\tilde{A}_{t_\varepsilon}, \Omega_{1,t}) \sim t^{n/2m} \hat{V}(a^{-\infty} - 1, R^n).$$

Furthermore, using (28.14) and the Hardy inequality, we obtain that $\forall s > 0$,

$$\mathcal{N}_0(A_{t_\varepsilon}, \Omega_{-c,t}) \leqslant \mathcal{N}_0(A^s_{Ct}, \Omega_{-c,t}),$$

where

$$A^s_{Ct} = \sum_{|\alpha|=m} D^\alpha \langle \cdot \rangle^{2k-2s} D^\alpha - Ct\langle \cdot \rangle^{2k'} + \langle \cdot \rangle^{2(k-m)-s}.$$

Let

$$q_{s,t}(x, \xi)^2 = \langle x \rangle^{2k-2s}|\xi|^{2m} + \langle x \rangle^{2(k-m)-s} + t\langle x \rangle^{2k'},$$
$$g_{s,t,x,\xi}(y, \eta) = (\langle x \rangle^{-2+s/m} + |\xi|^2)^{-1}|\eta|^2 + \langle x \rangle^{-2}|y|^2.$$

Then, uniformly with respect to $t \geqslant 1$, $g_{s,t}$ is a σ-temperate metric, $q_{s,t} \in O(C^1, g_{s,t})$, $A^s_t \in \mathcal{L}(g_{s,t}, q^2_{s,t})$ and $h_t(X) \leqslant C\langle X \rangle^{-c}$ for some $C, c > 0$.

Therefore, we can apply Theorem 7.2 and obtain the estimate

$$\mathcal{N}_0(A_{t_\varepsilon}, \Omega_{-c,t}) \leqslant C_1 \operatorname{mes}\{(x, \xi) \mid q_{s,t}(x, \xi)^2 \leqslant C_1 t\langle x \rangle^{2k'}, |x| > ct^\kappa\}$$
$$\leqslant C_2 t^{n/2m} \int\limits_{ct^\kappa}^{\infty} |x|^{n(k'-k+s)/m} dx = C_3 t^{n/2m-c},$$

where $c > 0$, if $k' - k + s < -m$.

Now estimate (28.15) follows from (28.16)–(28.18).

APPENDIX

BASIC VARIATIONAL THEOREMS

A.1. First, we cite the well-known facts about quadratic forms in a Hilbert space H (see, for example, Chapter 6 of Kato's book [1]).

Let a sesquilinear form $\mathcal{A}$ with a domain $D(\mathcal{A})$ satisfy the condition

$$\mathcal{A}[u, v] = \mathcal{A}\overline{[v, u]} \quad \forall u, v \in D(\mathcal{A}).$$

Then the form $\mathcal{A}$ is called a Hermitian form, and the function $\mathcal{A}[u] = \mathcal{A}[u, u]$ is called a quadratic form, associated with $\mathcal{A}[u, v]$.

The correspondence between Hermitian and quadratic forms is one-to-one. The forms $\mathcal{A}[u, v]$ and $\mathcal{A}[u]$ are simply called forms when confusion is impossible.

The form $\mathcal{A}$ is semi-bounded from below if there exists $c \in R$ such that $\mathcal{A}[u] \geqslant c\|u\|_H^2 \quad \forall u \in D(\mathcal{A})$. We say that a sequence $\{u_n\} \subset D(\mathcal{A})$ is $\mathcal{A}$-convergent (to $u \in H$) and denote $u_n \overset{\mathcal{A}}{\to} u$, if $u_n \to u$ in H and $\mathcal{A}[u_n - u_m] \to 0$ as $n, m \to \infty$. The form $\mathcal{A}$ is closed if $u_n \overset{\mathcal{A}}{\to} u$ implies $u \in D(\mathcal{A})$ and $\mathcal{A}[u_n - u] \to 0$.

Let $u_n \overset{\mathcal{A}}{\to} 0$ imply $\mathcal{A}[u_n] \to 0$. Then $\mathcal{A}$ has a closure $\widetilde{\mathcal{A}}$ which is defined as follows. $D(\widetilde{\mathcal{A}})$ is a set of all $u \in H$ for which there exists a sequence $\{u_n\} \subset D(\mathcal{A})$ such that $u_n \overset{\mathcal{A}}{\to} u$, and $\widetilde{\mathcal{A}}$ is defined by the equality

(A.1) $$\widetilde{\mathcal{A}}[u, v] = \lim_{n \to \infty} \mathcal{A}[u_n, v_n]$$

for any sequences $u_n \overset{\mathcal{A}}{\to} u$, $v_n \overset{\mathcal{A}}{\to} v$.

We say that the form $\mathcal{A}$ is closable.

Let $\mathcal{A}$ be a closable form, $D' \subset D(\mathcal{A})$ be a subspace, $\mathcal{A}'$ be a contraction of $\mathcal{A}$ on D', and $\mathcal{A}$ be a closure of the form $\mathcal{A}'$. Then D' is called a kernel of the form $\mathcal{A}$.

A.2. The following theorem is due to Friedrichs [1] (see also, for example, Riesz, Nagy [1]).

Theorem A.1.   *Let A be a densely defined closable form semi-bounded from below in H. Then there exists a self-adjoint operator A semi-bounded from below in H such that*

1) $D(A) \subset D(\tilde{A})$ *and*

$$\tilde{A}[u, v] = \langle Au, v \rangle_H \quad \forall u \in D(A), \ v \in D(\mathcal{A}).$$

2)

(A.2) $D(A)$ *is a kernel of the form* $\tilde{A}$.

Condition 1) uniquely defines the operator A. We say that A is an operator associated with the variational triple $\mathcal{A}$, $D(\mathcal{A})$, H.

If A_0 is a densely defined operator semi-bounded from below and if the form $\mathcal{A}$ is defined by the equality

$$\mathcal{A}[u, v] = \langle A_0 u, v \rangle_H \quad \forall u, v \in D(A_0) = D(\mathcal{A}),$$

then the form $\mathcal{A}$ is closable. The operator A associated with the variational triple $\mathcal{A}$, $D(\mathcal{A})$, H is called the Friedrichs extension of the operator A_0.

Throughout the following, the form $\mathcal{A}$ satisfies the conditions of Theorem A.1.

If the interval $(-\infty, t)$ contains points of the essential spectrum of the operator A, then we assume that $N(t, A) = \infty$; otherwise, $N(t, A)$ is the number of eigenvalues (counting multiplicity) of the self-adjoint operator A on $(-\infty, t)$.

The following Glazman's lemma (Glazman [1], see also Reed, Simon [1]) enables us to see whether the interval $(-\infty, t)$ contains points of the essential spectrum and to calculate $N(t, A)$.

Lemma A.1. $N(t, A) = \mathcal{N}(A - t, D(A)).$ ∎

The next lemma follows from (A.1), (A.2), and Lemma A.1:

Lemma A.2.   *Let A be an operator associated with the variational triple $\mathcal{A}$, $D(\mathcal{A})$, H and let $\mathcal{V}$ be a kernel of the closure of the form $\mathcal{A}$. Then $N(t, A) = \mathcal{N}(\tilde{A} - t\| \cdot \|_H^2, \mathcal{V})$.*

In particular, if A is the Friedrichs extension of the operator A_0, then

$$N(t, A) = \mathcal{N}(A_0 - t, D(A_0)).$$

∎

Similarly, we shall modify the well-known Courant–Weyl maximin principle (see, for example, Reed, Simon [1]):

Theorem A.2. *Let $\mathcal{A}, A$ be the same as in Theorem A.1. Let*

$$\mu_n(\tilde{\mathcal{A}}, \mathcal{V}) = \sup_{\substack{L \subset \mathcal{V} \\ \text{codim } L \leqslant n}} \inf_{0 \neq u \in L} \tilde{\mathcal{A}}[u]/\|u\|_H^2,$$

where $\mathcal{V}$ is a kernel of the form $\tilde{\mathcal{A}}$.

Then for any n, either there exist n eigenvalues (counting multiplicity) of the operator A lying below the essential spectrum and $\mu_n(\tilde{\mathcal{A}}, \mathcal{V})$ is the nth eigenvalue of the operator A, or $\mu_n(\tilde{\mathcal{A}}, \mathcal{V})$ is a lower bound of the essential spectrum and $\mu_n(\tilde{\mathcal{A}}, \mathcal{V}) = \mu_{n+1}(\tilde{\mathcal{A}}, \mathcal{V}) = \cdots$. ∎

Theorem A.2 and Lemma A.2 yield the following corollary:

Corollary A.1. *Let $D(\mathcal{A}_1) \subset D(\mathcal{A}_2)$, $\mathcal{A}_2[u] \leqslant \mathcal{A}_1[u]$ $\forall u \in D(\mathcal{A}_1)$, let A_j be an operator associated with the variational triple $\mathcal{A}_j$, $D(\mathcal{A}_j)$, H, and let $\lambda_k(A_j)$ be eigenvalues of the operator A_j that lie below the lower bound $\lambda^\infty(A_j)$ of its essential spectrum and are numbered in increasing order (counting multiplicity).*

Then

1) *if there exist $\lambda_k(A_1)$ and $\lambda_k(A_2)$, then $\lambda_k(A_1) \geqslant \lambda_k(A_2)$;*
2) *$\lambda^\infty(A_2) \leqslant \lambda^\infty(A_1)$;*
3) *for any λ, $N(\lambda, A_2) \geqslant N(\lambda, A_1)$.* ∎

A.3.

Lemma A.3. *Let H_i be Hilbert spaces, let $\mathcal{A}_i$ be closed forms in H_i with the closure $\tilde{\mathcal{A}}_i$, let $\mathcal{V}_i$ be kernels of the forms $\tilde{\mathcal{A}}_i$ $(i = 1, \ldots, m)$, and let $\mathcal{A}_0$ be a form in $H_0 = \bigoplus_{1 \leqslant i \leqslant m} H_i$ defined on $D(\mathcal{A}_0) = \bigoplus_{1 \leqslant i \leqslant m} D(\mathcal{A}_i)$ by the equality*

$$\mathcal{A}_0[u] = \sum_{1 \leqslant i \leqslant m} \mathcal{A}_i[u_i], \quad u = (u_i, \ldots, u_m) \in D(\mathcal{A}_0).$$

Then

$$(A.3) \qquad N(\tilde{\mathcal{A}}_0 - \lambda\| \cdot \|_{H_0}^2, \mathcal{V}_0) = \sum_{1 \leqslant i \leqslant m} N(\tilde{\mathcal{A}}_i - \lambda\| \cdot \|_{H_i}^2, \mathcal{V}_i)$$

for any kernel $\mathcal{V}_0$ of the form $\tilde{\mathcal{A}}_0$—in particular, for

$$\mathcal{V}_0 = \bigoplus_{1 \leqslant i \leqslant m} \mathcal{V}_i \qquad \mathcal{V}_0 = \bigoplus_{1 \leqslant i \leqslant m} D(\mathcal{A}_i).$$

Proof. Let A_i be the operators associated with the variational triple $\tilde{\mathcal{A}}_i$, $\mathcal{V}_i$, H_i. Then

$$A_0 = \bigoplus_{1 \leqslant i \leqslant m} A_i, \qquad N(\lambda, A_0) = \sum_{1 \leqslant i \leqslant m} N(\lambda, A_i),$$

and (A.3) follows from Lemma A.2.   ∎

Lemma A.4. *Let V be a kernel of the closed form A, and let $V' \subset V$ be a space of codimension m. Then*

(A.4) $$\mathcal{N}(A, V') \leqslant \mathcal{N}(A, V) \leqslant \mathcal{N}(A, V') + m.$$

Proof. The left inequality is obvious. We shall prove the right side. Let H' be a closure of V' in H. Then the contraction A' of the form A on V' is a closable form in H'. Therefore, the operator A' associated with the variational triple A', V', H' is defined. Due to Glazman's lemma, $\lambda^\infty(A') \geqslant \lambda^\infty(A)$. Furthermore,

(A.5) $$\mu_n(A', V') \leqslant \mu_{n+m}(A, V);$$

hence, $\lambda^\infty(A') \leqslant \lambda^\infty(A)$, due to Theorem A.2. Thus, $\lambda^\infty(A') = \lambda^\infty(A)$; and (A.4) holds, again due to Glazman's lemma, provided that $0 > \lambda^\infty(A)$ (each term of the inequality is equal to ∞).

Now let $0 \leqslant \lambda^\infty(A)$. If $n = \mathcal{N}(A', V')$, then (A.5) and Theorem A.2 imply that $\lambda_n(A') \leqslant \lambda_{n+m}(A)$; and (A.4) is proved.   ∎

Lemma A.5. *Let A_i be closable forms in H_i and $D(A_i) = V_i$, $i = 1, 2$, let $\ell : V_1 \to V_2$ be a linear operator and $\dim \operatorname{Ker} \ell = r < \infty$, and let*

$$A_1[u] \geqslant A_2[\ell u] \quad \forall u \in V_1.$$

Then

$$\mathcal{N}(A_1, V_1) \leqslant \mathcal{N}(A_2, V_2) + r.$$

Proof. Let $V' \oplus \operatorname{Ker} \ell = V_1$. Lemma A.4 implies $\mathcal{N}(A_1, V_1) \leqslant \mathcal{N}(A_1, V') + r$, and the inequality $\mathcal{N}(A_1, V') \leqslant \mathcal{N}(A_2, V_2)$ follows from the injectiveness of the operator $\ell : V' \to V_2$.   ∎

Corollary A.2. *If $\ell : V_1 \to V_2$ is an isomorphism, and $A_1[u] = A_2[\ell u]$ for all $u \in V_1$, then $\mathcal{N}(A_1, V_1) = \mathcal{N}(A_2, V_2)$.*   ∎

Corollary A.3. *If $V = D(A_1) = D(A_2)$, $c > 0$ and $A_1[u] \geqslant c A_2[u] \quad \forall u \in V' \subset V$, where $\operatorname{codim} V' = r$, then*

$$\mathcal{N}(A_1, V) \leqslant \mathcal{N}(A_2, V) + r.$$

                                                                            ∎

Lemma A.6. *Let $\mathcal{A}_i$, $i = 0, \ldots, m$ be the same as in Lemma A.3, let $\mathcal{A}$ be a closable form in H with domain $D(\mathcal{A})$, let $\ell : D(\mathcal{A}) \rightarrow D(\mathcal{A}_0)$ be a linear operator, and let $\dim \operatorname{Ker} \ell = r_1 < \infty$ and*

$$\mathcal{A}[u] \geqslant c \sum_{1 \leqslant i \leqslant m} \mathcal{A}_i[u_i], \qquad \ell u = (u_1, \ldots, u_m)$$

for all $u \in \mathcal{V} \subset D(\mathcal{A})$, where $\operatorname{codim} \mathcal{V} = r_2$, $c > 0$.
  Then

$$\mathcal{N}(\mathcal{A}, D(\mathcal{A})) \leqslant \sum_{1 \leqslant i \leqslant m} \mathcal{N}(\mathcal{A}_i, D(\mathcal{A}_i)) + r_1 + r_2.$$

Proof. This follows from an obvious combination of Lemmas A.3–A.5.

A BRIEF REVIEW OF THE BIBLIOGRAPHY

This bibliography and its review do not presume to be complete. Basically, we discuss the works which are closest to the text of this book. A more detailed review of the literature up to 1976 is given by Birman and Solomyak in [4]. Aspects of spectral asymptotics are also discussed in the books by Agmon [1], Shubin [1], and in lectures by Birman and Solomyak [5] and by Kostuchenko [1].

An extensive bibliography of works on asymptotics of the spectrum of operators on a compact manifold is contained in the monograph by Ivriĭ [2] and for all types of spectral problems in recent book by Rosenblum, Solomyak and Shubin [1].

On Chapter 1

PDOs in their modern form were first introduced by Kohn and Nirenberg [1]. Their original theory, as well as a number of the following ones, was essentially local (see the review of the literature by Beals [2]). Soon thereafter, Hörmander [1] proved boundedness in L_2 of PDO with symbols of the class $S^0_{\rho,\delta}$, $0 \leqslant \delta < \rho \leqslant 1$, and Calderon and Vaillancourt analyzed the boundary case $\delta = \rho < 1$. Then, in a number of studies (see review by Beals [2]), various versions of calculus, which allow for polynomial growth of the symbols with respect to x, were constructed. The PDO classes, constructed in these works, were generalized by Beals and Fefferman [1] and by Beals [1,2]. The next step was made by Hörmander [3, 5]; generalization of his calculus is constructed here. A calculus, close to Hörmander's, was constructed by Unterberger [1].

The construction of the scale of weight spaces, given here, is a modification of the scheme developed by Beals [1]; Beals [3] and Lerner [1] constructed the scale under weaker conditions, but the methods of their works are more complicated and, as it seems to me, are applicable only for the scalar case.

The reader can acquaint himself with theories of PDO on a compact manifold without boundary in the monographs by Hörmander [5], Shubin [1], Taylor [1], and Treves [1].

The Calculus of so-called double PDO, which allows one to analyze DO with coefficients rapidly increasing at infinity, was developed by Feigin [1,2]. Levendorskiĭ [1,2] generalized the calculus of Feigin and extended it to DO with coefficients which have singularities on a boundary.

On Chapter 2

The ASP method was first introduced by Tulovsky and Shubin [1] for the study of hypoelliptic PDO in R^n; the estimate for the remainder, within the frame of Tulovsky and Shubin's approach, was sharpened by Hörmander [4]. Roitburd [1], Feigin [4], Hörmander [4] and Levendorskiĭ [2] studied quasi-classical asymptotics (the last two works also studied asymptotics with respect to spectral parameters). Feigin [4] and Levendorskiĭ [2] also studied asymptotics with respect to a small parameter when the spectral parameter approaches the lower bound of the spectrum of the symbol.

An important modification of the original version of the ASP method (construction of ASP in the form $\mathcal{E}^2(3-2\mathcal{E})$) was offered by Feigin [2]; in the same work, the ASP method was applied to systems of equations for the first time. PDO on a closed manifold were studied by Bezyaev [1,2].

DO with coefficients, which grow rapidly at infinity or near the boundary of a domain, were studied by Feigin [1,2] and Levendorskiĭ [1,2]. Feigin [3,5] was the first to apply the ASP method to boundary value problems in an unbounded domain; his technique essentially utilized ellipticity and locality of an operator. Therefore, we use a different one here.

The proof of estimate (9.10) belongs to Feigin [4], and Lemma 10.2 to Hörmander [4]. Lemma 10.1 is a generalization of the lemma by Hörmander [4]; the construction of ASP in the proof of Theorem 7.4 is due to Feigin [2].

On Chapter 3

Sections 13 and 14: For regular boundary value problems, the principal term of the asymptotics of the spectrum is defined by the Weyl formula and does not depend on boundary conditions; this holds also for weaker constraints than in §14 on a domain of an operator (see Michaileč [5]; the conditions could also be weakened within the ASP method). For "nonclassical" boundary conditions (i.e., for arbitrary semi-bounded self-adjoint extensions of a given minimal operator) the asymptotics can be arbitrary (see Ilyin [1,2], Ilyin and Filippov [1], Gorbachuk [1], Michaileč [3], and the review of the literature by Birman an Solomyak [4]).

For a scalar differential operator A in a domain with a sufficiently smooth boundary, the principal term was obtained by Agmon [1,2] and Browder [1] for the condition $D(A^s) \subset H^{sm}(\Omega)$, where $m = ord A$, $s \geq 1$ is an integer. Birman and Solomyak [1,2] obtained the classical formula of asymptotics of the spectrum for linear pencils with principal elliptic operators under very weak conditions, which in a number of cases coincide with conditions of finiteness of integrals in asymptotic formulas (formulation of the problem is variational); they also calculated the principal term of asymptotics of spectrum of PDO of negative order (see [5]).

Agmon [3] obtained, for elliptic PDO with C^∞-coefficients, the estimate for the remainder

$$(B.1) \qquad N(t, A) = c_0 t^{n/m} + O(t^{(n-\delta)/m}),$$

where $\delta \in (0,1)$, if the higher coefficients are constant, and $\delta \in (0, 1/2)$, otherwise, they are arbitrary.

The following estimate was obtained by Courant [1] for the Laplace operator and by Brüning [1], Boimatov and Kostuchenko [3] for an arbitrary elliptic operator

$$N(t, A) = c_0 t^{n/m} + O(t^{(n-1)/m} \ln t).$$

This estimate is better than (B.1), but Agmon's technique extends to systems without assumption of constancy of multiplicities of eigenvalues of the principal symbol (see Kozlov [1]), where estimate (B.1), with $\delta \in (0, 1/2)$, is obtained for linear pencils with a principal elliptic operator which does not have a definite sign and an lower elliptic positive-definite operator.

Robert [1] obtained estimate (B.1), with $\delta \in (0, 1)$, for scalar PDO, which satisfy transmission condition, and Grubb [3] derived it, with $\delta \in (0, 1/2)$, for matrix elliptic PDO. The same estimate, for scalar DO with Lipschitz coefficients, was obtained by Metivier [1].

The sharp estimate of the remainder

$(B.2)$ $$N(t, A) = c_0 t^{n/m} + O(t^{(n-1)/m})$$

was obtained for scalar DO of the second order by Levitan [1] and Avacumovic [1]; examples of the Laplace operator on a sphere and in some domains on a sphere show that, in general, estimate (2) is not improvable (see, for instance, examples by Berard [1]).

Developing the idea of Levitan [1], Hörmander offered a general method of estimating a spectral function (which yields an estimate for $N(t, A)$), based on the technique of Fourier integral operators. He studied scalar PDO on a closed manifold (see [2]). A similar result for elliptic semi-bounded systems of first order was obtained by Levitan [2] with the assumption of constancy of multiplicity of eigenvalues of the symbol.

For elliptic scalar PDO on a closed manifold and under some additional conditions, expressed in terms of acorresponding Hamiltonian system, Duistermaat and Guillemin [1] obtained the formula of the type

$(B.3)$ $$N(t, A) = c_0 t^{n/m} + c_1 t^{(n-1)/m} + o(t^{(n-1)/m})$$

(for DO, always $c_1 = 0$). Volovoy [1] obtained some estimate of the remainder in formula (B.3).

For operators on a manifold with a boundary $\partial \Omega \in C^\infty$, formula (B.3) is proved, under additional assumptions of the geometry of bicharacteristics, by Ivriĭ [1,2], Melrose [1], Vasiliev [2,3,5]; Safarov [1–3] obtained (B.3) for diffraction problems and showed that formula (B.3) can be valid for them without the usual constraint on the geometry of bicharacteristics. Gureev and Safarov [1] and Safarov [4] obtained the second oscillating term of asymptotics for the case when the measure of recurrent points is not equal zero.

Metivier [5] succeeded in obtaining a sharp estimate for the remainder in (B.2) by using not the method of hyperbolic operator, but the resolvent technique.

Vasiliev [4] obtained a sharp estimate for the remainder (and, under additional conditions, the second term of asymptotics) for scalar operators of arbitrary order in domains with piecewise smooth boundary, and Ivriĭ [15] obtained this for the Laplace operator in domains with conic singularities of the boundary.

The principal term of the asymptotic function for linear pencils with degenerate subordinate operators was calculated by Pleyel [1]; Kozlov [2,3] an Andreev [8] obtained an estimate for the remainder, which in some cases is worse than the estimates of §§13 and 14, and in other cases is better. Estimates for the remainder of type (B.2) and the formulas of type (B.3) for linear sheaves were obtained by Ivriĭ [2,5,9] for the following cases: a) A, B differ little from natural exponents of the Beltrami–Laplace operator; b) A is an elliptic DO, B is either a PDO of first order, or an operator of multiplication by a regularly degenerating matrix.

For PDO of negative order, estimates for the remainder were obtained by Andreev [3,4,6]; in some cases they are better than the estimates of §13, in others they are worse.

Rosenblum [7–9] obtained full asymptotic expansion of the function $N(t, A)$ for systems of PDO on a circle under conditions of constancy of multiplicities of eigenvalues of principal symbols. Modifying his method, Agranovich [2] obtained a similar result for operators, close to normal.

The principal term of the asymptotics of a spectrum for Douglis–Nirenberg operators on a closed manifold was calculated by Kozhevnikov [1], and on a compact manifold with a boundary by Grubb [1]. An estimate of the remainder of the type (B.1) with $\delta \in (0, 1/2)$ for operators on a manifold without boundary was obtained by Grubb [2], and an estimate of the type (B.2) by Kozhevnikov [2]. We note that his method does not extend to operators on a manifold with a boundary.

Sections 15 and 16: The asymptotic function of eigenfrequencies of piezoactive bodies was calculated by Levendorskiĭ [6,14]; general boundary value problems with a resolvable constraint, which can be reduced to a problem with nonlocal subordinate operator, were studied by Alekseev and Birman in [1], where they also calculated the principal term of the asymptotic function of eigenvalues of the resonator filled with a medium with complicated discontinuities of ε, μ.

The full resonator was also studied by Weyl [3,4]; for other references see the beginning of Section 16.

Section 17: The principal term of the asymptotic function of a series which becomes dense at $+\infty$, was calculated by Grubb and Geymonat [1]; an estimate for the remainder, in a much more general situation then here (and using a different method), was obtained by Levendorskiĭ [5]. The asymptotic function of series which become dense at $\mu_j - 0$ was calculated by Andreev [7]; he also proved that eigenvalues do not become dense at $\mu_j + o$ and found out when μ_j is an eigenvalue of infinite multiplicity. In the cases of multiple

μ_j or nonsmooth boundary, Andreev's estimate is better than that obtained here.

Section 18: The principal term of the asymptotics of the spectrum of operators with an unresolvable constraint was calculated by Metivier [3], the estimate for the remainder was obtained by Levendorskiĭ [6,14]. A sharp estimate of a remainder of the type (B.2) for Navier–Stokes operators in a domain with a boundary $\partial\Omega \in C^\infty$ was obtained by Kozhevnikov [3] (the last condition is essential).

For other types of problems with constraints see, for example, Kozhevnikov [1], Birman and Solomyak [4,6,7].

Section 19: Asymptotics of fundamental frequencies of a thin shell in a vacuum was studied in a series of works by Lidskiĭ and his colleagues—see the summarizing monograph by Aslanyan and Lidskiĭ [1]—who obtained the estimate

$$(B.4) \qquad n_h(\lambda) = c_0 h^{-1} + O(h^{-1+\delta}),$$

$\forall \delta \in (0, 1/8)$. A similar result was obtained by Gulgazaryan and Lidskiĭ [1] for an anisotropic layered shell. Kozlov [2] obtained (B.4) for δ, depending on a type of degeneracy of the symbol and a shell with $\partial\Omega \in C^\infty$ (the last condition is essential). His best estimate is for $\delta = 1/4$, and not for $\delta \in (0, 1/3)$, as in §19, but in some cases Koslov's estimate is better.

The ASP method is also applicable to the study of a shell which contains liquid (see Levendorskiĭ [13]). This problem was considered by Aslanyan [1], Aslanyan, Vasiliev and Lidskiĭ [1]. We do not mention works which consider axially symmetric shells.

On Chapter 4

Section 20: Here we consider only a Schrödinger operator, but the ASP method allows us to obtain similar results for operators of higher orders—see Levendorskiĭ [3,13]—and for DO with coefficients which grow rapidly at infinity (Feigin [1,2,5], Levendorskiĭ [1,2]).

The sharp estimate for a remainder in the formulas of §20 is given for $\delta = 1$, but it is obtained only for operators in R^n and for fixed h (Tamura [5,6] for the Schrödinger operator, Helffer and Robert [1,2], Helffer [1], Ivriĭ [6,8], Feigin [6–8], and Mohamed [3] for general elliptic operators with weight) or λ (Helffer and Robert [3–6], Helffer [1], Ivriĭ [3]), and also under more rigid conditions than ours.

For a fixed h and an operator in R^n, Ivriĭ [6] and Feigin [7,8] also singled out, with additional assumptions, the second term of asymptotics.

For operators in an unbounded domain, the well-known works do not give a better estimate for the remainder than with $\delta \in (0, 1/2)$ (see for example, Robert [2], Boimatov [4,5], Boimatov and Kostuchenko [1,2]).

The quasi-classical asymptotic function for PDO on a closed manifold was studied by Vasiliev [1] and Ivriĭ [3,4]; the Schrödinger operator on a manifold with a boundary was studied by Ivriĭ [3]. They obtained a sharp estimate for the remainder.

The classical formula for the Schrödinger operator (for $h = 1$) was first obtained by Wett and Mandl [1]. Titchmarsh [1], Levitan [3], Rosenblum [4] analyzed and weakened constraints on the potential; Rosenblum [4], Otelbaev and Sultanaev [1] and Boimatov [1] showed that oscillation of the potential can perturb the classical asymptotics. Operators of high order were first analyzed by Kostuchenko [1] (see his bibliography).

Section 21: The principal term of asymptotics of the spectrum of the Schrödinger operator with rapidly decreasing potential was calculated by Birman and Solomyak [1,2], and that of the Dirac operator with slowly decreasing scalar potential (for $h = 1$) was calculated by Tamura [2]. A sharp estimate of the remainder was obtained by Tamura [3,4] only for $h = 1$ and operators in R^n (Schrödinger operator with non-degenerating potential) and Ivriĭ [6,7,10] (general operators of indefinite sign). Ivriĭ, under additional conditions, also obtained the second term of asymptotics [7,10]; in [7] he also analyzed multipartial Schrödinger operators. The case of $h = 1$ was also studied by Agmon [4] and Rosenblum [5,6].

Ivriĭ [12–14] gives sharp estimates for the remainder and, under additional assumptions of global character—a second term of the asymptotic function—for the Schrödinger operator with increasing and decreasing potentials; in the last case, he analyzes the asymptotic function with respect to both spectral and small parameters and also the joint asymptotics with respect to two parameters.

The potential can degenerate unregularly and have singularities; for the method of analysis of a Schrödinger operator with potential that has singularities, see Ivriĭ and Fydorova [1].

On Chapter 5

Section 22: The distribution of the spectrum of a Sturm–Liouville operator with an operator-valued potential was first studied by Kostuchenko and Levitan [1]. For further results in this direction see Michailec̆ [1] and the monograph by V. I. Gorbachuk and M. L. Gorbachuk [1].

In his series of works, Boimatov (see [4,5] and their bibliographies) studied DO and PDO with symbols of the type $A(x,\xi)+A_0$, where $A(x,\xi)$ are bounded operators for all (x,ξ). Classes of PDO studied by Karol [1,2] and Robert [3] are wider, but they considered only PDO in the entire space and on a closed manifold, respectively.

The ASP method was first applied to PDO with operator-valued symbols by Levendorskiĭ [11].

Section 23: For operators of the type $A(x, hD) + A_0$, where A is a scalar operator, the asymptotic function for fixed λ was studied by Lifschitz [1], where also a relatively simple resonator was analyzed.

On Chapter 6

Section 24: A hypothesis, replacing the Weyl formula, was suggested by Fefferman [1]; however, it is much more complicated and, as far as the author knows, had not yet been used for studying the asymptotics of a spectrum of concrete operators.

For some operators with constant coefficients, Andreev [5] obtained formulas similar to, but different from the Weyl formula.

The ideology and technique in §24 are applicable to the calculation of asymptotics of the spectrum of some integral Fourier operators; in particular, one can obtain results from the works by Grubb [5] and Laptev [1].

Section 25: Here we complement the results obtained by Solomyak [1]; however, in the case of a weak degeneracy and $n \geqslant 3$, his conditions are weaker. Earlier, the operator $-\Delta + |x|^k |y|^m$ on a plane was analyzed by Simon [1]; his result can be deduced from results by Robert [3], which analyzed the operator $-\Delta + \langle x \rangle^k |y|^m$.

The ASP method allows us also to analyze such operators as well as operators in cones and operators with potentials which vanish on transversally intersecting manifolds (see Levendorskiĭ [11]).

Sections 26 and 27: The three types of degeneracy were first singled out by Solomeshch [1,2], who calculated principal terms of asymptotics for the cases of weak and intermediate degeneracies and, for the case of strong degeneracy, obtained correct, with respect to order, two-sided estimates.

Birman and Solomyak [1,2] (under additional restrictions on the order of degeneracy) and Tashchiyan [1] obtained classical formulas also for the case of nonpower degeneracy. For DO with C^∞-coefficients weakly degenerating on a boundary, Berger [1] and Tashchiyan [2] obtained an almost sharp estimate for the remainder. For the model operator

$$-\partial_x (x^k \partial_x) - x^\ell \partial_y^2,$$

Matsuzawa and Shimakura [1] obtained, in a rectangle or in a semi-strip, sharp estimates for the remainder in some cases, and in two cases ($k = \ell = 1$ or $\ell = 2$, $k = 0$) which correspond to intermediate degeneracy—two terms of asymptotics. Their use of the method of separation of variables does not allow extension of results on operators on general-type domains.

The principal term of the asymptotic function for operators of second order was first calculated by Nordin [1] for the case of strong degeneracy (for one value of exponents, which defines the type of degeneracy) and by Vulis and Solomyak [1,2] (for all values of the exponents that correspond to a strong degeneracy). Operators of higher order were studied by Vulis [1,2], Pham The Lai [1,2], Metivier [2], Boimatov [2], Bolley, Camus, Pham The Lai [1], and Levendorskiĭ [10].

In the conclusion of the review of works on degenerating DO in a bounded domain, we mention the work by Metivier [4], where he showed that the principal term of asymptotics of the spectrum of the δ-Neumann problem is equally defined by a boundary and an interior of a domain (in §§26 and 27 and in all the quoted works, it is defined either by an interior or by a boundary), and the work by Karol and Solomyak [1], who considered operators similar to tensor products of DO in bounded domains; the asymptotics is similar to that in the case of strong degeneracy.

The ideology of §24 can be applied to the analysis of hypoelliptic PDO with multiple characteristics (see review, Levendorskiĭ [17]); such operators are similar in their properties to DO degenerating on a boundary (the principal term degenerates not on $\partial\Omega \times R^n$, but on a conic submanifold in $T^*\Omega$)—see Menikoff and Sjöstrand [1–3], Sjöstrand [1], and Mohamed [1,2].

In particular, one can also analyze a more general type of degeneracy on a smooth submanifold than those studied in the works mentioned above.

Section 28: DO degenerating in R^n were first studied by Skachek [1]. Gasimov [1] studied operators on a half-line and singled out the cases when discreteness of the spectrum and its asymptotics are defined by one of its leading coefficients. A wide class of systems in R^n was studied by Boimatov [3].

If domain Ω is a cone or the entire space, then the spectrum is discrete only under conditions of growth of some coefficients (sufficiently rapid, if the potential is bounded). If, on the other hand, Ω narrows at infinity, then the conditions on the growth of the coefficients can be weakened or removed (Clark [1], Birman and Solomyak [1,2], Rosenblum [1–3]); if Ω narrows slowly, then the spectrum is discrete, but the asymptotic function differs from the classical one (Rosenblum [3], Tamura [1], Fleckinger [1]).

We note that Rosenblum and Fleckinger considered operators in domains of an essentially more general type than here; however, our method also allows us to consider domains which were studied by Fleckinger (but not all domains studied by Rosenblum).

Appendix

All lemmas of Appendix are well-known and widely used in spectral theory (see, for example, Rosenblum, Solomyak and Shubin [1]).

BIBLIOGRAPHY

Agmon, S.,

[1] (1965). *Lectures on Elliptic Boundary Value Problems.* Princeton, N. J., 291.

[2] (1965). "On kernels, eigenvalues and eigenfunctions of operators related to elliptic problems." *Comm. Pure Appl. Math.* **18**(4), 627–663.

[3] (1968). "Asymptotic formulas with remainder estimates for eigenvalues of elliptic operators." *Arch. Rat. Mech. and Anal.* **28**(3), 165–183.

[4] (1975). "Spectral properties of Schrödinger operators and scattering theory." *Ann. Scuola Norm. Super. Pisa* 4(2,2), 151–218.

Agranovich, M. S.,

[1] (1977). "Spectral properties of diffraction problems." In: *Generalized Method of Fundamental Oscillations in Diffraction Theory* (Voitovich, N. N., Katzelenbaum, B. S., Sivov, A. N., eds.), Moscow: Nauka, pp. 289–416.

[2] (1984). "On elliptic pseudodifferential operators on a closed curve." *Works of the Mosc. Math. Soc.* **47**, 22–67.

Alekseev, A. B., and Birman, M. Sh.,

[1] (1976). "Asymptotics of the spectrum of elliptic boundary value problems with solvable constraints." *Dokl. AN USSR* **230**(3), 505–507.

Andreev, A. S.,

[1] (1982). "Asymptotics of the spectrum of Dirichlet problems for one class of pseudodifferential operators." In: *Problems of Mathematical Physics* **10**. Leningrad: LGU, 7–20.

[2] (1983). "Asymptotics of the spectrum of Dirichlet problem for differential operators with constant coefficients." *Funct. Anal. and its Appl.* **17**(3), 61–62.

[3] (1984). "Asymptotic estimates for limit phases of potential scattering." *Vestn. LGU* **4**, 117–119.

[4] (1984). " Spectral asymptotics of compact pseudodifferential operators with a "constant" symbol." Notes of scientific seminars, LOMI **138**, 3–7.

[5] (1984). "On the Weyl formula for spectral asymptotics for Dirichlet problem for non-elliptic operators." In: *Problems of Mathem. Physics* **9**. Leningrad: LGU, 3–18.

[6] (1985). "On estimates of a remainder in spectral asymptotics of pseudo-differential operators of negative order." In: *Problems of Math. Physics.* Leningrad: LGU, **11**, 31–46.

[7] (1985). "Spectral asymptotics for equations of multi-group diffusion model." *Vestnik LGU* **18**, 78–90.

[8] (1985). "Spectral acymptotics of semi-bounded Dirichlet problems for differential operators with constant coefficients." *Vestnik LGU* **15**, 82–87.

Aslanyan, A. G.,

[1] (1979). "Formula for a number of frequencies of free oscillations of a shell, filled with a liquid." *Funct. Anal. and its Appl.* **13**(4), 59–61.

Aslanyan, A. G., and Lidskiĭ, V. B.,

[1] (1974). "Distribution of eigenfrequencies of thin elastic membranes." Moscow: Nauka, 156.

Aslanyan, A. G., Vasiliev, D. G., and Lidskiĭ, V. B.,

[1] (1981). "Frequencies of free oscillations of a thin shell which interact with a liquid." *Funct. Anal. and its Appl.* **15**(3), 1–9.

Avakumovic, V. G.,

[1] (1956) "Über die Eigenfunktionen auf geschlossenen Riemannischen Mannigfaltigkeiten." *Math. Z.* **65**(4), 327–344.

Beals, R.,

[1] (1974). "Spatially inhomogeneous pseudodifferential operators. 2." *Comm. Pure Appl. Math.* **27**, 161– 205.

[2] (1975). " A general calculus of pseudodifferential operators." *Duke Math. J.* **42**(1), 1–41.

[3] (1981). "Weighted distribution spaces and pseudodifferential operators." *J. Anal. Math.* **39**, 131–187.

Beals, R., and Fefferman, C.,

[1] (1974). "Spatially inhomogeneous pseudodifferential operators." *Comm. Pure Appl. Math.* **27**, 1-24.

Belokon, A. V., and Vorovich, I. I.,

[1] (1979). "Some mathematical problems of the theory of electroelastic bodies." In: *Actual Problems of Mechanics of Deformable Medium.* Dnepropetrovsk: DGU, pp. 53–69.

Bérard, Pierre H.,

[1] (1983). "Remarques sur la conjecture de Weyl." *Comp. Math.* **48**(1), 35–53.

Berger, G.,

[1] (1979). "Asymptotics of a spectrum with estimates of a remainder for elliptic operators with weak degeneracy." *Vestnick LGU* **7**, 14–19.

Bezyaev, V. I.,

[1] (1979). "Asymptotics of eigenvalues of hypoelliptic operators on a closed manifold." *Dokl. AN SSSR* **244**(5), 1054–1057.
[2] (1982). —. *Mathem. Trans.* **111**(2), 161–180.

Birman, M. Sh.,

[1] (1980). "On quasiclassical spectral asymptotics of a class of integral operators." Notes of scientific seminar, LOMI, **98**, 22–32.
[2] (1980). "Quasiclassical asymptotics of a partial trace of a class of integral operators. *UMN* **35**(4), 162.
[3] (1986). "The Maxwell operator for a resonator with contained edges." *Vestnik LGU* **1**(3), 3–8.

Birman, M. Sh., and Solomyak, M. Z.,

[1] (1972). "Spectral asymptotics of non-smooth elliptic operators. I." *Works of the Moscow Math. Soc.* **27**, 3–52.
[2] (1973). "Spectral asymptotics of non-smooth elliptic operators. II." *Works of the Moscow Math. Soc.* **28**, 3–34.
[3] (1974). "Numerical analysis in Sobolev imbedding theorems and applications to spectral theory." In: *Tenth Mathem. School. Kiev*, 5–189.
[4] (1977). "Asymptotics of the spectrum of differential equations." In: *Mathematical Analysis* **14**, (Results of science and technology, VINITI). Moscow: Nauka, pp. 5–53.
[5] (1977). "Asymptotics of the spectrum of pseudodifferential operators with anisotropically-homogeneous symbols." *Vestnik LGU* **13**, 13–21.
[6] (1979). "Asymptotics of the spectrum of variational problems for solutions of elliptic equations." *Sibir. Mat. Zh.* **20**(1), 3–22.
[7] (1982). "Asymptotics of the spectrum of variational problems for solutions of elliptic systems." Notes of scientific seminar, LOMI **115**, 23–39.
[8] (1987). "The Maxwell operator in domains with non-smooth boundary." *Syb. Mathem. Journal* **28**(1), 23–36.
[9] (1987). "Weyl asymptotics of the Maxwell operator for domains with Lipschitz boundary." *Vestnik LGU* **8**, 3–11.
[10] (1987). "L_2-theory of the Maxwell operator in arbitrary domains." *UMN* **42**(6), 115.

Birman, M. Sh., and Yafaev, D. G.,

[1] (1982). "Asymptotics of limit phases for scattering on potential without spherical symmetry." *TMF* **51**(1), 44–53.
[2] (1981). "Asymptotics of the spectrum of scattering matrix." Notes of scientific seminar, LOMI **110**, 3–29.

Boimatov, K. Kh.,

[1] (1976). "Asymptotics of the spectrum of the Schrödinger operator with singular potential." *UMN* **31**(1), 241–242.

[2] (1979). "Distribution of eigenvalues of degenerating elliptic operators." *Dokl. AN SSSR* **248**(3), 521–523.

[3] (1978). "Asymptotics of the spectrum of an elliptic differential operator in the degenerate case." *Dokl. AN SSSR* **243**(6), 1369–1372.

[4] (1981). "Spectral asymptotics of differential and pseudodifferential operators. I." Works of I. G. Petrovsky seminar, *MGU* **5**, 50–100.

[5] (1983). "Spectral asymptotics of differential and pseudodifferential operators. II." Works of I. G. Petrovsky seminar, *MGU* **9**, 240–263.

Boimatov, K. Kh., and Kostuchenko, A. G.,

[1] (1976). "Distribution of eigenvalues of elliptic operators in the entire space." Works of I. G. Petrovsky seminar, *MGU* **2**, 113–143.

[2] (1984). "Distribution of eigenvalues of the equation $Au = \lambda Bu$ in the entire space." *Dokl. AN SSSR* **277**(6), 1292–1295.

[3] (1985). "Spectral function of elliptic operators in a bounded domain." *Funct. Anal. and its Appl.* **19**(2), 67–69.

Bolley, P., Camus, J., and Pham, The Lai,

[1] (1978). "Noyau, résolvante et valeurs propres d'une classe d'opérateurs elliptiques et dégénérées." *Lecture Notes Math.* **660**, 33–46.

Boutet de Monvel, L.,

[1] (1971). "Boundary problems for pseudodifferential operators." *Acta Math.* **126**, 11–51.

Browder, F. E.,

[1] (1965). "Asymptotic distribution of eigenvalues and eigenfunctions for non-local elliptic boundary value problems. 1." *Amer. J. Math.* **87**(1), 175–195.

Brüning, J.,

[1] (1974). "Zur Abschätzung der Spektralfunktion elliptischer Operatoren." *Math. Z.* **137**(1), 75–85.

Calderon, A. P., and Vaillancourt, R.,

[1] (1972). "A class of bounded pseudodifferential operators." *Proc. Nat. Acad. Sci. USA* **69**, 1185–1187.

Carleman, T.,

[1] (1935). "Propriétés asymptotiques des fonctions fondamentales des membranes vibrantes." *C. R. 8éme Cong. des Math. Scand. Stockholm, 1934, Lund. 1935*, 34–44.

Clark, C.,

[1] (1966). "An asymptotic formula for the eigenvalues of the Laplacian operator in an unbounded domain." *Bull. Amer. Math. Soc.* **72**(4), 709–712.

Cordes, H. O.,

[1] (1979). "Elliptic pseudodifferential operators – an abstract theory." *Lect. Notes Math.* **756**, 331p.

Courant, R., and Gilbert, D.,

[1] (1951). "Methods of mathematical physics. V. 1." Moscow: *Gos. Izd. Techn.-Teoret. lit.*, 476p.

Duistermaat, J. J., and Guillemin, V. W.,

[1] (1975). "The spectrum of positive elliptic operators and periodic bicharacteristics." *Invent. Math.* **29**(1), 39–79.

Eskin, G. I.,

[1] (1973). *Boundary Value Problems for Elliptic Pseudodifferential Equations*. Moscow: Nauka, 230 p.

Fefferman, C. L.,

[1] (1983). "The uncertainty principle." *Bull. Amer. Math. Soc.* **9**(2), 129–206.

Feigin, V. I.,

[1] (1978). "New classes of pseudodifferential operators in R^n and some applications." *Works of the Moscow Math. Soc.* **36**, 155–194.

[2] (1976). "Asymptotic distribution of eigenvalues for hypoelliptic systems in R^n." *Mathem. Trans.* **99**(4), 594–614.

[3] (1977). "On spectral asymptotics for boundary value problems and on asymptotics of a negative spectrum." *Dokl. AN SSSR* **232**(6), 1269–1272.

[4] (1979). "Asymptotic distribution of eigenvalues and Bohr–Sommerfeld-type formula." *Mathem. Trans.* **111**(1), 66–87.

[5] (1980). "Asymptotic distribution of eigenvalues for elliptic operators in unbounded domains." In: *Differential Equations with Partial Derivatives*. Works of S. L. Soboven Seminar, 1980, No. 2. Novosibirsk, pp. 116–151.

[6] (1982). "Sharp estimates of a remainder in spectral asymptotics for pseudodifferential operators in R^n." *Funct. Anal. and its Appl.* **16**(3), 88–89.

[7] (1982). "Sharp estimates in spectral asymptotics and Fourier integral operators of general type." *UMN* **37**(4), 99–100.

[8] (1984). "Spectral asymptotics for pseudodifferential operators in R^n." *Dokl. AN SSSR* **275**(3), 557–561.

Fleckinger, J.,

[1] (1981). "Asymptotic distribution of eigenvalues of elliptic operators on unbounded domains." *Lecture Notes in Math.* **846**, 119–128.

Friedrichs, K.,

[1] (1934). "Spektraltheorie halbbeschränkter Operatoren." *Math. Ann.* **109**, 465–487.

Gasimov, M. G.,

[1] (1969). "On the distribution of eigenvalues of a self-adjoint ordinary differential operator." *Dokl. AN SSSR* **186**(4), 753–756.

Glazman, I. M.,

[1] (1963). *Direct methods of qualitative spectral analysis of singular differential operators*. M. Physmatgiz, 339p.

Gorbachuk, M. L.,

[1] (1971). "Self-adjoint boundary value problems for a differential equation of second order with unbounded operator coefficient." *Funct. Anal. and its Appl.* **5**(1), 10–21.

Gorbachuk, V. I., and Gorbachuk, M. L.,

[1] (1984). "Boundary value problems for differential operator equations." Kiev: Naukova dumka, 284p.

Grisvard, P.,

[1] (1981). "Boundary value problems in non-smooth domains." France: Université de Nice.

Grubb, G.,

[1] (1977). "Spectral asymptotics for Douglis–Nirenberg elliptic and pseudodifferential boundary problems." *Comm. Part. Differ. Equat.* **2**, (11), 1071–1150.

[2] (1978). "Remainder estimates for eigenvalues and kernels of pseudodifferential elliptic systems." *Math. Scand.* **43**(2), 275–307.

[3] (1978). "Estimation du reste dans l'étude des valeurs propres des problèmes aux limites pseudodifférentiels auto-adjoints." *C. R. Acad. Sci.* **AB 287**(15), A1017–A1020.

[4] (1979). "On the coerciveness of Douglis–Nirenberg elliptic boundary value problems." *Boll. Unione mat. ital.* **B16**(3), 1049–1080.

[5] (1983). "Comportement asymptotique du spectre des opérateurs de Green singuliers." *C. R. Acad. Sci.* 1(296,1), 35–38.

Grubb, G., and Geymonat, G.,

[1] (1979). "Eigenvalue asymptotics for selfadjoint elliptic mixed-order systems with a nonempty essential spectrum." *Boll. Unione Mat. Ital.* **B 19**(3), 1032–1048.

Gulgazaryan, G. R., and Lidskiĭ, V. B.,

[1] (1982). "Density of frequencies of free oscillations of a thin anisotropic shell, consisting of anisotropic layers." *Isvestia AN SSSR, ser. MTT* **3**, 171–174.

Gureev, T. E., and Safarov, Yu. G.,

[1] (Preprint). "Precise asymptotics of the spectrum for the Laplace operator on manifolds with periodic geodesics." *LOMI* **E-1-86**, 33p.

Helffer, B.,

[1] (1984). "Théorie spectrale pour des opérateurs globalement elliptiques." *Asterisque* **112**, 197.

Helffer, B., and Robert, D.,

[1] (1981). "Comportement asymptotique précisé du spectre d'opérateurs globalement elliptiques dans R^n." *C. R. Acad. Sci., Sér., 1* **292**(6), 363–366.

[2] (1982). "Propriétés asymptotiques du spectre d'opérateurs pseudodifférentiels sur R^n." *Comm. Part. Differ. Equat.* **7**(7), 795–882.

[3] (1981). "Comportement semi-classique du spectre des hamiltoniens quantiques elliptiques." *Ann. Inst. Fourier* **31**(3), 169–223.

[4] (1981). "Comportement semi-classique du spectre des hamiltoniens quantiques hypoelliptiques." *C. R. Acad. Sci., Sér. 1*, **292**(1), 47–50.

[5] (1982). "Comportement semi-classique du spectre des hamiltoniens quantiques hypoelliptiques." *Ann. Scuola Norm. Super. Pisa. Cl. Sci* **9**(3), 405–431.

[6] (1983). "Calcul fonctionnel par la transformation de Melin et opérateurs admissibles." *J. Funct. Anal.* **53**(3), 246–268.

Hörmander, L.,

[1] (1966). "Pseudodifferential operators and hypoelliptic equations." *Proc. Symp. on Singular Integrals, Chicago, 1966, v. 10*. Providence, R.I.: Amer. Math. Soc., pp. 138–183.

[2] (1968). "The spectral function of an elliptic operator." *Acta Math.* **121**, 193–218.

[3] (1979). "The Weyl calculus of pseudodifférential operators." *Comm. Pure Appl. Math.* **32**,(3), 367– 443.

[4] (1979). "On the asymptotic distribution of the eigenvalues of pseudodifferential operators in R^n." *Ark. Mat.* **17**(2), 297–313.

[5] (1985). "The analysis of linear partial differential operators III." Berlin, New York, Heidelberg: Springer-Verlag.

Ilyin, V. A.,

[1] (1970). "On the characteristics of the spectrum of selfadjoint non-negative extensions of elliptic operators and about exact conditions of convergence and Riesz integrability of Fourier series for different classes of functions." *Mathem. Notes* **7**(4), 515–523.

[2] (1970). "On the expansion over eigenfunctions of arbitrary self-adjoint extensions of some elliptic operators." In: *International Congress of Mathematicians in Nice*. Moscow: Nauka, pp. 102–110.

Ilyin, V. A., and Filippov, A. F.,

[1] (1970). "On the characteristics of the spectrum of self-adjoint extensions of the Laplace operator in a bounded domain (fundamental systems of functions with arbitrary data of a sequence of fundamental numbers)." *Dokl. AN SSSR* **191**(2), 267–269.

Ivriĭ, V. Ya.,

[1] (1980). "On the second term of spectral asymptotics for the Laplace–Beltrami operator on manifolds with boundary and for elliptic operators on fibers." *Dokl. AN SSSR* **250**(6), 1300–1302.

[2] (1984). "Precise spectral asymptotics for elliptic operators." *Lect. Notes in Math.* **1100**, 238p.

[3] (1982). "On quasiclassical spectral asymptotics for the Schrödinger operator on manifolds with boundary and for h-pseudodifferential operators on fibers." *Dokl. AN SSSR* **266**(1), 14-18.

[4] (1983). "On the exact quasi-classical spectral asymptotics for h-pseudodifferential operators on fibers." In: *Imbedding Theorems and Their Applications to Problems of Mathem. Physics.* Novosibirsk, pp. 30–54.

[5] (1983). "On the asymptotics of a spectral problem connected with the Laplace–Beltrami operator on a manifold with boundary." *Funct. Analysis and its Appl.* **17**(1), 71–72.

[6] (1984). "On the exact asymptotics of eigenvalues of two classes of differential operators in R^n." *Dokl. AN SSSR* **276**(2), 268–270.

[7] (1984). "On the asymptotics of the discrete spectrum of the Schrödinger operator for a system of n particles." *Dokl. AN SSSR* **277**(4), 785–788.

[8] (1984). "Global and partially global operators, propagation of singularities and spectral asymptotics." *Contemp. Math.* **27**, 119–125.

[9] (1984). "Exact classical and quasi-classical asymptotics of eigenvalues for spectral problems on manifolds with boundary." *Dokl. AN SSSR* **277**(6), 1307–1310.

[10] (1985). "On the asymptotics of the discrete spectrum for some operators in R^n." *Funct. Anal. and its Appl.* **19**(1), 73–74.

[11] (1986). "Precise eigenvalue asymptotics for transversally elliptic operators." *Current Topics in Part. Differ. Equat.*, 55–62.

[12] (1986). "Estimation pour le nombre de valeurs propres négatives de l'opérateur de Schrödinger avec potentiels singuliers." *C. R. Acad. Sci.* **302**(1, 13), 467–470.

[13] "Estimation pour le nombre de valeurs propres négatives de l'opérateur de Schrödinger avec potentiels singuliers et application au comportement asymptotique des grandes valeurs propres." *C. R. Acad. Sci.* **302**(14), 491–494.

[14] "Estimation pour le nombre de valeurs propres négatives de l'opérateur de Schrödinger avec potentiels singuliers et application au comportement asymtotique des valeurs propres s'accumulant vers – 0, aux asymptotique

à deux paramètres et à la densité des états." *C. R. Acad. Sci.* **1**(15), 535–538.

[15] (1986). "Weyl asymptotic formula for the Laplace–Beltrami operator in Riemannian polyhedrons and in domains with conic singularities of boundary." *Dokl. AN SSSR* **288**(1), 35-38.

Ivriĭ, V. Ya., and Fedorova, S. I.,

[1] (1986). "Dilatations and asymptotics of eigenvalues of spectral problems with singularities." *Funct. Anal. and its Appl.* **20**(4), 29–34.

Karol, A. I.,

[1] (1981). "On the ζ-function of the Dirichlet degenerate elliptic boundary value problem." *Dokl. AN SSSR* **260**(1), 20–22.

[2] (1983). "Operator-valued pseudodifferential operators and resolvent of degenerate elliptic operators." *Mathem. Trans.* **121**(4), 562–575.

Karol, A. I., and Solomyak, M. Z.,

[1] (1982). "On a class of differential operators with non-standard behaviour of the spectrum of the Dirichlet problem." *Mathem. Notes* **31**(5), 747–751.

Kato, T.,

[1] (1966). *Perturbation Theory for Linear Operators.* Berlin, Heidelberg, New York: Springer-Verlag.

Kohn, J. J., and Nirenberg, L.,

[1] (1965). "An algebra of pseudodifferential operators." *Comm. Pure Appl. Math.* **18**, 269–305.

Kostuchenko, A. G.,

[1] (1968). "The asymptotic behavior of the spectral function of self-adjoint elliptic operators." In: *Forth Mathem. School.* Kiev, pp. 42–117.

Kostuchenko, A. G., and Levitan, B. M.,

[1] (1967). "On the asymptotic behavior of eigenvalues of the Sturm–Liouville operator problem." *Funct. Anal. and its Appl.* **1**(1), 86–90.

Kozhevnikov, A. N.,

[1] (1973). "Spectral problems for pseudodifferential systems, which are Douglis–Nirenberg elliptic, and their applications." *Mathem. Trans.* **92**(1), 60–88.

[2] (1981). "Remainder estimates for eigenvalues and complex powers of the Douglis–Nirenberg elliptic systems." *Comm. Part. Differ. Equat.* **6**(10), 1111–1136.

[3] (1983). "On the remainder term in the asymptotics of the spectrum of Stokes' problem." *Dokl. AN SSSR* **272**(2), 294–296.

Kozlov, V. A.,

[1] (1979). "On the estimate of a remainder in the formula for the asymptotics of the spectrum of non-semi-bounded elliptic systems." *Vestnik LGU* **19**, 112–114.

[2] (1983). "Remainder estimates in formulas for asymptotics of the spectrum for linear operator pencils." *Funct. Anal. and its Appl.* **17**(2), 80–81.

[3] (1984). "On the estimates of a remainder in formulas of asymptotics of the spectrum for linear operator pencils." In: *Problems of Mathem. Anal.* **9**. Leningrad: LGU, 34–56.

Laptev, A. A.,

[1] (1981). "Spectral asymptotics of a class of Fourier integral operators." *Tr. MMO* **43**, 92–115.

Levendorskiĭ, S. Z.,

[1] (1979). "Algebras of pseudodifferential operators with discontinuous symbols." *Dokl. AN SSSR* **248**(4), 777–779.

[2] (1982). "Asymptotic distribution of eigenvalues." *Izv. AN SSSR* **46**(4), 810–852.

[3] (1983). "Distribution of eigenvalues of systems of the type $Au = tBu$." *Funct. Anal. and its Appl.* **17**(2), 82–83.

[4] (1983). "Approximate spectral projection method as a general method of substantiation of the classical formula of asymptotics of the spectrum." *Dokl. AN SSSR* **271**(2), 287–291.

[5] ((1983). "Spectral asymptotics of non-linear operator pencils and operators with a non-empty essential spectrum." *Dokl. AN SSSR* **272**(6), 1314–1317.

[6] (1984). "Asymptotics of the spectrum for problems with constraints." *Dokl. AN SSSR* **276**(2), 282–285.

[7] (1984). "Asymptotics of the spectrum of linear operator pencils." *Math. Trans.* **124**(2), 251–271.

[8] (1984). "Asymptotics of eigenvalues of problems of the type $Au = tBu$ for Douglis–Nirenberg elliptic operators." *Funct. Anal. and its Appl.* **18**(3), 84–85.

[9] (1985). "Asymptotics of the spectrum of degenerate elliptic systems in unbounded domains." *Funct. Anal. and its Appl.* **19**(2), 80–81.

[10] (1985). "Asymptotics of the spectrum of degenerate systems." *Dokl. AN SSSR* **280**(4), 793–796.

[11] "Asymptotics of the spectrum of differential operators with operator-valued coefficients and some applications." *Dokl. AN SSSR* **280**(6), 1303–1306.

[12] (1985). "On the asymptotics of the spectrum of linear pseudodifferential pencils in the case of degeneracy." *Dokl. AN SSSR* **285**(6), 1321–1324.

[13] (1985). "Approximate spectral projection method." *Izvestia AN SSSR* **49**(6), 1177–1229.

[14] (1986). "Asymptotics of the spectrum of problems with constraints."
    Math. Trans. 129(1), 73–89.

[15] (1986). Asymptotics of the spectrum of weakly elliptic operators." Sib.
    Math. Journal 27(1), 111–122.

[16] (1981). "Quasi-classical asymptotics of eigenvalues of a system of pseu-
    dodifferential operators." Dokl. AN SSSR 256(1), 32–36.

[17] (1988). "Non-classical spectral asymptotics." Russian Mathematical
    Surveys 43(1), 149-192.

Lerner, N.,

[1] (1979). "Sur les espaces de Sobolev généraux associés aux classes récentes
    d'opérateurs pseudodifférentiels." C. R. Acad. Sci. 289(A), 663–666.

Levitan, B. M.,

[1] (1954). "On the eigenfunctions expansion for Laplace operator." Math.
    Trans. 35(2), 267–316.

[2] (1971). "Asymptotic behaviour of spectral function of elliptic equation."
    UMN 26(6), 151–212.

[3] (1955). "On the eigenfunctions expansion for the Schrödinger operator in
    the case of an unboundedly growing potential." Dokl. AN SSSR 103(2),
    191–194.

Lifshitz, A. E.,

[1] (1980). "On the oscillation of anisotropic resonators." Funct. Anal. and
    its Appl. 14(2), 63–64.

Lions, J. L., and Magenes, E.,

[1] (1968). "Problèmes aux limites non homogènes et applications." Paris:
    Dunod.

Markus, A. S., and Matsaev, V. I.,

[1] (1982). "Theorems of comparison of the spectrums of linear operators and
    spectral asymptotics." Works of the Mosc. Math. Soc. 45, 133–181.

[2] (1979). "On the asymptotics of the spectrum of operators, close to nor-
    mal." Funct. Anal. and its Appl. 13, 93–94.

[3] (1981). "Operators, generated by sesquilinear forms and their spectral
    asymptotics." Mathem. Research, Kishinev 61, 86–103.

Matsuzawa, and Shimakura, N.,

[1] (1984). "Valeurs propres d'une classe d'opérateurs elliptiques dégénérés."
    J. Math. Pures et Appl. 63(1), 15–36.

Melrose, R.,

[1] (1980). "Weyl's conjecture for manifolds with concave boundary." Proc.
    Symp. on Pure Math, Bull. Amer. Math. Soc. 36, 257–273.

Menikoff, A., and Sjöstrand, J.,

[1] (1978). "On the eigenvalues of a class of hypoelliptic operators." *Math. Ann.* **235**(1), 55–85.

[2] (1979). "On the eigenvalues of a class of hypoelliptic operators. 2." *Lecture Notes in Math.* **755**, 201–247.

[3] (1979). "On the eigenvalues of a class of hypoelliptic operators. 3. The non-semibounded case." *J. Anal. Math.* **35**, 123–150.

Metivier, G.,

[1] (1977). " Valeurs propres de problémes aux limites elliptiques irréguliers." *Bull. Math. Soc. France* **51–52**, 125–219.

[2] (1976). "Comportement asymptotique des valeurs propres d'opérateurs elliptiques dégénérés." *Asterisque* **34–35**, 121–149.

[3] (1978). "Valeurs propres d'opérateurs definis par la restriction de systèmes variationnels à des sous-espaces." *J. Math. Pures et Appl.* **57**(2), 133–156.

[4] (1981). "Spectral asymptotics for the $\overline{\partial}$-Neumann problem." *Duke Math. J.* **48**(4), 779–806.

[5] (1982). "Estimation du reste en théorie spectrale." *Equat. aux derivées partielles, Saint-Jean-De-Montes*, Exp. **1**.

Michaileč, V. A.,

[1] (1977). "Distribution of eigenvalues of the Sturm–Liouville operator equation." *Izv. AN SSSR, Ser. Mathem.* **41**(3), 607–619.

[2] (1978). "Sharpening of asymptotic formulas for the spectrum of Laplace operators in domains of the type $G_1 \times G_2$." *UMN* **33**(4), 219–220.

[3] (1979). "Self-adjoint extension of operators with a given asymptotics of the spectrum." *Dokl. AN SSSR* **A**(5), 338–341.

[4] (1981). "Distribution of eigenvalues of operators which are close to being self-adjoint." *Funct. Anal. and its Appl.* **15**(1), 78–79.

[5] (1982). "Asymptotics of the spectrum of elliptic operators and boundary conditions." *Dokl. AN SSSR* **266**(5), 1059–1062.

Mohamed, A.,

[1] (1982). "Etude spectrale d'opérateurs hypoelliptiques à caractéristiques multiples. 1." *Ann. Inst. Fourier* **32**(3), 39–90.

[2] (1983). "Etude spectrale d'opérateurs hypoelliptiques à caractéristiques multiples. 2." *Comm. Part. Differ. Equat.* **8**(3), 235–297.

[3] (1984). "Comportement asymptotique, avec estimation du reste, des valeurs propres d'une classe d'opérateurs pseudodifferentiels sur R^n." *C. R. Acad. Sci.* **1**(299, 6), 177–180.

Nordin, C.,

[1] (1972). "Asymptotic distribution of the eigenvalues of a degenerate elliptic operator." *Ark. Math.* **10**(1), 9–21.

Otelbaev, M., and Sultanaev, Ya. T.,

[1] (1974). "On formulas of the distribution of eigenvalues of singular differential operators." *Mathem. Notes* **14**(3), 361–368.

Pham, The Lai,

[1] (1976). "Comportement asymptotique du noyau de la résolvante et des valeurs propres d'une classe d'opérateurs elliptiques dégénérés non necessairement autoadjoint." *J. Math. Pures et Appl.* **55**(4), 379–420.

[2] (1976). "Estimation du reste dans la théorie spectrale d'une classe de problèmes d'opérateurs elliptiques dégénérés." *Astérisque* **33–35**, 263–277.

Pleyel, A.,

[1] (1961). "Certain indefinite differential eigenvalue problems—the asymptotic distribution of their eigenfunctions." *Part. Differ. Equat. and Cont. Mech.* Madison: Wisconsin Press, pp. 19–37.

Reed, M., and Simon, B.,

[1] (1978). "Methods of modern mathematical physics. V.4." New York, San Francisco, London: Academic Press.

Riesz, F., and Sz.-Nagy, B.,

[1] (1972). "Leçons d'analyse fonctionnelle." Budapest: Akadémiai Kiado.

Robert, D.,

[1] (1978). "Développement asymptotique du noyau résolvant d'opérateurs elliptiques." *Osaka J. Math.* [**15**, 233–245.

[2] (1978). "Propriétés spectrales d'opérateurs pseudodifferentiels." *Comm. Part. Differ. Equat.* **3**(9), 755–826.

[3] (1982). "Comportement asymptotique des valeurs propres d'opérateurs du type Schrödinger à potentiel 'dégénérés'." *J. Math. Pures et Appl.* **61**(3), 275–300.

Roitburd, V. L.,

[1] (1976). "On the quasi-classical asymptotics of the spectrum of a pseudodifferential operator." *UMN* **31**(4), 275–276.

Rosenblum, G. V.,

[1] (1971). "On the distribution of eigenvalues of the first boundary value problem in unbounded domains." *Dokl. AN SSSR* **200**(5), 1034–1036.

[2] (1972). "Distribution of the discrete spectrum of singular differential operators." *Dokl. AN SSSR* **202**(5), 1012–1015.

[3] (1972). " On eigenvalues of the first boundary value problem in unbounded domains." *Mathem. Trans.* **89**(2), 234–247.

[4] (1974). "Asymptotics of eigenvalues of the Schrödinger operator." *Math. Trans.* **93**(3), 346–367.

[5] (1975). "On estimates of the spectrum of the Schrödinger operator." In: *Problems of Mathem. Analysis, No. 5.* Leningrad: LGU, pp. 151–166.

[6] (1977). "Asymptotics of the negative discrete spectrum of the Schrödinger operator." *Mathem. Notes* **21**(3), 399–407.

[7] (1975). "Almost-similarity of operators and spectral asymptotics of pseudodifferential operators on a circle." *Dokl. AN SSSR* **223**(4), 569–571.

[8] (1978). "Almost-similarity of operators and spectral asymptotics of pseudodifferential operators on a circle." *Works of the Mosc. Math. Soc.* **36**, 59–84.

[9] (1977). "Almost-similarity of pseudodifferential systems on a circle." In: *Probl. of Mathem. Analysis, No. 6*. Leningrad: LGU, pp. 143–157.

[10] (1990). "Encyclopedia of contemporary Mathematics. V.64. Spectral Theory of Differential Operators". Springer-Verlag.

Safarov, Yu. G.,

[1] (1984). "On the asymptotics of eigenvalues of diffraction problems." *Preprint LOMI*.

[2] (1984). "Asymptotics of the spectrum of diffraction problems." *Notes of Scient. Seminar, LOMI* **138**, 137–145.

[3] (1985). "On the asymptotics of eigenvalues of diffraction problems." *Dokl. AN SSSR* **281**(5), 1058–1061.

[4] (1986). "Asymptotics of the spectrum of a pseudodifferential equation with periodic bicharacteristics." *Notes of Scient. Seminar LOMI* **152**, 93–104.

[5] (1983). "On the asymptotics of the spectrum of the Maxwell operator." *Notes of Scient. Seminar LOMI* **51**, 52–61.

Schwartz, L.,

[1] (1957). "Distribution a valeurs vectorielles. 1." *Ann. Inst. Fourier* **7**, 1–141.

[2] (1958). "Distribution a valeurs vectorielles. 2." *Ann. Inst. Fourier* **8**, 1–209.

Shubin, M. A.,

[1] (1978). *Pseudodifferential operators and spectral theory*. Moscow: Nauka, 280 p.

Simon, B.,

[1] (1983). "Nonclassical eigenvalue asymptotics." *J. Funct. Anal.* **53**(1), 84–98.

Sjöstrand, J.,

[1] (1980). "On the eigenvalues of a class of hypoelliptic operators. 4." *Ann. Inst. Fourier* **30**(2), 109–169.

Skachek, B. Ya.,

[1] (1966). "On the asymptotic behaviour of the eigenvalues of singular diffraction operators." *Function Theory, Funct. Anal. and its Applic., Republ. Scient. Trans.* **3**, 110–116.

Solomeshch, I. A.,

[1] (1961). "On the eigenvalues of some degenerate elliptic equations." *Math. Trans.* **54**(2), 295–310.

[2] (1962). "On the asymptotics of eigenvalues of bilinear forms connected with some elliptic equations degenerating on a boundary." *Dokl. AN SSSR* **144**(3), 727–729.

Solomyak, M. Z.,

[1] (1984). "On the asymptotics of the spectrum of the Schrödinger operator with irregular homogeneous potential." *Dokl. AN SSSR* **278**(2), 291–295.

[2] (1985). "Asymptotics of the spectrum of the Schrödinger operator with irregular homogeneous potential." *Mathem. Trans.* **127**(1), 21–40.

Stein, E. M.,

[1] (1970). *Singular Integrals and Differentiability Properties of Functions.* Princeton, N.J.: Princeton University Press.

Tamura, H.,

[1] (1976). "The asymptotic distribution of eigenvalues of the Laplace operator in an unbounded domain." *Nagoya Math. J.* **60**, 7–33.

[2] (1976). "The asymptotic distribution of discrete eigenvalues for the Dirac operator." *J. Fac. Sci. Univ. Tokyo* **1A**(23,1), 167–197.

[3] (1981). "Asymptotic formulas with sharp remainder estimates for bound states for Schrödinger operators. 1." *J. Anal. Math.* **40**, 166–182.

[4] (1982). "Asymptotic formulas with sharp remainder estimates for bound states of Schrödinger operators. 2." *J. Anal. Math.* **41**, 85–108.

[5] (1982). "Asymptotic formulas with remainder estimates for eigenvalues of Schrödinger operators." *CPDE* **7**(1), 1–53.

[6] (1982). "Asymptotic formulas with sharp remainder estimates for eigenvalues of elliptic operators of second order." *Duke Math. J.* **49**(1), 87–119.

Tashchiyan, G. M.,

[1] (1981). "Classical formula for asymptotics of the spectrum of elliptic equations which degenerate on a boundary of a domain." *Mathem. Notes* **30**(6), 871–880.

[2] "On the remainder estimate in the classical formula for spectral asymptotics for degenerate elliptic equations." In: *Problems of Mathem. Analysis, No. 8* . Leningrad: LGU, pp. 181–188.

Taylor Michael, E.,

[1] (1981) *Pseudodifferential Operators.* Princeton, N.J.: Princeton University Press.

Titchmarsh, E. C.,

[1] (1954). "On the asymptotic distribution of eigenvalues." *Quat. J. Math.* **5**(19), 228–240.

Treves, F.,

[1] (1982). *Introduction to Pseudodifferential and Fourier Integral Operators.* New York and London: Plenum Press.

Tulovsky, V. N., and Shubin, M. A.,

[1] (1973). "On the asymptotic distribution of eigenvalues of pseudodifferential operators in R^n." *Mathem. Trans.* **92**(4), 571–588.

Unterberger, A.,

[1] (1979). "Oscillator harmonique et opérateurs pseudodifférentiels." *Ann. Inst. Fourier* **29**, 201–221.

Vasiliev, D. G.,

[1] (1980). "Asymptotics of the distribution function of the spectrum of pseudodifferential operators with parameters." *Funct. Anal. and its Appl.* **14**(3), 67–68.
[2] (1983). "Binomial asymptotics of the spectrum of a boundary value problem." *Funct. Anal. and its Appl.* **17**(4), 79–81.
[3] (1984). "Binomial asymptotics of the spectrum of a boundary value problem for intrinsic mapping of general type." *Funct. Anal. and its Appl.,* **18**(4), 1–13.
[4] (1986). "Binomial asymptotics of the spectrum of the boundary value problem in case of a piecewise smooth boundary." *Dokl. AN SSSR* **290**(2), 270–274.
[5] (1986). "Asymptotics of the spectrum of the boundary value problem." *Works of the Mosc. Math. Soc.* **49**, 167–237.

Vasiliev, D. G., and Lidskiĭ, V. B.,

[1] (1982). "Oscillations of thin shells which interact with a liquid." *UMN* **37**(5), 224–225.

Volovoy, A. V.,

[1] (1987). "Improved binomial asymptotic distribution of eigenvalues of elliptic operators on a compact manifold." *Dokl. AN SSSR* **294**(5), 1037–1041.

Vulis, I. L.,

[1] (1976). "Spectral asymptotics of elliptic operators of arbitrary order with strong degeneracy on a boundary of domain." In: *Problems of Mathem. Physics.* Leningrad: LGU, **8**, 56–69.
[2] (1977). "Spectral asymptotics of a class of degenerate elliptic operators" In: *Problems of Mathem. Analysis.* Leningrad: *LGU* **6**, 18–30.

Vulis, I. L., and Solomyak, M. Z.,

[1] (1972). "Spectral asymptotics of degenerate elliptic operators." *Dokl. AN SSSR* **207**(2), 262–265.
[2] (1974). "Spectral asymptotics of degenerate elliptic operators of second order." *Isvestia AN SSSR*, ser. mathem. **38**(6), 1362–1392.

Wett, J. S., and Mandl, F.,

[1] (1950). "On the asymptotic distribution of eigenvalues." *Proc. Roy. Soc.* **A200**, 572–580.

Weyl, H.,

[1] (1912). "Das asymptotische Verteilungsgesetz der Eigenwerte linearer partieller Differentialgleichungen." *Math. Ann.* **71**, 441–479.

[2] (1912). "Über die Abhängigkeit der Eigenschwingungen einer Membran von der Begrenzung." *J. Reine Angew. Math.* **141**, 1–11.

[3] (1912). "Über das Spectrum der Höhlraumstralung." *J. Reine Angew. Mth.* **141**, 163–181.

[4] (1913). "Über die Randwertaufgabe der Stralungstheorie und asymptotische Spectralgesetze." *J. Reine Angew. Math.* **143**, 177–202.

Zweifel, P. F.,

[1] (1973). *Reactor Physics*. New York: McGraw–Hill.

Index of Notation

SUBJECT INDEX